1/2012

Climate Change and National Security

Climate Change and National Security

A Country-Level Analysis

Daniel Moran, Editor

Georgetown University Press / Washington, D.C.

Georgetown University Press, Washington, D.C. www.press.georgetown.edu

Library of Congress Cataloging-in-Publication Data

Climate change and national security : a country-level analysis / Daniel Moran, editor.
    p. cm.
  Includes bibliographical references and index.
  ISBN 978-1-58901-741-2 (pbk. : alk. paper)
    1. Global environmental change. 2. Climatic changes—Social aspects.
  3. International relations—Forecasting.
  I. Moran, Daniel.
  GE149.C55 2011
  355′.033—dc22

                                       2010036728

15 14 13 12 11   9 8 7 6 5 4 3 2 First printing

Printed in the United States of America

Contents

Illustrations

Acknowledgments

The contents of this volume originated as presentations at a workshop sponsored by the National Intelligence Council, in support of the National Intelligence Assessment on Climate Change that it prepared for the U.S. House of Representatives in June 2008.

It is a pleasure to be able to thank Mat Burrows and Paul Herman, of the National Intelligence Council's Long-Range Analysis Unit, and Rich Engel, of its Office of Science and Technology, for their support. Everyone involved owes an enormous debt to Marc Levy, deputy directory of the Center for International Earth Science Information Network at Columbia University, whose team assembled much of the country-level climate data used by the contributors to this project. The Peterson Institute for International Economics was also kind enough to allow chapter 5 of William R. Cline's *Global Warming and Agriculture: Impact Estimates by Country* (2007) to be distributed to all the conference participants. I am also glad for the opportunity to thank my colleagues James Russell and Michael Malley for their assistance in assembling the team of experts whose work is presented here. It goes without saying that all the contents of this volume reflect the judgment of the individual chapter authors. Nothing in it has been reviewed or approved by any department or agency of the United States government.

Abbreviations

The following abbreviations and short titles are used throughout.

CIESIN	Center for International Earth Science Information Network at Columbia University
GDP	gross domestic product
IPCC	United Nations Intergovernmental Panel on Climate Change
IPCC, *Climate Change 2007: Impacts*	Intergovernmental Panel on Climate Change, *Climate Change 2007: Impacts, Adaptation and Vulnerability.* Contribution of Working Group II to the Fourth Assessment Report of the Intergovernmental Panel on Climate Change, edited by M. L. Parry, O. F. Canziani, J. P. Palutikof, P. J. van der Linden, and C. E. Hanson (Cambridge: Cambridge University Press, 2007).
IPCC, *Climate Change 2007: Science*	Intergovernmental Panel on Climate Change, *Climate Change 2007: The Physical Science Basis.* Contribution of Working Group I to the Fourth Assessment Report of the Intergovernmental Panel on Climate Change, edited by S. Solomon, D. Qin, M. Manning, Z. Chen, M. Marquis, K. B. Averyt, M. Tignor, and H. L. Miller (Cambridge: Cambridge University Press, 2007).
UNDP	United Nations Development Program
UNFCCC	United Nations Framework Convention on Climate Change

Unless otherwise indicated, all monetary figures are in U.S. dollars.

1

Introduction

Climate Science and Climate Politics

Daniel Moran

This book seeks to appraise the intermediate-term security risks that climate change may pose to the United States, its allies, and to regional and global order. It is intended to be a contribution to the growing literature on "environmental security," a phrase that encompasses a wide range of policy problems. For many environmental security is chiefly about addressing the challenges that climate change may present to humanity and its institutions. Security in this context is to be sought through measures designed to mitigate or adapt to changes in the Earth's ecology, which may some day, and perhaps quite soon, make current social and economic practices unsustainable.[1] Others have sought to interpret climate change less as a direct threat than as an additional source of stress on the sinews of public life, which may cause fragile governments to fail, or may provide a new impetus for a range of violent outcomes, ranging from social upheaval to aggressive war.[2]

The present work is of the latter kind. It does not seek to comment on the likelihood that the daunting environmental changes foreseen by current earth science will come to pass, nor to evaluate the policies that might be chosen in response to them. It is, instead, an attempt to lay the problems hypothesized by science on top of the known or anticipated challenges of international life and to consider what might change as a consequence. It seeks to do this in a relatively precise and disciplined way, however, and it is useful to begin by considering why, in the field of environmental security, precision and discipline can be hard to achieve.

The greatest difficulty arises from the different speeds at which politics and the natural environment operate, or more precisely the rates at which they change. It must be admitted that the phrase "intermediate-term security risks" that was employed a few sentences ago to describe this project's goals was chosen for no better reason than because it is equally unsatisfactory to scientists and statesmen alike. As has already been mentioned in the acknowledgments, this volume originated in a project directly connected to real-world policy-making, and it was intended to produce a result of practical value in that context. To that end, all the contributors were asked to evaluate the projected security implications of climate change until 2030. In the world of politics, twenty years is not the intermediate term. It is a long time—the far horizon within which serious strategic planning has traditionally been done.[3] To policymakers and planners, risks that are thought to be a generation away are long-term risks.

For earth scientists, on the other hand, twenty years, if not quite the blink of an eye, is certainly a brief span of time, across which the models by which they seek to anticipate our environmental future are barely able to generate useful results. The mainstream of credible work on climate change, as exemplified by the reports from the United Nations Intergovernmental Panel on Climate Change (IPCC), has tended in recent years to look further outward, toward 2050 or 2100, in presenting its conclusions.[4] Of necessity, models that are intended to anticipate conditions in fifty or a hundred years lose precision when asked to render judgments over significantly shorter (or longer) periods.

If we imagine the models of the earth scientist as a kind of lens through which the future may be viewed, then we must accept that, like real lenses, those models have an optimum focal range, on either side of which the image becomes progressively blurry. The same can be said of the assumptions, experiences, and predilections of policymakers, which are the tools by which they seek to envision the future. The central challenge in the field of environmental security arises from the fact that these two lenses have such different focal lengths that their useful fields of vision barely overlap. The time frame of this study was chosen not because it is optimal in terms of either politics or science but because it is at least credible in both contexts, and sufficiently so to impose a reasonable restraint on the range of conditions to be considered. Twenty years may be a long time in politics, but not so long as to be unimaginable. It may be a short time in the life of the Earth, but not so short that the changes anticipated by today's scientists will not have become subject to preliminary "out-of-sample" testing, which will in turn strengthen (or not) the credibility of their vision of the longer-term future.

In the same way that scientists and policymakers diverge in their sense of how quickly time passes, they also differ with respect to the maps of the world they carry in their heads. The earth scientist's map is drawn by natural forces on a planetary scale, which may be affected by human actions in the aggregate but which nevertheless operate without regard to human institutions. Its crucial boundaries are drawn by nature, in the form of seacoasts, mountain ranges, forests, deserts, glaciers, and so on. These things matter to the policymaker, too, of course, but their influence is often trumped by other lines, drawn by human hands, that divide one sovereign state from another. It is within these lines that the political choices that govern humanity's response to climate change will be made.

As the contents of this volume demonstrate, there is good reason to believe that those choices will vary a good deal, even among polities confronted with substantially similar forms of environmental stress. This project appraises the security risks that climate change poses when its most readily anticipated effects are applied to the known political and social conditions of states whose fortunes are of particular importance to the United States. At its heart lies a body of path-breaking country-level data prepared by Columbia University's Center for International Earth Science Information Network (CIESIN), which allows such issues to be addressed comprehensively for the first time. CIESIN's work originally extended across 179 countries, in practice all those for which reliable demographic and scientific information is available. From these a secondary list was chosen, including states for which the scientific data suggest that anticipated climate effects will be especially severe, and others that may face less grave challenges but whose place in the world system makes them especially important to the United States.

The process of selection, it is worth emphasizing, was driven by concrete considerations of policy as judged by those tasked with making it, and not by social-scientific criteria aspiring to general theoretical validity. The resulting list, as reflected in the table of contents and the appendixes to this volume, could easily have been extended to almost any length. It also omits some countries whose policy relevance would appear to demand their inclusion. A number of nations of intense current interest—including North Korea, Afghanistan, Iran, and the states of the Horn of Africa—have not been included because their politics are at present so opaque or unsettled that it did not seem possible to disentangle and evaluate the likely consequences of their exposure to climate change, however grave that exposure might appear to be. Nevertheless, the resulting study is still the most comprehensive of its kind, and it is broadly representative of the security challenges that climate change may pose during the next few decades.

CIESIN's data on temperature change, freshwater availability, and sea-level rise are summarized in appendixes A and B, which provide additional detail on the methods by which the data were gathered. Those data were presented to all the contributors, along with other selected materials chosen to reflect the mainstream of current climate science. These included the "Technical Summary" and relevant regional sections from the report of the IPCC's Working Group II, some maps showing current and anticipated future changes in the global distribution of freshwater,[5] and selections from William Cline's *Global Warming and Agriculture: Impact Estimates by Country*, some of whose findings are also incorporated into appendix A.[6] All this material reflected what might be described as down-the-middle assumptions about how anthropogenic climate change will unfold during the next few decades—that is, it has been assumed that basic trends, both human and natural, will continue in their current directions and at their current pace.

Climate science is all about building scenarios and then modeling projected results across future time periods. Because this project considers a relatively short period in environmental terms, albeit an extended one in political terms, it seemed most reasonable to employ scientific data based on models that assume the world economy will continue to develop more or less as it has recently: that the large-scale distribution of goods will continue to be governed by international markets; that market participants will continue to seek economic growth as a social good; and that no remarkable changes for better or worse will occur in the consumption of fossil fuels, other that those that may arise as a consequence of rising costs in the marketplace.

Other sets of assumptions are possible, and they have often been explored by earth scientists as a way of testing the range of outcomes their models may generate under varying patterns of human activity—increasing or decreasing economic growth, static or increasing efforts at environmental mitigation, and so on.[7] Because the focus of this project is on human activity itself, however, it was essential to hold the earth science constant, and also to avoid analyses predicated upon "worst-case scenarios"—or "best-case" ones, for that matter. All the contributors were free to supplement the common body of data and information originally presented to them with whatever other evidence they deemed appropriate. Because the aim was to cast light on political conditions that may pose security risks in the future, however, no attempt has been made to challenge, evaluate, or extend the scientific hypotheses by which that future has been envisioned in environmental terms. The contributors to this volume were invited to participate because they are policy experts, not climate experts (though a few are both). One purpose of the project, indeed, was to broaden the range of scholars participating in the academic conversation about climate politics.

The area in which the "no-worst-case-scenarios" rule proved most difficult to enforce was with respect to rising sea level, a particularly contentious area of climate science. Many well-founded efforts to estimate sea-level rise, including the work summarized by the IPCC, have been revealed as too conservative in light of subsequent experience, owing to the currently rapid melting of polar and glacial ice, whose pace has outstripped the scientific consensus.[8] CIESIN data on sea-level rise are pegged to "low-elevation coastal zones" of 1 and 3 meters, which, within our time frame, imply a rate of sea-level rise significantly faster than that foreseen by the IPCC even a few years ago. On balance, and assuming the persistence of current conditions, a 1-meter rise would lie at the far end of the range of what is currently thought possible by 2030.[9] A 3-meter rise would lie outside that range, and it has been incorporated as a (relatively conservative) proxy for the kind of tidal surge that severe weather events like tropical cyclones routinely produce, and whose incidence is expected to increase as a consequence of global warming.[10] The tidal surge produced by Hurricane Katrina, the most famous such episode in recent American history, exceeded 7 meters.[11]

The chapters in this volume are organized around six primary questions that the contributors were asked to address in whatever form seemed best to them. They were free to ignore questions that did not warrant consideration in relation to their countries or regions, and also to integrate whatever additional forms of analysis they deemed necessary to produce an accurate appraisal of emerging security risks. Although this may seem an artificially constrained starting point, it offers the advantage of strengthening the internal coherence and comparability of analyses that span a very wide range of social and political conditions. As will become apparent, an excess of uniformity is scarcely the dominant impression that has resulted.

Here are the six questions as originally posed:

1. Your country or region may already be susceptible to unruly sociopolitical change due to existing traditional risk factors. Are the climate impacts hypothesized for 2030 likely to be an appreciable additional factor in triggering disruptive change, internally or externally?
2. The most basic negative responses to environmental stress are either fight (civil conflict or external aggression) or flight (internal or external migration). For the people of your country or region, which response would you consider more likely? Please provide what detail you can with respect to the political parties, social or ethnic groups, or local areas where the actions you anticipate may be most serious or most likely to occur.
3. Considering the disruptive possibilities you have described in response to the questions above, how would you assess the risk that the net result will be complete failure of the state?
4. Conversely, might your country or region possess latent reserves of social resilience and ingenuity, or of institutional capital, which will allow it to meet the challenges of climate change successfully? Are there important social or other interest groups in your country or region that might benefit from climate change? Or expect to?
5. Would you anticipate that climate change might make your state an object of aggressive war to control scarce resources? Might it become a destination for migrants and refugees from neighboring countries?
6. What do the behaviors and developments that you have foreseen in your state suggest about the outlook and tone of official foreign policy? Would you expect it to become more or less open to engagement with the larger world? More or less amenable to Western and American influence and interests?

These are not quite the questions that scholars, left to themselves, would have formulated. They reflect the perennial need of policymakers for firm, "actionable" answers to straightforward questions, and as a consequence they may sometimes trample upon analytic nuances that are worth preserving in other contexts. It is apparent, for instance, that the choices available to a population confronted by climate stress go beyond "fight" or "flight." The most likely choice, and the one most thoroughly explored in the chapters that follow, is adaptation in place—a possibility that is effectively acknowledged in the subsequent question about ingenuity and social resourcefulness. Similarly, though it is possible that climate change may produce "winners" and "losers" within given societies, it is more likely to produce complex patterns of differential loss, thus increasing inequality while depressing economic performance and degrading social conditions for everyone.

From the outset, the governing assumption of this project has been that realistic arguments about environmental security must be conducted within a politically relevant time frame, and with reference to real states and the particular societies they govern. Its outstanding inference, in turn, is that easy generalizations about the security implications of climate change will be hard to come by. Nevertheless, there is one general point that is worth emphasizing: Given the power of the natural forces that appear to be bearing down upon humankind, it is easy to lose sight of human agency and to assume that all essential questions will eventually be decided by phenomena whose scale dwarfs all human endeavor. That may be true in the long run. It certainly will be if the worst anticipated effects of climate change come to pass. But it is not the case during the period that concerns the contributors to this book.

On the basis of the evidence presented here, it is clear that the most significant consequences of climate change during the next few decades are likely to arise from, or be substantially amplified by, human responses to natural phenomena whose immediate effects may appear to be relatively modest. Climate politics is likely to be hard politics because increasing numbers of people are already concluding that the modest appearance is deceiving. Not the least of the stresses that climate change presents to politics will arise not from any actual experience of destabilizing climate change but from increasing concern that such change is in the offing. In this sense climate politics is like all other politics; it is partly about what is real and partly, sometimes predominantly, about what is believed, expected, and feared. The success with which governments are able to allay fears and manage expectations may prove as important to their countries' survival as their capacity to master the new material conditions that climate change will create.

Notes

1. For a representative selection of recent scholarship in this vein, see the papers presented at the Workshop on Human Security and Climate Change, Oslo, June 21–23, 2005, collectively available at www.gechs .org/2005/06/24/holmen_workshop/. It goes without saying that traditional concepts of national and international security, and newer ones of human and environmental security, are not mutually exclusive. See, e.g., Jon Barnett and W. Neil Adger, "Climate Change, Human Security, and Violent Conflict," *Political Geography* 26, no. 6 (August 2007): 639–55. An interesting example of how the rhetorical strategies of Cold War security studies can be adapted to the challenges of reducing dependence on fossil fuels and cutting greenhouse gas emissions is given by Sharon Burke and Christine Parthemore, eds., *A Strategy for American Power: Energy, Climate, and National Security*, Solarium Strategy Series (Washington, DC: Center for a New American Security, 2008).

2. Climate change was first injected into the mainstream literature of security in the 1990s as a consequence of a number of challenging works hypothesizing that, in the wake of the Cold War's unwinding, casus belli should henceforth be expected mainly from competition over resources, which climate change would exacerbate. See, e.g., Thomas Homer-Dixon, *Environment, Scarcity, and Violence* (Princeton, NJ: Princeton University Press, 1999); and Michael T. Klare, *Resource Wars: The New Landscape of Global Conflict* (New

York: Metropolitan Books, 2001). More recent work has tended to back away from such neo-Malthusian scenarios, primarily because it has proven difficult to specify the nature of the causal links required to connect environmental stress and international violence. Yet it remains apparent that such linkages are possible and even likely in given circumstances. See, e.g., Idean Salehyan, "From Climate Change to Conflict? No Consensus Yet," *Journal of Peace Research* 45, no. 3 (2008): 315–26; Halvard Buhaug, Nils Petter Gleditsch, and Ole Magnus Theisen, "Implications of Climate Change for Armed Conflict," paper presented at the World Bank Workshop on the Social Dimensions of Climate Change, Washington, March 5–6, 2008, http://site resources.worldbank.org/INTRANETSOCIALDEVELOPMENT/Resources/SDCCWorkingPaper_Conflict .pdf; and the articles assembled by Ragnhild Nordås and Nils Petter Gleditsch in *Political Geography* 26, no. 6, Special Issue on Climate Change and Conflict (August 2007): 627–735. Nordås and Gleditsch begin their volume with a useful survey of the evolution of the literature on environmental security (627–38), which may be compared to an earlier one by Gleditsch—"Armed Conflict and the Environment: A Critique of the Literature," *Journal of Peace Research* 35, no. 3 (1998): 381–400—and a more recent one by Geoffrey D. Dabelko, "Planning for Climate Change: The Security Community's Precautionary Principle," *Climate Change* 96 (2009): 13–21, www.springerlink.com/content/gn2h652887023576.

3. The current top-line statement of American military doctrine is *Joint Vision 2020*, May 2000, www .fs.fed.us/fire/doctrine/genesis_and_evolution/source_materials/joint_vision_2020.pdf. It will eventually be succeed by a new statement, called *Joint Vision 2030*, which is now under development. Also see the series of long-range planning documents created by the National Intelligence Council: *Global Trends 2010* (1997); *Global Trends 2015* (2000); *Mapping the Global Future 2020* (2004), and the current statement, *Global Trends 2025: A Transformed World*, www.dni.gov/nic/PDF_2025/2025_Global_Trends_Final_Report.pdf, which appeared in 2008. And see NATO, "Multiple Futures Project: Trends and Challenges in Global Security to 2030," www.act.nato.int/multiplefutures/ACT-RUSI%20Roundtable%20Report.pdf.

4. IPCC WGII, *Climate Change 2007*. See also Nick Mabey, *Delivering Climate Security: International Security Responses to a Climate Changed World*, Royal United Services Institute Whitehall Paper 69 (London: Routledge, 2008).

5. These maps are given by Marc A. Levy, Bridget Anderson, Melanie Brickman, Chris Cromer, Brian Falk, Balazs Fekete, Pamela Green, et al., *Assessment of Climate Change Impacts on Select US Security Interests*, Center for International Earth Science Information Network Working Paper (New York: Columbia University, 2008), 48–50, www.ciesin.columbia.edu/documents/Climate_Security_CIESIN_July_2008 _v1_0.ed070208_000.pdf.

6. William Cline, *Global Warming and Agriculture: Impact Estimates by Country* (Washington, DC: Peterson Institute for International Economics, 2007).

7. An influential set of alternative future scenarios for use in this context was presented in the *IPCC Special Report on Emission Scenarios*, 2000 (www.grida.no/publications/other/ipcc_sr/?src=/climate/ipcc/ emission/), which sought to envision future conditions varying with relation to economic practices, regional and global integration, and increasing or static emphasis of environmental remediation. Within the IPCC framework, the assumptions of this project are consistent with scenario A1B, the one most frequently discussed in subsequent IPCC reports. Scenario A1B hypothesizes that economic and demographic trends will continue along current lines, and that energy consumption will remain balanced among multiple sources, rather than shifting decisively away from fossil fuels. An example of a study that seeks to consider global security across a range of climate scenarios—roughly the mirror image of the present project's structure—is *The Age of Consequences: The Foreign Policy and National Security Implications of Global Climate Change*, by Kurt M. Campbell, Jay Gulledge, J. R. McNeill, John Podesta, Peter Ogden, Leon Fuerth, R. James Woolsey, et al. (Washington, DC: Center for Strategic and International Studies, 2007), www.csis.org/media/csis/ pubs/071105_ageofconsequences.pdf.

8. See, e.g., Susmita Dasgupta, Benoit Laplante, Craig Meisner, David Wheeler, and Jianping Yan, *The Impact of Sea-Level Rise on Developing Countries: A Comparative Analysis*, World Bank Policy Research Working Paper 4136 (Washington, DC: World Bank, 2007), www-wds.worldbank.org/external/default/WDSContentServer/IW3P/IB/2007/02/09/000016406_20070209161430/Rendered/PDF/wps4136.pdf; James Hansen, "Huge Sea-Level Rises Are Coming—Unless We Act Now," *NewScientist Environment*, July 25, 2007, www .newscientist.com/article/mg19526141.600; Alun Anderson, "The Great Melt: The Coming Transformation

of the Arctic," *World Policy Journal* 26, no. 4 (December 2009): 53–65; and Orville Schell, "The Message from the Glaciers," *New York Review of Books*, May 27, 2010, 46–50. The melting of Arctic sea ice will not contribute to sea-level rise, though it is symptomatic of the same environmental trends. It has lately emerged as a security issue in its own right, because of the opportunities for resource competition and other forms of national rivalry that it may incite. See Scott G. Borgerson, "Arctic Meltdown: The Economic and Security Implications of Global Warming," *Foreign Affairs* 87, no. 2 (March–April 2008): 65–68; and Klaus Dodds, "The Arctic: From Frozen Desert to Open Polar Sea?" in *Maritime Strategy and Global Order*, edited by Daniel Moran and James A. Russell (Washington, DC: Georgetown University Press, in press).

9. A 1-meter low-elevation coastal zone is the lowest level at which effects upon population can be calculated with any precision. This is owed in part to significant discrepancies between relevant geographic and demographic data. See Levy et al., *Assessment of Climate Change Impacts*, 2–29; and appendix B in this volume.

10. On the significance of extreme weather events as an aspect of environmental security, see Joshua W. Busby, *Climate Change and National Security: An Agenda for Action*, Special Report 23 (New York: Council on Foreign Relations, 2007); Joshua W. Busby, "Who Cares about the Weather? Climate Change and U.S. National Security," *Security Studies* 17 (2008): 468–504; and Sven Harmeling, *Global Climate Risk Index 2010: Who Is Most Vulnerable? Weather-Related Loss Events since 1990 and How Copenhagen Needs to Respond*, Germanwatch Briefing Paper, December 2009, www.germanwatch.org/klima/cri2010.pdf.

11. Richard D. Knabb, Jamie R. Rhome, and Daniel P. Brown, *Tropical Cyclone Report: Hurricane Katrina, 23–30 August 2005*, updated August 10, 2006, National Hurricane Center, www.nhc.noaa.gov/pdf/TCR-AL122005_Katrina.pdf.

2

China

Joanna I. Lewis

Global climate change will increasingly alter the environment. Although its causes are generated around the world, its impacts can be highly variable at the local level. Some parts of the world may be spared dramatic environmental shifts, but many others may face major climatic changes that could create new social and economic challenges.

China's role in the climate change problem has increasingly become a topic of international attention. Because China is now the world's largest emitter of greenhouse gases measured on an annual basis, albeit with relatively low emissions per capita (table 2.1), the country can no longer ignore its contribution to this challenge. Though China's leadership has held firm against international pressures to curb its emissions, a clearer realization of the impacts that climate change may have within the country is giving rise to growing concern, and it may in fact be this concern that acts as the driving force behind any future climate change mitigation strategy the country adopts.

Given that China's domestic realities inform its international policy choices, an understanding of how climate change may affect its population and natural resources is critical to global climate stabilization efforts. The country's size, geography, regional politics, resource endowment, and role in international trade will each play a role in determining how climate change affects it, and how these impacts will in turn affect the rest of the world. This chapter explores how China will fare under our changing climate, and it examines

China

Table 2.1 China's Greenhouse Gas Footprint

Source	Annual Share of Global Emissions (%)	Annual Per Capital Emissions (tons)
China	24	5.1
United States	21	19.4
EU-15	12	8.6
India	8	1.8
Russian Federation	6	11.8
Rest of the World	29	—

Note: EU-15 = the European Union at the time it had fifteen member states. The data are for 2007. The data include carbon dioxide emissions from fossil fuels only.

Source: Netherlands Environmental Assessment Agency and British Petroleum, 2008, www.mnp.nl/en/service/pressreleases/2008/20080613ChinacontributingtwothirdstoincreaseinCO2emissions.html.

how the impacts of climate change facing the country, and its response, may drive security challenges domestically, within the greater Asian region, and around the world.

Climate Change Impacts and China

As the scientific understanding of anthropogenic climate change continues to improve, so does the understanding of the impacts that climate change will have on specific regions. China, like many parts of the world, is expected to experience significant climate-related impacts.

The State of the Science

We now know that human activity is altering the Earth's climate. Driven primarily by a century and a half of fossil fuel combustion, carbon dioxide concentrations in the atmosphere had reached 379 parts per million by 2005, 35 percent higher than preindustrial levels.[1] Average global temperatures have risen by 0.76 degree Celsius since the late 1800s, and the impacts are evident in extreme weather events, changed weather patterns, floods, droughts, glacial and Arctic ice melt, rising sea levels, and reduced biodiversity.[2] Average temperatures are projected to increase by another 3 degrees Celsius upon a doubling of carbon dioxide concentrations.[3] In China, the observed data show that the nationwide mean surface temperature has risen between 0.5 and 0.8 degree Celsius during the past hundred years, and it will likely increase further by 3 to 4 degrees Celsius by the end of this century.[4] Even if all emissions were to stop today, the greenhouse gases already accumulated in the atmosphere will remain there for decades to come, which will result in more warming and stronger climate impacts.

Climate Impacts in China

A synthesis report compiled by China's leading climate change scientists stated: "It is very likely that future climate change would cause significant adverse impacts on the ecosystems, agriculture, water resources, and coastal zones in China."[5] The impacts already being observed in China include extended drought in the north, extreme weather events and flooding in the south, glacial melting in the Himalayas, declining crop yields, and rising seas along heavily populated coastlines.[6] Such reports are increasingly acknowledged by China's leaders. For example, China's former special ambassador for climate change, Yu Qingtai, recently stated that "climate change . . . is in fact a comprehensive question with scientific, environmental, and development implications and involves the security of agriculture and food, water resource, energy, ecology, and public health and economic competitiveness," and "if the climate changes dramatically, the survival of mankind and the future of Earth might be impacted."[7] Such statements, however, have not yet influenced China's international negotiating positions on climate change, or its domestic policies, in any tangible way.

This lack of response is worrisome, because the science is ever more clearly showing that climate change is expected to wreak havoc on China through the following channels:

- *Decreased precipitation.* Studies predict that precipitation may decline by as much as 30 percent in the Huai, Liao, and Hai river regions in the second half of this century.[8] Climate change could decrease river runoff in northern China, where water scarcity is already a problem, and increase it in southern China, where flooding and heavy rains are already a problem. Declining runoff to the six largest rivers in China has been observed since the 1950s, with some rivers even facing intermittent flow.[9] Decreased precipitation is also tied to increased desertification. In China it is estimated that about 150 square kilometers of cultivated land is lost annually as a result of desertification.[10]
- *Severe weather events.* Climate change is expected to increase the frequency and severity of storm surges, droughts, and other extreme climate events, particularly in coastal areas. Currently, the Yellow River Delta, the Yangtze River Delta, and the Pearl River Delta are the most vulnerable coastal regions in China. Climate change could also increase the frequency and intensity of heat waves, resulting in higher mortality and morbidity from heat-related weather events. Hotter temperatures over time could increase the occurrence and the transmission of infectious diseases, including malaria and dengue fever. Malaria is still one of the most significant vector-borne diseases in parts of southern China, and climate change could expand its geographical range into the temperate and arid parts of the country.[11]
- *Sea-level rise.* Since the 1950s, sea-level rise has been observed along China's coastline at a rate of 0.0014 to 0.0032 meter per year.[12] Estimates of future sea-level rise along China's coastline range between 0.01 and 0.16 meter by 2030, and between 0.4 and 1 meter by 2050.[13] Higher sea levels magnify the possibility of flooding and intensified storm surges, and exacerbate coastal erosion and saltwater intrusion. A 1-meter rise in sea level would submerge an area the size of Portugal along China's eastern seaboard;[14] the majority of Shanghai—China's largest city—is less than 2 meters above sea level.[15] China's twelve coastal provinces contain 42 percent of its population and contribute 73 percent of its gross domestic product (GDP).[16]
- *Glacial melt.* The effect of climate change on the glaciers on China's Tibetan Plateau will have severe repercussions for the country's lakes and river systems. The total area of its western glaciers is projected to decrease 27.2 percent by 2050. During the same period, glacier thawing will increase water discharge by 20 to 30 percent per year until water levels peak between 2030 and 2050. This increased water discharge would then decline as the glaciers disappear.[17] The mountain and highland lakes that rely on inland glaciers for recharge, such as the lakes on the Tibetan Plateau and Pamir Plateau, could initially enlarge as a result of glacier melting but will eventually shrink as the glaciers are reduced over time. The Yellow and Yangtze rivers, which support the richest agricultural regions of the country and derive much of their water from the Tibetan glaciers, will initially experience floods as the glaciers melt, and then drought, once the glacial runoff is gone.[18]

As a result of all these trends, China's agricultural system, trade system, economic development, and human livelihoods will all face new threats in a warming world.

Environment, Resources, and Security

Post–Cold War security scholars have pushed for the escalating threats to the global environment to be recognized in the concept of national security.[19] Empirical studies have supported a neo-Malthusian model of scarcity-driven conflict, where population pressures and high levels of resource consumption lead to competition and eventually to violent confrontation.[20] In a survey of conflicts, Thomas Homer-Dixon found that whereas environmental scarcity is unlikely to be the only factor to cause the outbreak of armed conflict, more often than not it will be a contributing factor. He further predicted that environmental pressures in China, including its unequal distribution of resources and its large population, may eventually cause the country's fragmentation.[21]

Lester Brown argued in his book *Who Will Feed China?* that China's demand for resources would have not just local but also dire global implications, stating that "China's rising food prices will become the world's rising food prices. China's land scarcity will become everyone's land scarcity. And water scarcity in China will affect the entire world."[22] Though he did not go as far as to suggest that this scarcity would incite armed conflict, he did suggest that China's demand for resources could lead to a global redefinition of national security away from military preparedness and toward maintaining adequate food supplies.

A study published in the *China Science Bulletin* examined the connection between climate changes and historic societal developments in China by interrelating paleoclimatic records with historical data of wars, social unrest, and dynastic transitions.[23] Looking at the period from the late Tang Dynasty to the Qing Dynasty in ancient China, the authors found that climate change was one of the most important factors in determining the dynastic cycle and alternation of war and peace. Furthermore, they found that 70 to 80 percent of peak war activity, and most dynastic transitions, took place during "cold phases," a phenomenon they attributed to diminishing thermal energy input and the resulting decline in the productivity of land, which strained resources and human livelihood and, as a result, drove conflict.

Other studies have found the link between resource conflict and international military conflict tenuous at best, because countries will try many other solutions before resorting to military action. For example, Daniel Deudney has written that resource wars between countries are unlikely, especially for developed countries, and that although free-rider problems may generate severe conflict, it is doubtful that states will find military instruments to be the most useful tools for coercion and compliance.[24] He has further stated that "if, for example, China decided not to join a global climate agreement, it seems unlikely that the other major countries would really go to war with China over this. In general, any state sufficiently industrialized to be a major contributor to the carbon dioxide problem will also present a very poor target for military coercion."[25]

The issue of whether to cast climate change as an issue of national security is itself somewhat controversial. For example, though many in the U.S. environmental policy community believe that the "securitization" of the climate change issue will raise its level of importance on the national policy agenda, others fear that a security frame will lead to militarization at the expense of cooperation.[26] Few would dispute, however, that it is appropriate to view climate change as a "threat multiplier" that is likely to exacerbate many existing environmental and resource challenges, including those in some of the most volatile regions of the world.[27] For example, Robert Kaplan has argued that resource depletion, along with urbanization and population increase, are undermining already-fragile governments in the developing world.[28]

Although an international war with China over climate change is unlikely, domestic instability within China is probable if current trajectories continue. If not adequately managed, problems within China would probably reverberate across international borders and markets. These challenges are discussed further below.

Domestic Security Risks

The predicted changes to China's climate will have social and economic impacts across the country. Particularly at risk are its agricultural system, its ability to maintain strong economic development and foreign trade, and the livelihoods of its population.

Agriculture System Vulnerabilities

A commonly cited statistic is that China feeds more than 20 percent of the world's population with 7 percent of the world's arable land.[29] Climate change is expected to decrease the stability of agricultural production, causing larger variations in crop productivity. Scientists predict a 5 to 10 percent decline in overall crop productivity in China by 2030 as a result of climate change, and a decline of up to 37 percent in rice, maize, and wheat yields after 2050.[30] This marks a serious challenge for the country's long-term food security. If this decline in supply were to result in global scarcity and elevated food prices, it could have particularly severe impacts in Africa, where food insecurity is a grave threat.[31]

Although higher carbon dioxide levels in the atmosphere may increase crop growth and yield, mainly through their effect on photosynthetic processes, higher temperatures generally decrease yield by speeding up a plant's development so that it matures sooner, while also increasing its demand for water. Higher temperatures also affect the prevalence of pests, diseases, and weeds. The impacts will differ across regions within China, being most severe in regions where water is already scarce, such as the northwest. Some model simulations indicate that water shortages in this region could reach about 20 billion cubic meters per year between 2010 and 2030.[32]

Encroaching desertification also threatens China's agricultural system. Desertification in China has been primarily caused by climate change, and particularly by strong wind regimes (with high sand transport potential), but it is exacerbated by development and urbanization. The area affected by desertification is approximately 2.6 million square kilometers, or 27 percent of the total land territory. At present, more than 60 percent of the country's arid and semiarid areas (usually in regions above 35 degrees north latitude and having an annual rainfall of less than 450 millimeters) is being managed using traditional pastoral and agricultural systems that could be seriously endangered, which would jeopardize the existence of nearly 200 million people.[33] In addition, as more and more of China's land is lost to urbanization, agricultural land is at a premium, and losses in yield due to climate change could have even more of an impact.

Economic Development and Trade Impacts

A fundamental target of the current development plan articulated by China's leadership is an annual economic growth rate of 7.5 percent.[34] Many have argued that continued rapid economic growth in China is critical to the Communist Party's legitimacy. As noted above,

Table 2.2 Projected Sea-Level Rise in China's Eastern Coastal River Deltas

Region	Nearby Economic Centers	Sea Level Rise by 2050 (meters)
Yellow River Delta (Huanghe)	Tianjin, Beijing	0.7–0.9
Yangtze River Delta (Changjiang)	Shanghai	0.5–0.7
Pearl River Delta (Zhujiang)	Guangzhou, Shenzhen, Hong Kong, Zhuhai	0.4–0.6

Sources: Fan Daidu, Li Congxian, "Complexities of China's Coast in Response to Climate Change," *Advances in Climate Change Research* 2 (Suppl. 1) (2006); C. X. Li, D. D. Fan, B. Deng, and D. J. Wang, "Some Problems of Vulnerability Assessment in the Coastal Zone of China," in *Global Change and Asian Pacific Coasts*, Proceedings of APN/SURVAS/LOICZ Joint Conference on Coastal Impacts of Climate Change and Adaptation in the Asia-Pacific Region, ed. N. Mimura and H. Yokoki (Kobe, 2001), 49–56; and Lu Xianfu, "Climate Change Impacts, Vulnerability and Adaptation in Asia and the Pacific," GEF Training Workshop on Environmental Finance and GEF Portfolio Management, May 22–25, 2007, http://ncsp.va-network.org/UserFiles/File/PDFs/Resource%20Center/Asia/BackgroundPaper.pdf.

China's eastern coastal provinces are responsible for the lion's share of that growth, about three-quarters of GDP. As a result, the fact that China's core economic development infrastructure is directly vulnerable to the impacts of climate change poses a great risk to its leadership.

China contains roughly 144,000 square kilometers of low-lying coastal lands with an elevation no greater than 5 meters, mainly distributed across the deltas of the Yellow, Yangtze, and Pearl rivers, and including the major industrial centers of Tianjin, Shanghai, and Shenzhen/Guangzhou.[35] This is a land area roughly equivalent to the entire country of Bangladesh. China's low-lying coastal area is also its most densely populated, and its population is growing twice as quickly as the national average; furthermore, the urban sections are growing at three times the national rate.[36] China is not only attracting more people to the coast at the present time; it is also establishing an urban migration structure, which means that the population will continue to move to the coast far into the future. The projected sea-level rises in the three delta regions on China's eastern coast, which are also core economic centers, are listed in table 2.2.

Extreme climate events are the main causes of natural disasters in low-lying coastal regions. In China, both the frequency and intensity of tropical storm surges have increased since the 1960s.[37] Studies predict ever more frequent exceptional floods in eastern China. The South China region, which includes the low-lying Pearl River Delta, is susceptible to sea-level rise. Sea-level rise threatens the continued rapid economic growth of these low-elevation coastal regions, which not only surround the economic centers of Shanghai, Tianjin, and Guangzhou but are also home to key ports through which imports and exports pass, connecting China to the rest of the world. About 14 percent of China's freight goes through Shanghai and 8 percent through Tianjin, whereas 29 percent of its foreign trade income comes from Guangzhou.[38]

Extreme weather events have already taken a toll on China's eastern coastal region, which perhaps indicates what is to come. In 2006 such disasters cost China more than $25 billion in damage. Much of this damage is related to the infrastructure in these densely populated areas. Other economic impacts related to climate change might stem from decreased access to hydropower resources, which currently generate 15 percent of China's electricity.[39] The annual direct economic loss caused by desertification is approximately $6.5 billion.[40] In addition to the impacts on agriculture mentioned above, declining yields of rice and other grains could have implications for revenue from trade.[41] China is the world's largest producer of rice and second largest producer of wheat and coarse grains, and it is among the principal sources of tobacco, soybeans, peanuts, and cotton.[42] Exports of rice and maize amounted to $417 million and $420 million, respectively, in 2006.[43] The total value of exported agricultural products in 2004 was about $17 billion, and today it is significantly higher.[44]

Threats to Human Livelihood

Climate change increases the possibility of disease incidence and transmission, greatly influencing the distribution and the potential danger of vector-borne infectious diseases. The risk of climate-related disease will more than double globally by 2030. Although there is limited evidence of the impact that future climate change will have on specific diseases, climate cycles have been shown to cause inter-annual variations in disease incidence.[45]

A greater incidence of ill health is likely to result from the spread of vectors of tropical diseases such as malaria, dengue, Chagas's disease, and perhaps others such as the Ebola virus, into previously temperate regions of the world.[46] The IPCC has concluded that the changes in environmental temperature and precipitation could expand the geographical range of malaria in the temperate and arid parts of Asia. A Chinese study suggests that the lethal H5N1 virus (avian flu) will spread as climate change shifts the habitats and migratory patterns of birds.[47] Disease can also be spread by stagnant water following flooding disasters, which are expected to become more frequent as a result of climate change. Studies have shown that after a flood, the incidents of infectious diarrhea diseases such as cholera, dysentery, typhoid and paratyphoid often increase.[48] Heat itself is an issue, especially among the elderly and very young, and in densely populated megacities. In addition to heat stroke, the incidence of cardiovascular and respiratory disease is affected by temperature.

Another threat to human livelihood that could be exacerbated by climate change relates to large numbers of refugees. Climate change could trigger civil unrest as a result of human migrations to avoid climate impacts. In the short term, populations from regions where water is scarce may move toward northwest and southwest China, where rainfall and glacier runoff may result in increased water availability. The movement of Han Chinese into Muslim Uighur and ethnic Tibetan areas may aggravate existing ethnic tensions in these regions, many of which are caused by ongoing resource extraction from the "west" for the "east."[49] As glaciers melt and inhabitants of the Tibetan Plateau region must deal with more water scarcity, they may migrate toward central or eastern China. A similar phenomenon may occur as Inner Mongolia faces more desertification as a result of climate change. A population migration from Tibet and Inner Mongolia into predominantly ethnic Chinese regions could further ethnic tensions, and even threaten regional stability.

Agriculture, economic development, and human livelihood are all interrelated, and while the exact nature of the interactions may be hard to predict, feedback between these systems is highly likely. Changing weather and precipitation patterns will affect agricultural productivity, which at a macro level will affect economic output and trade, and at a micro level will have an impact on farmers' livelihoods. In addition, the rising sea level will affect economic development on China's eastern coastline—the key region for overall economic productivity—which in turn will affect global trade. Populations will need to migrate inland to flee the rising sea level and storms, which will change agricultural patterns and potentially cause regional conflicts. Rising temperatures increase the prevalence of certain diseases that affect both humans and ecosystems and agricultural productivity.

Increased desertification in northern China will exacerbate water scarcity, reducing the viability of hydropower for energy, and also creating human migration from areas of minority ethnic population such as Inner Mongolia, Tibet, and Xinjiang, which will increase the risk of ethnic conflict in areas that currently lack such minorities.

It is evident that climate impacts on China will vary markedly across regions. The four maps that make up figure 2.1 illustrate different aspects of China's vulnerability to climate change. Map 1 shows where China's population is located, as well as the percentage of minority population in each province. Map 2 illustrates each province's contribution to GDP, as well as the provinces that depend heavily on agriculture (defined as provinces where more than 25 percent of GDP comes from farming). Map 3 illustrates per capita water resource availability by province, as well as water consumption per capita, highlighting those provinces with a water deficit (i.e., where use exceeds available resources within the province). Map 4 portrays the provincial distribution of China's coal and hydropower resources.

Looking at the four maps together, several distributional trends become clear. First, eastern China is the most densely populated region, and it currently contains a very small percentage of minorities (map 1). This makes it vulnerable to immigration-related tensions, including overpopulation and ethnic conflict. Eastern China is also the economic center of China (map 2). The high-value economic activity, combined with population density, makes the coastal economic centers particularly vulnerable to the costly impacts of sea-level rise and extreme weather events. The central eastern and southeastern provinces are heavily reliant on agriculture, which results in high water consumption, despite relatively low water resource availability (map 1). Western China, particularly Tibet and Qinghai, are very rich in water resources (map 3), but the rivers, lakes, and glaciers in these provinces are under threat in a warming world. Extreme water scarcity is already a problem in two major population centers, the greater Beijing and Shanghai regions, as well as in the northern provinces (map 3). Eastern China also contains minimal energy resources (map 4), with both hydropower and coal being scarce. This makes these provinces heavily dependent on imported resources and power from other provinces. Hydropower resources are richest in the western provinces of Tibet, Sichuan, and Yunnan, though the impact of climate change on China's rivers that flow from the Tibetan Plateau region will directly threaten the continued availability of water for electricity.

Regional and Global Security Risks

Expanding the scope of assessment beyond China's borders highlights myriad security implications stemming both from the impacts the country faces and from its likely response to a warming world.

Impacts on the Tibetan Plateau

One of the most severe threats concerns the Tibetan Plateau region, which is potentially a hot spot with respect to both climate and conflict. The Tibetan Plateau, which has been called both "the reservoir at the top of the world" and "the water tower of Asia," is the source of most of Asia's major rivers. As global temperature rises, Tibet's glaciers are melting and grassland permafrost is thawing at an alarming rate. The region's warming climate is causing glaciers to recede at a rate faster than anywhere else in the world, and in some regions of Tibet by 3 feet per year.[50] These changes will have an impact on the millions living downstream in China, India, Pakistan, Afghanistan, Nepal, Vietnam, and other countries that are dependent for their water supply on the waters of such rivers as the

Yangtze, Yellow, Mekong, Salween, Irrawaddy, Indus, Sutlej, and Brahmaputra. It has been estimated that almost half the world's population lives in the watersheds of rivers whose sources lie on the Tibetan Plateau, and at least 500 million people in Asia and 250 million people in China are at risk from declining glacial flows on that plateau.[51]

Given these statistics, and Tibet's historic capacity to store more freshwater than any other place on Earth except the North and South poles, it is not surprising that Tibet is becoming an ever more crucial strategic territory. The impacts related to climate change are exacerbated by a host of serious environmental challenges to the quantity and quality of Tibet's freshwater reserves caused by industrial activities. These include large-scale erosion and siltation resulting from increased deforestation, along with air and water pollution stemming from mining, manufacturing, and other human activities.

Under a scenario in which the melting of the Tibetan glaciers reducers river flow across East and Southeast Asia, there could be significant conflicts over scarce water resources. Water scarcity could cause China to divert these rivers from flowing into surrounding countries, which could cause tensions with neighbors such as Vietnam, Myanmar, Laos, Pakistan, Nepal, and India. Water scarcity in the region would raise concerns about water available not only for consumption and agriculture but also for electricity generation, because hydropower is heavily relied upon among many countries in the region, including China.

If China wins this resource battle, it could become a destination for migrants or refugees from the countries that lose out. It could also become a destination for climate refugees from neighboring Asian countries, such as Central Asian states severely affected by drought, North Korea, Pakistan, and Afghanistan.[52] Migration into China by other ethnic groups could cause social unrest and ethnic conflict outside its borders.

All in all, the Tibetan Plateau represents a concentration of potential political tensions and climate change impacts that could combine to create civil unrest and threaten political stability.[53]

Regional Economic and Trade Impacts

In a warming world, changes within China will generate many regional economic and trade-related impacts. As outlined above, climate change is projected to have a considerable impact on China's current agricultural system, and China is a major player in international agricultural commodity markets. The lower yields projected under a warming climate in parts of China could mean reduced exports of key crops such as rice, maize, and wheat from China to other countries. Because trade in agricultural products is a valuable part of China's economy, currently accounting for about 15 percent of its GDP, this would pose a risk to economic development.[54]

China's future energy consumption will also be influenced by the domestic impacts of climate change, and because of the size of China's energy markets, its consumption patterns will have international economic repercussions. Increased water scarcity may lead to reduced hydropower resources, which in turn may mean heavier reliance on fossil fuels. Hydropower scarcity, combined with a scenario under which China takes no major policy actions to reduce carbon emissions by moving away from fossil fuels, may mean increased Chinese reliance on imported energy resources, which would create continued pressure on world oil markets. If China were able to implement a massive shift to other forms of domestic renewable energy and nuclear power, or to achieve widespread implementation of carbon capture technologies, this pressure on oil markets could be averted—but such a path-changing scenario is unlikely in the near future.

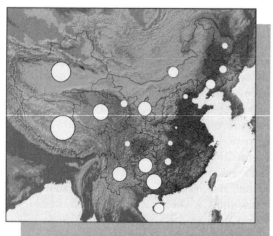

Map 1. Population Indicators

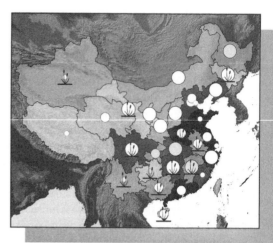

Map 2. Economic Indicators

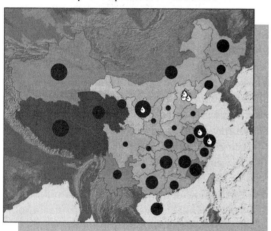

Map 3. Water Resource and Usage

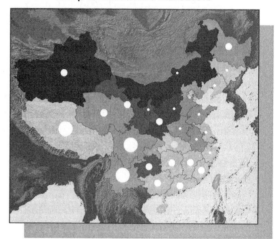

Map 4. Energy Resources

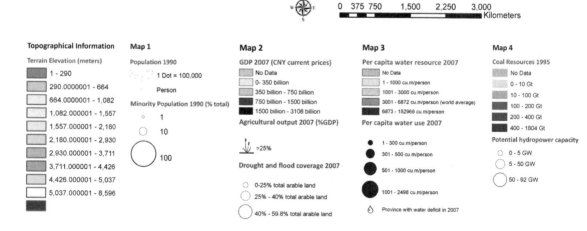

0   375 750   1,500   2,250   3,000
                                                    Kilometers

Topographical Information	Map 1	Map 2	Map 3	Map 4
Terrain Elevation (meters)	**Population 1990**	**GDP 2007 (CNY current prices)**	**Per capita water resource 2007**	**Coal Resources 1995**
1 - 290	1 Dot = 100,000	No Data	No Data	No Data
290.0000001 - 664	Person	0- 350 billion	1 - 1000 cu.m/person	0 - 10 Gt
664.0000001 - 1,082		350 billion - 750 billion	1001 - 3000 cu.m/person	10 - 100 Gt
1,082.000001 - 1,557	**Minority Population 1990 (% total)**	750 billion - 1500 billion	3001 - 6872 cu.m/person (world average)	100 - 200 Gt
1,557.000001 - 2,180	1	1500 billion - 3108 billion	6873 - 152969 cu.m/person	200 - 400 Gt
2,180.000001 - 2,930	10	**Agricultural output 2007 (%GDP)**	**Per capita water use 2007**	400 - 1804 Gt
2,930.000001 - 3,711	100	>25%	1 - 300 cu.m/person	**Potential hydropower capacity**
3,711.000001 - 4,426			301 - 500 cu.m/person	0 - 5 GW
4,426.000001 - 5,037		**Drought and flood coverage 2007**	501 - 1000 cu.m/person	5 - 50 GW
5,037.000001 - 8,596		0-25% total arable land	1001 - 2498 cu.m/person	50 - 92 GW
		25% - 40% total arable land		
		40% - 59.8% total arable land	Province with water deficit in 2007	

The potential impacts of climate change on crucial economic infrastructure, particularly in China's eastern coastal provinces, also pose a threat to the country's trade partners. The three major eastern coastal river deltas, which serve as crucial economic and trade centers, cause the highest concern because of their vulnerability to sea-level rise, storm surges, flooding, and typhoons. As the current global financial crisis has illustrated, China's economic well-being is inherently tied to the economies of the United States, the European Union, and other major economic markets. As a result, the economic impact of climate change on China has potential implications for all major economies connected to that of China. For example, the European Union and the United States are the top two trading partners of Shanghai, with 2008 trade volumes amounting to $71.9 billion for the EU and $53.4 billion for the United States.[55]

International Responses to China's Inaction

Another risk confronting China in conditions of global warming is the high probability of international retaliation if it fails to engage proactively with the international community to forge a solution to this problem in whose creation it has played a significant role. Should the developed world move ahead with climate change mitigation actions and China fail to follow suit, it is highly likely that international pressure on the Chinese will mount.

It is uncertain how China will respond politically either to the climate change threat or to its role in the challenge. Although dramatic, path-changing action is unlikely in the near future, it is possible that gradual changes will be instituted that begin to address domestic greenhouse gas emissions. These changes are most likely to occur, however, if they simultaneously address other pressing issues. For example, the Chinese government is already

---

Figure 2.1 Climate Change Factors in China

Sources: Map 1: Topographical and population data from China Historical GIS, Harvard University, www.fas.harvard.edu/~chgis/data/chgis/downloads/v4. Minority population data for regions with minority populations greater than 1 million. From *Geography of Chinese Nationalities*, ed. Li Zhihua (Shanghai: Shanghai Educational Publishing House, 1997).

Map 2: *China Statistical Yearbook 2008*. Agricultural output data are from 2007. Agriculture, animal husbandry, and fishery data were calculated as a percentage of total gross domestic product by dividing agricultural outputs by the total gross domestic product for each region.

Map 3: *China Statistical Yearbook 2008*. Drought and flood coverage refers to the land covered by flood and drought, as a percentage of the total cultivated land, at the end of 2007. The area covered by flood and drought is defined as those areas that have lost more than 10 percent of cultivation land (compared with previous years) due to droughts or floods. If several disasters occurred on the same land, only the event in which the largest area of land was lost is counted.

Map 4: Per capita water resource and regions with water deficit are from *China Statistical Yearbook 2008*. Note that the per capita water resource world average, according to the World Bank, is 6,872 cubic meters per person. The water deficit is calculated by subtracting the water use per capita from the water resource per capita. The potential hydropower capacity by region (1980) and coal resources (1995) are from Lawrence Berkeley National Laboratory, *China Energy Databook*, volume 7.0 (October 2008). The potential hydropower capacity for each region is estimated on the basis of the sites that can accommodate power plants ≥500 kW of installed capacity. The figure for Hebei Province includes Beijing and Tianjin. The figure for Jiangsu includes Shanghai.

Maps drawn by Tian Tian, Georgetown University.

under domestic pressure to address severe local air and water pollution, and reducing these pollutants by using less coal could also be a climate change mitigation strategy. Likewise, current policies to promote energy efficiency and renewable energy are in line with domestic priorities and are also crucial to reducing greenhouse gas emissions. Several proposals are being discussed in the context of the post-2012 United Nations climate negotiations that would require China's commitment to the international community to be expressed in the form of policy actions, or sector-specific actions, rather than national mandatory emissions limits.[56] Other studies have attempted to quantify the potential for mitigating greenhouse gas emissions from China through such ongoing policy activities.[57]

It is possible, however, that these win–win actions, or mitigation actions with other associated benefits, may not be sufficient in the eyes of the international community as we approach 2030. Although Europe is thus far the only global region to have implemented mandatory national measures to reduce greenhouse gas emissions, several countries are expected to soon follow suit, including Australia, New Zealand, and Japan. Many experts believe that under the Barack Obama administration, the United States is likely to adopt mandatory climate policies sometime during the next few years; at the global climate change conference in Copenhagen in 2009, the United States pledged to reduce emissions 17 percent below 2005 levels by 2020, "in conformity with anticipated U.S. energy and climate legislation."[58] In addition, several U.S. states have already adopted mandatory greenhouse gas reductions.[59] If both the United States and the EU have adopted mandatory action to combat climate change, it is primarily from these quarters that the international pressure on China is likely to come.

Historically, the developing countries have for the most part remained unified in their approach to the international climate change negotiations, representing their positions in the context of the Group of 77 (G-77).[60] Additional pressure could mount if other Non–Annex I countries (i.e., the developing countries that are currently exempted from binding mitigation commitments under the UN Framework Convention on Climate Change, or UNFCCC) opt to take on mitigation commitments in the current round of negotiations for the post-2012 period. For example, Brazil and Mexico have already signaled their willingness to pledge national actions within an international framework.[61] This puts pressure on China, and it creates an environment for a unified, multilateral approach against China, led by the United States and the EU.

Despite the fact that the United States does not yet have mandatory greenhouse gas emissions limits in place at the federal level, several draft proposals for legislation in Congress include trade measures aimed at large developing countries, primarily China.[62] In the 110th Congress (2007) these included S 1766, the "Low Carbon Economy Act" introduced by Senators Jeff Bingaman and Arlen Specter, and S 2191, "America's Climate Security Act," introduced by Senators Joseph Lieberman and Mark Warner. With a stated purpose of protecting the United States against foreign countries' undermining a U.S. objective of reducing greenhouse gas emissions, both bills stipulate that U.S. importers of certain goods from certain countries must buy international reserve allowances to offset the lower energy costs of manufacturing those goods. The requirement for a purchase of international reserve allowances amounts to a carbon allocation associated with the amount of carbon embedded in the imported product on a per unit basis. These "border adjustments" specifically target greenhouse gas–intensive products, including iron, steel, aluminum, cement, bulk glass, and paper. Although some least-developed countries are excluded from this requirement, it applies to developing countries representing greater than 0.5 percent of global greenhouse gas emissions, unless they have taken policy action at home deemed comparable to

U.S. action.[63] Although the impact of such measures on leveling the carbon playing field between the United States and China has been questioned, it is widely believed that U.S. legislation will contain some form of "China provision" aimed at appeasing labor interests, which have helped shape the provision described here and widely support it.[64]

It is possible that trade measures beyond those currently being proposed in draft legislation could escalate, if the imbalance between countries that "act" and countries that do not increases. A unified approach to climate change that includes the Group of Eight and possibly other developing countries but does not include China could result in a multilateral effort to coerce China into taking action. Under such circumstances it is easy to imagine the development of a mutually reinforcing cycle of intensifying economic or political sanctions, whose impacts on broader patterns of international relations would be difficult to anticipate. It seems certain that perceptions of relative justice and burden sharing will become increasingly central to climate politics, to the point where they may become an independent source of pressure for action by governments.

China and other developing countries have already voiced some concern about trade measures in proposed U.S. legislation.[65] The EU has not yet proposed similar measures, though many EU leaders are generally supportive of the idea as a means of encouraging action on climate change. At the March 2008 summit, a statement by EU leaders warned that "if international negotiations fail, appropriate measures can be taken," with concerns particularly focused on protecting European industry. The French president, Nicolas Sarkozy, went so far as to say that "our main concern is to set up a mechanism that would allow us to strike against the imports of countries that don't play by the rules of the game on environmental protection."[66]

China's Response to Climate Change

Many of the climate changes that are beginning to affect China, and will become more severe in the coming decades, could initiate disruptive change, both internally and externally. Although there are already signs of modest government responses to these threats, measures are likely to become bolder as impacts worsen.

The Chinese leadership will act to address the impacts of climate change if they are deemed to represent a serious threat to national stability. In examining the likely implications of climate change for China, three threats rise above the others: (1) severe impacts on minority regions, in particular Tibet, leading to migrations and potential regional tensions; (2) severe impacts on the core economic development zones in China's eastern coastal deltas; and (3) international pressure upon China as the developed world moves ahead with climate change mitigation actions and China does not follow suit.

Although all these risks are significant, the Chinese Communist Party has shown extreme resilience in facing both internal and external threats. Thus state failure is not likely, but when the threats from climate change are compounded with others, in particular those that threaten economic and therefore social well-being, it is possible.

China possesses the ingenuity and institutional capital to meet certain elements of the climate change challenge better than others. It has the technical, engineering, and innovation capacity that many other developing countries lack. China serves to gain from developing many of the technologies that will be crucial to dealing with climate change, from renewable energy to fossil fuel capture and sequestration technologies, which would allow the country to continue to rely on fossil fuels while mitigating some of the most severe impacts of climate change.

China is already involved in several efforts to develop and demonstrate low- or zero-emission coal technologies. These include GreenGen, a 400-megawatt-scale integrated-gasification, combined-cycle plant to which carbon capture and storage will be added by 2020;[67] and the Near Zero Emission Coal partnership between China, the EU, and the United Kingdom, whose goal is to have a coal plant with capture and storage online by 2020.[68] In addition, China's Huaneng Power Company, in cooperation with the Australian Commonwealth Scientific and Industrial Research Organization, has launched a postcombustion carbon capture project (without storage).[69]

China is also poised to become the world leader in renewable energy technology development. The country's wind power capacity has been growing at more than 100 percent per year, expanding from 1,266 megawatts in 2005 to more than 25,000 megawatts in 2009. An increasing number of these wind turbines are being manufactured domestically by Chinese companies. China also has moved aggressively into solar energy and has become the world's largest manufacturer of photovoltaic cells, currently accounting for about half the global market. It also both manufactures and utilizes more solar water heaters than the rest of the world combined. Its entry into the renewable energy technology marketplace is beginning to drive down costs due to the increased production scale, access to lower wages, and increased technological learning.

It is at least possible, then, that China may develop sufficient expertise in and commitment to next-generation low-carbon energy technologies to be able to benefit from climate change mitigation rather than merely tolerating it. Such a scenario would meet the precondition of China's political leaders, who have stated that climate change "can only be solved through development."[70] This is one of the few scenarios under which China is a winner in the climate change challenge, and under which the rest of the world stands to benefit as well.

Conclusions

This chapter has examined the impacts of climate change facing China, and the implications of these impacts and of China's response to them, for both China and the world. It finds that the impacts on China will be significant, and they may have sizable adverse economic implications, particularly on the nation's vulnerable eastern coastal economic centers. Water scarcity, a problem that is already challenging the country's leadership, will be exacerbated under projected climate impacts. China is likely to maintain its reliance on coal at least up to 2030, the end of the period on which this study focuses. This will increase the environmental and health impacts of coal use locally, and it will perpetuate a major source of carbon dioxide emissions globally. Although the global impact is likely to attract more attention around the world, the local impacts are particularly important, because domestic concerns are likely to be the primary motivation behind any actions China may take to curb greenhouse gas emissions.

Because of China's great size, whatever impacts it suffers and whatever actions it takes as a consequence of climate change are certain to have at least regional implications. A major regional threat that may be aggravated by climate change is that of political instability in Tibet. The crucial role of the Tibetan Plateau as the major watershed of Asia can only heighten the interest of outsiders in developments there, while perhaps emboldening both the Tibetan and Chinese leaderships in unpredictable ways. Other important climate security hot spots include the low-lying eastern coastal economic centers around the Yangtze, Pearl, and Yellow river deltas, which are particularly vulnerable to sea-level rise, storm surges, and typhoons. Finally, China faces the risk of international retaliation if it fails to undertake serious greenhouse gas mitigation actions. This is likely to take the form of trade sanctions or other constraints on China's exports imposed by importing countries. For a country whose economic prospects will remain closely tied to its overseas exports for the foreseeable future, this is a serious risk.

It is fair to say that if China were to stake its future exclusively on a policy of high growth fueled by high greenhouse gas emissions, showing only defiance or indifference in the face of deteriorating climatic conditions both locally and worldwide, its national stability would eventually come under threat. Yet China is not without options. It is already poised to become a leader in the low-carbon technology revolution. It has also come to perceive itself, and wishes to be perceived by others, as one of the world's leading nations, and not simply its most populous. Although it is unrealistic to expect China to sacrifice its drive for economic growth in order mitigate the impact of climate change, it is possible that the two goals could converge sufficiently to avert any overt clash with either its regional neighbors or its trading partners in the West.

Notes

The author would like to acknowledge the invaluable research assistance of Lynn Kirshbaum, and the GIS assistance of Tian Tian, both of Georgetown University's School of Foreign Service. An earlier version of this chapter appeared as an article, "Climate Change and Security: Examining China's Challenges in a Warming World," *International Affairs* 85, no. 6 (October 2009): 1195–1213.

1. "Summary for Policymakers," in IPCC, *Climate Change 2007: Science,* 1–17.

2. Ibid.

3. Ibid.

4. Lin Erda, Xu Yinlong, Wu Shaohong, Ju Hui, and Ma Shiming, "Synopsis of China National Climate Change Assessment Report (II): Climate Change Impacts and Adaptation," *Advances in Climate Change Research* 3 (Suppl.) (2007): 6–11; Lin Erda, Xiong Wei, Ju Hui, Xu Yinlong, Li Yue, Bai Liping, and Xie Liyong, "Climate Change Impacts on Crop Yield and Quality with CO_2 Fertilization in China," *Philosophical Transactions: Biological Sciences* 360, no. 1463 (2005): 2149–54.

5. Lin et al., "Synopsis," 1.

6. "Summary for Policymakers."

7. Yu Qingtai, "Special Representative for Climate Change Negotiations of the Ministry of Foreign Affairs Yu Qingtai Receives Interview of the Media," September 22, 2007, www.chinaembassy.org.in/eng/zgbd/t366696.

8. Elizabeth Economy, "China vs. Earth," *The Nation,* May 7, 2007, www.thenation.com/doc/20070507/economy.

9. Lin et al., "Synopsis," 7.

10. Eun-Shik Kim, Dong Kyun Park, Xueyong Zhao, Sun Kee Hong, Kang Suk Koh, Min Hwan Shu, and Young Sun Kim, "Sustainable Management of Grassland Ecosystems for Controlling Asian Dusts and Desertification in Asian Continent and a Suggestion of Eco-village Study in China," *Ecological Research* 21, no. 6 (November 2006): 907–11.

11. World Health Organization, *Climate Change and Human Health: Risks and Responses* (Geneva: World Health Organization, 2003).

12. Lin et al., "Synopsis," 7.

13. Fan Daidu and Li Congxian, "Complexities of China's Coast in Response to Climate Change," *Advances in Climate Change Research* 2 (Suppl. 1) (2006): 54–58; C. X. Li, D. D. Fan, B. Deng, and D. J. Wang, "Some Problems of Vulnerability Assessment in the Coastal Zone of China," in *Global Change and Asian Pacific Coasts,* Proceedings of APN/SURVAS/LOICZ Joint Conference on Coastal Impacts of Climate Change and Adaptation in the Asia-Pacific Region, ed. N. Mimura and H. Yokoki (Kobe, 2001): 49–56.

14. Economy, "China vs. Earth."

15. J. G. Titus, "Greenhouse Effect, Sea Level Rise, and Land Use," *Land Use Policy* 7, no. 2 (April 1990): 138–53. www.epa.gov/climatechange/impacts/downloads/landuse.pdf.

16. Li et al., "Some Problems of Vulnerability Assessment."

17. Several errors in the data surrounding Himalayan glaciers have been found in the IPCC report, which could mean that some of these projections are overstated or inaccurate. See, e.g., Damian Carrington, "IPCC Officials Admit Mistake over Melting Himalayan Glaciers," *The Guardian*, January 20, 2010, www.guardian .co.uk/environment/2010/jan/20/ipcc-himalayan-glaciers-mistake.

18. Economy, "China vs. Earth."

19. Jessica Tuchman Mathews, "Redefining Security," *Foreign Affairs* 68, no. 2 (Spring 1989): 162–77.

20. United Nations Environment Program, *Understanding Environment: Conflict and Cooperation* (Nairobi: United Nations Environment Program, 2004), www.unep.org/pdf/ECC.pdf.

21. Thomas F. Homer-Dixon, "Environmental Scarcities and Violent Conflict: Evidence from Cases," *International Security* 19, no. 1 (Summer 1994): 5–40.

22. Lester R. Brown, *Who Will Feed China: Wake-Up Call for a Small Planet* (Washington, DC: Worldwatch Institute, 1995).

23. Zhang Dian, Jim Chiyung, and Lin Chusehng, "Climate Change, Social Unrest and Dynastic Transition in Ancient China," *Chinese Science Bulletin* 50, no. 2 (2005): 137.

24. Daniel Deudney, "The Case against Linking Environmental Degradation and National Security," *Millennium Journal of International Studies* 19, no. 3 (1990): 474.

25. Ibid.

26. Avilash Roul, "Beyond Tradition: Securitization of Climate Change," *Society for the Study of Peace and Conflict* 112 (2007), http://sspconline.org/article_details.asp?artid=art124; Michael Renner, "Security Council Discussion of Climate Change Raises Concerns about 'Securitization' of Environment," *Worldwatch Perspective*, April 30, 2007, www.worldwatch.org/node/5049.

27. Center for Naval Analysis, *National Security and the Threat of Climate Change* (Alexandria, VA: Center for Naval Analysis, 2008), http://securityandclimate.cna.org/report/.

28. Robert Kaplan, *The Coming Anarchy: Shattering the Dreams of the Post Cold War* (New York: Vintage Books, 2000).

29. See, e.g., Wu Jiao, "Changing Diet Offers More Food for Thought," *China Daily* (Beijing), June 24, 2006, www.chinadaily.com.cn/china/2008-06/24/content_6790223.htm.

30. Lin et al., "Synopsis."

31. World Economic Forum, *Africa@Risk 2008*, www.weforum.org/pdf/Africa2008/Africa_RiskRe port_08.pdf.

32. Lin et al., "Synopsis," 9.

33. Xunming Wang, Fahu Chen, Eerdun Hasi, and Jinchang Li, "Desertification in China: An Assessment," *Earth Science Reviews* 88, nos. 3–4 (June 2008): 188–206.

34. "Facts and Figures: China's Main Targets for 2006–2010," *Gov.cn* [Chinese Government Official Web Portal], March 6, 2006, www.gov.cn/english/2006-03/06/content_219504.htm.

35. Fan and Li, "Complexities of China's Coast."

36. Gordon McGranahan, Deborah Balk, and Bridget Anderson, "The Rising Tide: Assessing the Risks of Climate Change and How Human Settlements in Low-Elevation Coastal Zones," *Environment & Urbanization* 19, no. 1 (April 2007): 17–37, http://sedac.ciesin.columbia.edu/gpw/docs/McGranahan2007.pdf.

37. Fan and Li, "Complexities of China's Coast."

38. China National Bureau of Statistics, *China Statistical Yearbook 2007* (Beijing: National Statistics Press, 2008), table 16-33; "Guangdong Reports 20% Growth in Foreign Trade in 2007," *SINA English*, January 11, 2008, http://english.sina.com/business/1/2008/0111/141171.html.

39. Lawrence Berkeley National Laboratory, *China Energy Databook* [CD-Rom] (Berkeley: Lawrence Berkeley National Laboratory, University of California, 2008), table 2A.4.2; available at http://china.lbl.gov/databook.

40. Secretariat of the China National Committee for the Implementation of the United Nations Convention to Combat Desertification (CCICCD), *China National Report on the Implementation of United Nations Convention to Combat Desertification and National Action Programme to Combat Desertification* (Beijing: CCICCD, 2000), www.unccd.int/cop/reports/asia/national/2000/china-eng.pdf.

41. Peng Peng, Jianling Huang, John E. Sheehy, Rebecca C. Laza, Romeo M. Visperas, Xuhua Zhong, Grace S. Centeno, Gurdev S. Khush, and Kenneth G. Cassman, "Rice Yields Decline with Higher Night Temperature from Global Warming," *Proceedings of the National Academy of Sciences* 101 (2004): 9971–75.

42. U.S. Department of Agriculture, "World Agricultural Supply and Demand Estimates," WASDE-466, January 12, 2009, www.usda.gov/oce/commodity/wasde/latest.pdf.

43. China National Bureau of Statistics, *China Statistical Yearbook 2007*, tables 18-9 and 18-10.

44. Dong Fengxia and Helen H. Jensen, "Challenges for China's Agricultural Exports: Compliance with Sanitary and Phytosanitary Measures," *Choices* 22, no. 1 (2007), 19–24, www.choicesmagazine.org/2007-1/foodchains/2007-1-04.htm; and *The Food and Agricultural Organization of the United Nations Database*, http://faostat.fao.org.

45. M. Ezzati, A. D. Lopez, A. Rodgers, and C. J. L. Murray, eds., *Comparative Quantification of Health Risks: The Global and Regional Burden of Disease Attributable to Selected Major Risk Factors*, 3 vols. [CD-Rom] (Geneva: World Health Organization, 2004), vols. 1 and 2.

46. Robert G. Arnold, David O. Carpenter, Donald Kirk, David Koh, Margaret-Ann Armour, Mariano Cebrian, Luis Cifuentes, et al., "Meeting Report: Threats to Human Health and Environmental Sustainability in the Pacific Basin," *Environmental Health Perspectives* 115, no. 12 (December 2007): 1770–75.

47. Economy, "China vs. Earth."

48. University of Wisconsin–Madison, "Climate Change and Disease," press release, November 16, 2006, reprinted in *Environment* 48, no. 2 (March 2006): 4–5.

49. John Podesta and Peter Ogden, "Security Implications of Climate Scenario 1: Expected Climate Change over the Next 30 Years," in *The Age of Consequences: The Foreign Policy and National Security Implications of Global Climate Change* (Washington, DC: Center for Strategic and International Studies, 2007), chap. 3.

50. "Summary for Policymakers," 2007.

51. Rajendra K. Pachauri, chairman of the IPCC, in an interview with Circle of Blue, May 8, 2008, www.circleofblue.org/waternews/world/china-tibet-and-the-strategic-power-of-water/.

52. Minxin Pei, "The Impact of Global Climate Change on China," memorandum prepared for Workshop on Climate Change and Regional Security, Long Range Analysis Unit, National Intelligence Council, December 2007.

53. Others have discussed how sea-level rise could cause China to need to resettle its own eastern coastal population in other countries in Europe or Russia, including neighboring Siberia. See R. James Woolsey, "The Security Implications of Climate Change Scenario 3," in *Age of Consequences*, 81–92.

54. Food & Fertilizer Technology Center of the Asian and Pacific Region, table 21, www.agnet.org/situationer/stats/21.html. Data are for 2005.

55. "Shanghai Foreign Trade Volume Rises 13.8% in 2008," Xinhuanet.com, January 15, 2009, http://news.xinhuanet.com/english/2009-01/25/content_10717893.htm.

56. Joanna Lewis and Elliot Diringer, *Policy Based Commitments in a Post-2012 International Climate Framework*, Pew Center on Global Climate Change Working Paper (Arlington, VA: Pew Center on Global Climate Change, 2007); Jake Schmidt, Ned Helme, Jim Lee, and Mark Houdashelt, "Sector-Based Approach to the Post-2012 Climate Change Policy Architecture," *Climate Policy* 8 (2008): 494–515.

57. Jiang Lin, Nan Zhou, Mark Levine, and David Fridley, "Taking Out 1 Billion Tons of CO_2: The Magic of China's 11th Five-Year Plan?" *Energy Policy* 36 (2008): 954–70; Center for Clean Air Policy, *Greenhouse Gas Mitigation in China, Brazil and Mexico: Recent Efforts and Implications* (Washington, DC: Center for Clean Air Policy, 2007).

58. Letter from the U.S. Special Envoy for Climate Change to the UNFCCC Executive Secretary, January 28, 2010, http://unfccc.int/files/meetings/application/pdf/unitedstatescphaccord_app.1.pdf.

59. Regional Greenhouse Gas Initiative, http://rggi.org/home; California Assembly Bill 32 (AB32), www.arb.ca.gov/cc/ab32/ab32.htm.

60. The Group of 77 (G-77) was established on June 15, 1964, by seventy-seven developing countries, which became signatories of the "Joint Declaration of the Seventy-Seven Countries" issued at the end of the first session of the United Nations Conference on Trade and Development in Geneva. Although membership in the G-77 has increased to 130 countries, the original name was retained because of its historic significance. See www.g77.org/doc/.

61. Government of Brazil, Interministerial Committee on Climate Change, *National Plan on Climate Change*, Executive Summary, November 21, 2007), www.mma.gov.br/estruturas/imprensa/_arquivos/96_11122008040728.pdf; Government of Mexico, Intersecretarial Commission on Climate Change, *Na-*

tional Strategy on Climate Change: Mexico, Executive Summary, 2007, wwwupdate.un.org/ga/president/61/
follow-up/climatechange/Nal_Strategy_MEX_eng.pdf.

62. "Trade Sanctions Emerge as Tool to Force China and India to Curb Emissions," *Greenwire*, March 21,
2007, www.wbcsd.org/plugins/DocSearch/details.asp?MenuId=1&ClickMenu=&doOpen=1&type=DocDet
&ObjectId=MjM1OTg.

63. As illustrated in table 2.1, China far exceeds this cutoff at 24 percent of global emissions.

64. Trevor Houser, Rob Bradley, Britt Childs, Jacob Werksman, and Robert Heilmayr, *Leveling the Carbon Playing Field: International Competition and US Climate Policy Design* (Washington, DC: Peterson Institute for International Economics and World Resources Institute, 2008). Also see memorandum from
Abraham Breehey, International Brotherhood of Boilermakers, to the Obama-Biden transition team, http://
otrans.3cdn.net/115767b67250063465_jhm6b9e65.pdf.

65. "Trade Plan Opposed by China, Brazil, and Mexico," *Greenwire*, September 26, 2007, www.earthportal
.org/news/?p=507.

66. Paul Ames, "EU Leaders Urge Trade Sanctions on U.S., China," *Seattlepi*, March 14, 2008, http://
seattlepi.nwsource.com/national/355174_eusummit15.html.

67. Xu Shisen, "Green Coal-Based Power Generation for Tomorrow's Power," *Presentation to the APEC
Energy Working Group: Expert Group on Clean Fossil Energy,* Thermal Power Research Institute, Lampang,
Thailand, February 24, 2006.

68. European Commission, "EU-China Summit: Joint Statement," September 5, 2005, http://ec.europa
.eu/comm/external_relations/china/summit_0905/index.htm; U.K. Department of Environmental, Food
and Rural Affairs, 2005, www.defra.gov.uk/environment/climatechange/internat/devcountry/china.htm.

69. "Carbon Capture Milestone in China," *Science Daily*, August 4, 2008, www.sciencedaily.com/releases/
2008/07/080731135924.htm.

70. Yu, "Special Representative for Climate Change Negotiations." He went on to say that "if the developing countries cannot maintain economic and social progress, eliminate poverty or raise people's living standards, the material foundation for coping with climate change will not exist, not mentioning the capacity to
fight against climate change."

3

Vietnam

Carlyle A. Thayer

The Socialist Republic of Vietnam is the world's thirteenth-most-populous country. Since 1986 Vietnam has undergone a remarkable transformation from a Soviet-style centrally planned economy to a market-led economy. And during the past two decades the country has achieved success across the board on most major indicators of development.[1] For example, it has sustained among the highest economic growth rates in East Asia for more than a decade.[2] And it recently overtook India to become the world's second-largest exporter of rice. Vietnam has attracted a windfall of foreign direct investment.

Vietnam

At the same time, Vietnam has achieved remarkable success in lowering the incidence of poverty from 58 percent in 1993 to 16 percent in 2006. The country is also expected to attain its UN Millennium Development Goals by reducing infant mortality and improving maternal health and overall national nutrition.[3] It is also poised to become a middle-income country by 2012. It has set its sights on becoming a modern and industrial country by 2020. To meet projected shortfalls in domestic sources of energy so that it can keep up the pace of industrial development, it is turning to nuclear power.[4]

Vietnam's successful economic transformation has been mirrored by a similar transformation in its external relations. The country is no longer one of the least-developed members of the Cold War–era Council for Mutual Economic Assistance. And it is also no longer economically and militarily dependent on the Soviet Union. Since the 1990s, Vietnam has looked outward toward its region and further afield to become a member of the Association of Southeast Asian Nations (ASEAN), the ASEAN Regional Forum, and the Asia-Pacific Economic Cooperation forum (APEC). In 2007 Vietnam became the 150th member of the World Trade Organization. It was also unanimously selected by the Asia bloc as its candidate for a nonpermanent seat on the United Nations Security Council. When Vietnam's nomination was put before the General Assembly in October 2007, it received 183 of the 192 votes cast.

Vietnam's potential to become a major regional power is being increasingly recognized by strategic analysts. For example, in 2007 Richard Armitage and Joseph Nye coauthored an article in which they wrote: "Perhaps the greatest opportunity over the next fifteen years lies in Vietnam, where political reforms needs to complement the economic reforms already well under way if the nation is to reach its potential and contribute more to the overall effectiveness of ASEAN."[5] Ashton Carter and William Perry have even suggested that the United States seek out Vietnam as a potential military partner.[6]

Any analysis of Vietnam's strategic future would not be complete, however, without a consideration of how it will cope with the impact of global climate change. According to the World Conservation Union, climate change is a "critical issue for Vietnam in the medium and long term" because "as the water level increases, inundation from the ocean may occur more frequently until permanent inundation occurs."[7] Vietnam would be threatened by crop failures, biodiversity loss, and damage to wetlands, coral reefs, and other critical ecosystems, particularly in coastal areas. In sum, global climate change has now emerged as one of—if not the most—serious security issues Vietnam will face in the next two decades and beyond. Global climate change threatens to undermine not only the country's remarkable economic and social progress but also possibly the unity and cohesion of the state itself.

This chapter considers the impact of global climate change on the Vietnamese state and society through 2030. The analysis that follows is divided into seven sections. The first sets out the scientific data on the likely impact of climate change on Vietnam. The second section considers the likely evolution of the Vietnamese state over the next two decades. The third section examines Vietnam's likely response to environmental stress. The fourth section examines the possibility of state failure in Vietnam. The fifth section discusses the resilience of Vietnamese state and society. The sixth section analyzes the likely impact of climate change on Vietnam's relations with its neighbors. And the seventh section explores Vietnam's likely external response as a consequence of global climate change. A concluding section offers a net assessment of the impact of climate change on Vietnam.

Global Climate Change and Vietnam

Vietnam is already experiencing the impact of global climate change. A study by the country's Institute of Meteorology, Hydrology, and Environment reports that during the past decade, sea levels have risen 2.5 to 3 centimeters per year and annual average temperature has risen 0.1 degree Celsius.[8] These developments have contributed to altered weather patterns, including increasingly violent and more frequent storms along the coast, the intrusion of saline water into the Mekong Delta, extremes of heat and cold, and the onset of desertification further inland.[9]

In 2007, according to Vietnam's Ministry of Agriculture and Rural Development (MARD), nearly five hundred people died due to flooding and 215,000 hectares of agricultural land was submerged.[10] Low-lying fields were inundated with salt water that destroyed crops. Rising temperatures have encouraged a plague of pests, including the brown plant hopper, which has been particularly destructive of Vietnam's rice and cashew crops. Four indicators provide an indication of Vietnam's current vulnerability to climate change: temperature rise, access to freshwater, agricultural productivity, and sea-level rise. Of the three Southeast Asian countries considered in this study, Vietnam appears to be in the most advantageous position with regard to the first three indicators but not the fourth, sea-level rise (see appendix A).

According to CIESIN data summarized in appendix A in this volume, Vietnam's aggregate temperature vulnerability score of 0.52 is less than that for Indonesia (0.55) and the Philippines (0.59) and well below the global average of 0.80. Vietnam's relative temperature vulnerability is rated average, the same for Indonesia and the Philippines. The country's temperature vulnerability in degrees Celsius, 0.54, is only marginally higher than that of the Philippines (0.53) but below Indonesia's (0.59). All three countries rank below the global average of 0.77. Vietnam's agricultural productivity impact is assessed as negligible, the same as Indonesia's, but better than that of the Philippines, where the rating is moderate. At present, Vietnam scores well on the third indicator, freshwater availability as measured by the percentage of population with access to less than 1,000 cubic meters of freshwater per capita. In 2000 Vietnam (11.5 percent) ranked below the Philippines (12.7 percent) and Indonesia (14.5 percent), and well below the global average of 30.8 percent. By 2030, however, the situation will be reversed, when the proportion of the Vietnamese population with access to less than 1,000 cubic meters of freshwater per capita will rise to 25.7 percent, above that of Indonesia (20.3 percent) and the Philippines (10.4 percent). In the meantime, increased salinity in the southern delta threatens water for irrigation and human consumption.[11] A separate report by the Vietnam–Sweden Cooperation Program forecast that Vietnam will be among sixty countries that will experience serious water shortages by 2030 because of population pressures, urbanization, and industrialization.[12]

However, as noted above, Vietnam scores less well on the fourth indicator, sea-level rise, in its coastal zones. The country has a 3,200-kilometer-long coastline and two of the largest low-lying river deltas in the world. It thus faces very serious problems due to projected sea-level rise. A recent *United Nations Human Development Report* found that "sea-level rise projected for 2030 would expose around 45 percent of the [Mekong] Delta's land area to extreme salinization and crop damage through flooding." This same report found that a projected sea-level rise of 1 meter by 2100 would lead to the displacement of up to 22 million people and a loss of up to 10 percent of GDP.[13] A rise of 1 meter in low-elevation coastal zones will affect 9.0 percent of the population, more than double for the Philippines (4 percent) and well above Indonesia (1.1 percent) (see appendix B).

A separate study by Vietnam's MARD estimates that if sea levels were to rise by 1 meter, 5,000 square kilometers of the Red River Delta and 20,000 square kilometers of the Mekong Delta would be inundated, affecting between 11 and 23 percent of Vietnam's population if no adaptive measures were taken.[14] The intrusion of seawater would exacerbate an already-existing problem of saline water intrusion caused by upland deforestation and dam building that is reducing Vietnam's supply of freshwater. In addition, according to a research study commissioned by the former Ministry of Irrigation, coastal plains would sink by nearly a meter and many major urban areas, including Ho Chi Minh City, would be flooded during high tides.[15]

Sea-level rise as high as 3 meters is unlikely within the time frame of this study, but possible in the longer run, and a reasonable proxy for extreme coastal weather events, whose incidence has recently been increasing. Such a rise would appear to be devastating, with 10.9 percent of the land area and 29.5 percent of the population affected. The main impact of sea-level rise would fall on Vietnam's rice-growing delta regions, which are densely populated and contain much of the country's industrial infrastructure. In 2007 agriculture contributed 15 percent of Vietnam's GDP.[16] The projected impact on GDP is severe—15.7 percent, by far the highest of the countries included in this study (see appendix B). At even more extreme levels, a 5-meter rise would have an impact on 16 percent of Vietnam's land area, 35 percent of its population, and 35 percent of its GDP. The greatest impact would be felt in the Mekong River and Red River deltas.[17]

A separate study by the World Bank placed Vietnam at the top of its list of eighty-four countries surveyed for the impact of sea-level rise. It calculated that 10.8 percent of Vietnam's population would be displaced with a 1-meter sea-level rise, with disproportionately high effects in the Mekong River and Red River deltas (even worse than the projected impact on Egypt's Nile River Delta). The World Bank report concluded, "At the country level, results are extremely skewed, with severe impacts limited to a relatively small number of countries. For these countries (e.g., Vietnam, Egypt, and the Bahamas), however, the consequences of [sea-level rise] are potentially catastrophic."[18]

Sea-level rise would not only have an impact on the population, industry, aquaculture, fishing, transport, and coastal energy sectors but, most important, would have impact on agriculture. Approximately 80 percent of Vietnam's population is rural. More than half its total population resides in low-lying delta areas. It is currently among the top three rice-exporting countries in the world. Rice depends on controlled irrigation and is highly vulnerable to fluctuations in water levels. Thus sea-level rise and saline intrusion would threaten not only Vietnam's food security but the food security of other states that are dependent on its exports. MARD concluded that a 1-meter rise in sea level would reduce the nation's food output by 12 percent, or 5 million tons per year.[19]

Global climate change brings with it altered weather patterns. According to Dao Xuan Hoc, deputy minister of MARD, an increase in the intensity and frequency of typhoons, for example, is likely to breach Vietnam's river and sea dike system, causing extensive flooding and loss of agricultural productivity. In addition, extreme weather could contribute to the outbreak of human and animal diseases.[20]

A recent study by the Organization for Economic Cooperation and Development of 130 key port cities worldwide analyzed the exposure of people, property, and infrastructure to a 101-year flood event.[21] In its estimate of the impact of climate change, the study assumed a mean sea-level rise of 0.5 meter by 2070, a hypothesis that is by no means extreme. In 2005, Vietnam ranked fifth in terms of exposed population. When the risk of exposure was projected out to 2070, Ho Chi Minh City still occupied fifth position, but the northern port city of Hai Phong rose to tenth position.[22]

In sum, Vietnam is slated to become one of the countries most adversely affected by global climate change, comparable to Bangladesh and small island states. The World Bank study of eighty-four countries cited above reveals that Vietnam ranks among the top ten countries that will feel the adverse impact of global climate change in six major areas: land, population, GDP, urban extent, agriculture extent, and wetlands. Vietnam ranks first on four of these indicators and second on the remaining two.[23]

Vietnam and Political Change to 2030

The history of Vietnam from the end of World War II to the present has been marked by prolonged periods of conflict and other manifestations of traditional risks. North Vietnam was struck by a devastating famine at the end of World War II. Vietnam then experienced an eight-year war of resistance to French colonialism (1946–54) that resulted in a political settlement partitioning Vietnam at the seventeenth parallel. In 1960 Communist North Vietnam initiated a "war of national liberation" to reunify Vietnam. What began as a guerrilla war in the early 1960s steadily escalated with the arrival of northern military units in the south. In 1965 the United States introduced combat forces to engage North Vietnamese regulars. Ground combat displaced hundreds of thousands of civilians. At the same time the United States conducted an air war over the north that destroyed much of its industrial and transport infrastructure. The northern air war was suspended in 1968 to encourage peace negotiations but was resumed in December 1972 to force a final settlement. In 1973 the United States withdrew all its combat forces and ground combat resumed between Communist and South Vietnamese armed forces. In 1975, the Communist side prevailed and the Republic of Vietnam surrendered unconditionally. In the immediate aftermath of the war and in subsequent years, more than a million South Vietnamese fled the country by boat to neighboring countries.

In 1977 and 1978 Khmer Rouge forces attacked population settlements along Vietnam's southwestern border. Vietnam responded by invading Cambodia in late 1978. China retaliated by attacking Vietnam's northern border in February–March 1979 and maintained hostilities until 1987.[24] Vietnam occupied Cambodia for a decade before withdrawing in September 1989. Finally, Chinese and Vietnamese naval vessels clashed in the South China Sea in 1988, and in 1992 China began to occupy features in the Spratly Archipelago.[25]

Developments in Vietnam since the end of the Cold War have seen a marked turn for the better. Southeast Asia has remained immune from state-versus-state conflicts, and Vietnam has reached a peaceful modus vivendi with its three neighbors, China, Laos, and Cambodia. And the potential for territorial conflict in the South China Sea region has diminished since the signing of the Declaration on Conduct of Parties in the South China Sea by China and the ten members of ASEAN, including Vietnam, in 2002.

On the domestic front, Vietnam has regularly experienced flooding, draught, typhoons, and famine. The birth of the Democratic Republic of Vietnam in September 1945 coincided with one of the worst famines in Vietnam's recorded history, with an estimated one million persons perishing.[26] Since reunification, Vietnam's central coast has been lashed by increasingly destructive tropical storms. In 2007 "six major storms from the South China Sea, a global typhoon hotspot, battered Vietnam, killing more than 400 people and leaving large areas in central Vietnam inundated for months."[27] In late November 2007 sea tides rose to 1.5 meters in Ho Chi Minh City, the highest in forty-eight years.

In precolonial times it was the duty of the emperor to protect his people from natural disasters. Vietnam has constructed 5,000 kilometers of river dikes, including an extensive system in the Red River Delta, and an additional 3,000 kilometers of sea dikes. The dikes in the Red River Delta are built above farming levels. If the dikes were to fall into disrepair or be breached, extensive flooding would result. According to Vietnamese tradition, natural disasters are a portent that the emperor has lost his "mandate from heaven" and thus justify rebellion or "uprising in a just cause."[28]

Today, it is as if the traditional Vietnamese political culture continues to cast a shadow over the country's future. The Vietnamese people expect their leaders to protect them from storm, flooding, drought, and pestilence. In recent interviews with officials and opinion makers in Hanoi, respondents noted that the current leadership deserves high marks for the priority it has set on mitigating the effects of natural disasters.[29]

Vietnam's current leadership has established a Central Committee for Storm and Flood Control as the body with responsibility for national disaster preparedness and mitigation. This committee comprises representatives from all the relevant government ministries as well as the Vietnam Red Cross, the Hydro-Meteorological Service, and the Department of Dike Management, Flood, and Storm Control in MARD.[30] The Central Committee for Storm and Flood Control has been quick to mobilize government resources in response to natural disasters. To deal with natural disasters, the government regularly calls on the resources of its large regular armed forces (with 484,000 soldiers) and extensive reserves, as well as the mobilization of young people.

An analysis of the impact of climate change on Vietnam until 2030 must consider likely changes in the country's society and political system between now and then. Currently, Vietnam is relatively stable compared with Indonesia and the Philippines, the two other Southeast Asian case studies considered in this volume. The Philippines continues to suffer from Communist insurgency, Islamic-inspired armed separatism, and attempted military coups. Indonesia was wracked by communal conflict in its outer islands in the late 1990s, and it continues to suffer from sporadic outbursts of communalism, terrorism, and armed separatism. By contrast, Vietnam has experienced only momentary political turbulence in the form of peasant demonstrations over land issues, ethnic unrest, and labor strikes.

In trying to make a forecast two decades into the future, it is important to take a look back at what Vietnam, Indonesia, and the Philippines were like twenty years ago and how their political futures changed over the course of the subsequent years. In the early 1980s, Indonesia, the Philippines, and Vietnam were considered more or less stable authoritarian regimes. Yet the Ferdinand Marcos regime collapsed in 1986, and New Order Indonesia fell twelve years later. Both countries then embarked on a process of democratic transition and consolidation. Few observers could have imagined that Communist Vietnam would begin to dismantle its socialist central planning system in late 1986, join ASEAN in 1995, and integrate with the global economy through membership in the World Trade Organization in 2007.

What will Vietnam's current soft-authoritarian, one-party state look like in 2030? As the above examples indicate, straight-line extrapolations of contemporary trends may be misleading and can map out only one of several plausible futures. The experiences of Taiwan and South Korea, in addition to those of Indonesia and the Philippines, indicate that authoritarian regimes are capable of liberalizing political change. There is more than one likely scenario. The Vietnamese Communist Party could continue in power or could gradually transform itself into a more pluralistic organization. It could also split into progressive and conservative factions. Other possibilities include political instability, as the party attempts to fend off challenges by prodemocratic forces or southern regionalism. Finally, Vietnam could muddle through after a period of political turbulence as it embarks on democratic transformation.

Perhaps the least likely possibility, given the general social and economic dynamism that has marked Vietnam's recent history, is that things will simply stay as they are. This analysis makes the assumption that Vietnam's current one-party political system will be substantially transformed by 2030, and that the initiative for this transformation will come from members of the Communist Party itself. They will seek to broaden the party's present elitist base by reaching out to politically active non–party members in society.[31] In sum, progressive and pragmatic members of the party are most likely to remain in power in the future and successfully manage Vietnam's adaptation to the security challenges posed by global climate change.

Vietnam's Response to Environmental Stress

How do societies react to environmental stress? The question examined in this section is whether the Vietnamese response will be fight, flight, or some mixed alternative. Fight refers to major internal instability leading to armed conflict, while flight refers to emigration. Both can occur simultaneously.

The Vietnamese people originated in what is today called the Red River Delta. As these people mastered wet rice cultivation, population density increased to such an extent that demographic pressures spurred a southward migration known as *nam tien* (southern advance). It is entirely realistic to portray this human movement from the Red River Delta south along the narrow coastline to the Mekong Delta as a product of environmental stress due to overpopulation. This push factor has been advanced to explain the history of the wars between Vietnam and the Cham Empire in northern-central Vietnam, which led to the demise of the latter.[32]

During the French colonial era, several hundred thousand Vietnamese fled to neighboring Laos and Cambodia, and also to Thailand. Some were motivated to escape conflict while others moved across the land border because of the opportunities for employment. When Vietnam was partitioned in 1954–56, nearly a million Vietnamese civilians fled south. Subsequently, the governments in both Hanoi and Saigon pursued population resettlement programs to shift ethnic Vietnamese from the overpopulated deltas to the highlands, which were inhabited by ethnic minority groups.

During the Vietnam War (1965–75) large numbers of rural South Vietnamese sought safety from the fighting by fleeing to district towns, provincial centers, and Saigon itself. On the other hand, in North Vietnam the residents of urban areas, particularly Hanoi, were evacuated to the countryside. When the war ended in 1975, more than a hundred thousand Vietnamese fled South Vietnam. After a brief decline, the numbers of "boat people" shot up two years later.[33] Although fear of political repression was a motive for early refugees, economic issues and the impact of major natural disasters in the south in 1977–78 were increasingly the most salient push factors.[34] Nearly a quarter million ethnic Chinese or Hoa people fled to southern China when relations deteriorated between Hanoi and Beijing in the late 1970s.

In the precolonial past, the political turmoil in China that accompanied the decline and fall of dynastic rule produced waves of refugees who fled south into Vietnam. To the extent that southern China is affected by global climate change, Vietnam could well come under pressure from a similar wave of migration. The arrival of large numbers of ethnic Chinese fleeing the impact of global climate change would most certainly arouse nationalist sentiment, especially in times of domestic socioeconomic crisis. This would undoubtedly produce friction between the governments of Vietnam and China. In 2009, for example, an anti-China backlash arose in Vietnam when it was revealed that Chinese workers had taken jobs in the bauxite mining industry in the Central Highlands while Vietnamese workers remained unemployed. General Vo Nguyen Giap even termed the presence of Chinese workers a threat to national security.

Vietnam's Communist rulers do not countenance political pluralism, organized opposition, or a multiparty system. In recent years, Vietnam's one-party system has come under challenge by ethnic minorities in the Central Highlands and a nascent prodemocracy movement in urban areas. The United Buddhist Church of Vietnam, for instance, has long resisted incorporation into Vietnam's "mono-organizational" political system.[35] Prodemocracy groups can be expected to be vocal critics of the government if it should fail to deal effectively with the challenges posed by climate change. These social forces are unlikely to mount a credible challenge to Communist Party rule in the coming decades, however.[36] There is some evidence that Vietnam's well-off citizens are now critical of the government's inability to satisfactorily address inflation, pollution, corruption, and traffic jams. Their solution is not to overturn the present system but to replace officials who are perceived as ineffective with more dynamic leaders from within the party.

These past patterns of history suggest that as adverse climatic effects are felt in Vietnam, some Vietnamese will respond by flight. Lowland Vietnamese settlers in provinces located along the central coastline can be expected to push into the Highlands and spark various forms of resistance by indigenous ethnic minorities. This resistance is likely to be subdued and result in the flight of ethnic minorities into neighboring Cambodia and Laos. Larger numbers of Mekong Delta farmers are likely to push into Cambodia's eastern provinces. Vietnamese living in the former South Vietnam may seek to emigrate in order to join relatives living abroad.

It seems unlikely that future adverse weather conditions will spark an exodus comparable to the waves of boat people that engulfed Southeast Asia in the 1970s. First, the option of political asylum or guaranteed resettlement in the United States is no longer on the table. Second, Vietnam's economic reforms have generated prosperity and a stake in the future of the country. This contrasts with the 1970s, when many South Vietnamese felt marginalized at a time when the economy was deteriorating. However, adverse climatic conditions could well inspire smaller numbers of desperate northern "boat people" to flee to Hong Kong and Hainan Island, while southerners are likely to head for the Philippines, Malaysia, and Indonesia.

Vietnam: A Failed State?

The discussion above indicates that global climate change is likely to increase the probability of disruptive change. The question examined in this section is whether the net effects of disruptive change will result in a breakdown or failure of the Vietnamese state.[37]

Vietnam has already recognized the severity of the threat posed by global climate change and is taking steps to mitigate its adverse impact.[38] Vietnam has a national body to coordinate data gathering, issue weather alerts, monitor flood and storm events, and coordinate disaster response and mitigation measures. The structure of the Central Committee for Storm and Flood Control extends from the national to province, district, and commune levels. Local storm and flood control committees work with local people's committees to implement appropriate protective and mitigation measures. Moreover, Vietnam has in place a Second National Strategy and Action Plan for Disaster Mitigation and Management for the period 2001–20.

A report by the Southern Institute for Water Resources Planning revealed serious erosion of the coastal area between Quang Ngai Province in the center to Binh Thuan

Province further south.[39] MARD is attempting to raise $676 million to strengthen dikes in fifteen of the most vulnerable areas. A feasibility study is being conducted on strengthening dikes from Quang Ngai Province to Kien Giang Province. MARD is also drawing up plans to upgrade the national irrigation system and to promote forestation.[40] Ho Chi Minh City has set up a Center for Flood Prevention and Control, and plans are set to reduce flooding by constructing eight major drains. In 2008, Vietnam's National Assembly approved a national program, submitted by the Ministry of Natural Resources, to cope with global climate change. The prime minister also set up a $1,436 million fund (2009–15) to devise strategies to respond to climate change.[41]

Vietnam is unlikely to experience a complete failure of the state as a result of adverse global climatic conditions. Vietnam's soft-authoritarian regime has proven resilient in the contemporary era in meeting perhaps even more challenging circumstances. More important, Vietnamese society has been resilient throughout its history, especially at the village and commune levels.[42]

Since unification the Vietnamese state has responded well to challenges from below, such as land disputes in the Mekong Delta in the 1980s, peasant unrest in Thai Binh Province in the late 1990s, ethnic minority resistance in the Central Highlands in 2001–4, and a wave of labor strikes in 2006–8. Vietnamese authorities have invariably responded swiftly and pragmatically to address real grievances. But the Vietnamese state has also shown its authoritarian nature by taking repressive action against those it has identified as ringleaders.

Instead of state failure, adverse climate effects would most likely result in the weakening of the state vis-à-vis the provinces. This could take on a regional dimension as well, with Ho Chi Minh City and the southern Mekong Delta seeking greater autonomy from the authorities in Hanoi to address pressing socioeconomic problems engendered by climate change.

The Resilience of State and Society

This section considers the question of whether or not the Vietnamese state possesses the latent reserves of social resilience, ingenuity, and institutional capital that would enable it to meet the challenges of global climate change successfully.

The Vietnamese state is composed of party elites occupying dual roles in the central administrative apparatus, provincial government, military, security services, and mass organizations. The state retains a considerable capacity for communication, mass mobilization, and the maintenance of public order. Vietnam's mass organizations respond well in times of adversity. These are national bodies with extensive networks at the local level, such as the multi-million-member Vietnam Women's Union, Vietnam Youth Union, and Vietnam Peasant's Union. Vietnam also has a tradition and a current policy of "compulsory labor" for youth.

A quick overview of Vietnamese history reveals a prolonged period of civil war between rival dynasties in the north and south, massive peasant rebellions (most notably the Tay Son rebellion and the Nghe-Tinh soviet movement), and national partition. The trajectory of these developments, however, has been in the direction of unifying the Vietnamese state, not its fragmentation.

The Vietnamese state is also strong because the country's provinces are robust units sometimes likened to autonomous kingdoms. During the air war period in the north, for example, the Hanoi regime ceased implementing a socialist central plan and permitted the provinces a great deal of autarky.[43] Vietnam also has a strong tradition of village communal solidarity, especially in the northern provinces.[44]

In May 2008, President Nguyen Minh Triet addressed the nation on the occasion of the sixty-second national day for dike management, flood prevention, and the control of storms and disasters. Triet urged all government officials from the national to the local levels to educate the public about the challenge of global climate change and to develop programs for dealing with its effects.

In May 2008 the government convened an online meeting of leaders from thirty-four coastal provinces to make preparation for an estimated six to eight major storms forecast to hit Vietnam during the year.[45] Deputy Prime Minister Hoang Trung Hai, who addressed the meeting, ordered provincial officials to focus on evacuation, stockpiling of food, and medicine and rescue. A representative from the Central Committee for Storm and Flood Control urged the Mekong Delta provincial authorities to speed up the fortification of dikes, houses, bridges, and roads so they could withstand major storms. The Ministry of Construction was directed by the government to ensure safety for schools, hospitals, apartment buildings, and other public structures. And the National Committee for Search and Rescue was given approval to increase the standards of its search-and-rescue force as well as to procure more vessels.

In summary, the present regime has given priority to developing the state's capacity to respond to such natural disasters as tropical storms, flooding, and forest fires, and to mitigating their impact. Vietnam would resort to the use of its armed forces and mass mobilization to deal with natural disasters and any attendant civil unrest.

Climate Change and Vietnam's Neighbors

This section shifts focus from the internal to the external and considers the question of whether global climate change might make Vietnam a destination for migrants and refugees. It also considers the question of whether Vietnam's neighbors might resort to aggression to seize resources and living space at a time of environmental stress.

Vietnam shares land borders with only three countries—Cambodia, Laos, and China. As noted above, Chinese citizens might flee south in the face of adverse conditions in their homeland in the decades after 2030.[46] But this would be a more or less spontaneous affair, not a result of a decision by the central government. The Chinese government could facilitate this migration by acquiescing and not imposing strict border controls. Neither Cambodia nor Laos has or is likely to develop the military capacity to pose a serious challenge to Vietnam in the future. But nationalist anti-Vietnamese sentiment lies just below the surface in Cambodia, and an aggressive response is possible as a consequence.

Although China has the capacity for military action, it also retains a historical memory of failed military campaigns against the Vietnamese, most recently in the 1979 border war.[47] Though the balance of power has shifted in China's favor since 1979, the regional context has also changed. China and the regional economies of Southeast Asia are and will become increasingly interdependent as the ASEAN–China Free Trade Agreement takes effect. Vietnam is not a lone actor but a member of ASEAN. China would have to carefully weigh the costs and benefits of aggressive action. It is more likely that the threat of economic sanctions, directed against the web of economic linkages connecting China with its regional trading partners, would act as a powerful constraint on China.

The main exception to this assessment is likely to be in the South China Sea region, where Vietnam and China have overlapping claims that touch on the sensitive issue of national sovereignty. Both occupy numerous features, rocks, and reefs. With few exceptions, these features do not support human habitation. Most are barely above the surface at high tide. Both China and Vietnam have built structures on them in order to physically occupy them. A rise in sea level could hamper the ability of both states to maintain these tenuous territorial claims, because the territory in question might literally disappear. At the same time, if Vietnam were distracted while dealing with the effects of global climate change on its mainland, China might be emboldened to become more assertive in the South China Sea, as demonstrated in 2009 when Chinese fishing vessels aggressively enforced a unilateral ban against Vietnamese fishermen.

Vietnam's External Relations

This section continues the focus on external relations. It considers the impact of global climate change on Vietnam's foreign policy in two dimensions: whether Vietnam would be more or less open to engagement with external states, and whether Vietnam would be more or less amenable to American influence and interests.

In November 1994, Vietnam ratified the UNFCCC. Nguyen Huu Ninh, one of Vietnam's foremost climate change scientists, participated on the IPCC as one of the lead authors in Working Group 2, which prepared the volume *Impacts, Adaptation and Vulnerability* released in 1995.[48] Vietnam ratified the Kyoto Protocol in September 2002. Vietnam is currently elevating global climate change to a key priority among the nontraditional security threats that must be dealt with by itself and the international community. In 2008 the Vietnamese cabinet adopted an Action Plan for 2007–10 to implement the Kyoto Protocol. The plan includes provisions on renewable energy, energy conservation, reforestation, and forest protection.[49]

In February 2008 Vietnam dispatched Tran The Ngoc, its deputy minister of natural resources and environment, to the UN General Assembly to participate in the debate on climate change. During the proceedings Ngoc stated that "[Vietnam is] looking forward to joining efforts undertaken by the international community to achieve the ultimate objective of the UN Framework Convention on Climate Change as well as global sustainable development."[50]

In addition to the self-help measures described above, Vietnam will pursue a layered multilateral strategy that will include engagement with the United Nations, World Bank, International Monetary Fund, Asian Development Bank, ASEAN, ASEAN Regional Forum, APEC, ASEAN Plus 3 (China, Japan, and South Korea), East Asian Summit, and any other relevant multinational institution, including those that may emerge in the next two decades. In October 2009, for example, Vietnam met with thirty international donor organizations to discuss measures and funding to cope with the impact of climate change.

There is no question that Vietnam will place great (though not exclusive) emphasis on ASEAN as its prime multilateral vehicle. To the extent that East Asia's regional architecture has been built up, Vietnam will also seek solutions through forums such as APEC, ASEAN Plus Three, and the East Asian Summit. Vietnam naturally sees itself as a key actor among developing countries and the nonaligned movement, and it will seek expressions of political solidarity if not support from this quarter. Vietnam can be expected to push multilateral buttons in order to apply pressure on the major powers in dealing with global climate change.

Vietnam also will be very open to engagement with the outside world, including the United States, the European Union, and other powers such as Japan and Australia. Vietnam will not have any qualms about working with and accepting U.S. leadership on this global issue—but only if Washington is seen as part of the solution and not part of the problem. If Vietnam faces pressure from China, as noted above, it will expect other outside parties, like the United States, to provide countervailing leverage. Vietnam will also be leery about any endeavor that is not all-inclusive in its approach to dealing with the adverse impact of global climate change.

Conclusion

Vietnam is poised to become a major regional actor in East Asia in the next two decades as its economy recovers from the global financial crisis and continues to develop on an upward trajectory. But Vietnam's national objective of becoming a modern, industrialized country by 2020 could be put in jeopardy by the effects of global climate change. The international community was quick to recognize Vietnam's rise as a so-called little dragon in the past. It has now come to realize that Vietnam is likely to be one of the countries most affected by global climate change.

Vietnam's location and topography make it one of the most disaster-prone countries in the world. It has a coastline of 3,200 kilometers and is regularly lashed by typhoons, which produce flooding on a large scale.[51] It is currently experiencing the effects of global warming and climate change. Its weather patterns are altering, and tropical storms are increasing in frequency and destructiveness.

This chapter has examined six major questions related to the hypothesized impact of global climate change on Vietnam and reached the following conclusions. First, Vietnam's one-party state is likely to transform itself over the next two decades as a result of reform from within the Vietnamese Communist Party itself. Second, Vietnamese society may respond to environmental stress by flight to neighboring Cambodia and Laos, and by sea in boats to the countries with the shortest landfall. Third, though Vietnam is unlikely to become a failed state, the stresses caused by global climate change could weaken the control exercised by the national government in Hanoi, resulting in enhanced autonomy for provinces and the southern region. Fourth, global climate change will put enormous strain on Vietnam's political system, state, and society. But it is unlikely to trigger disruptive internal unrest as long as the Vietnamese state demonstrates competence in mitigating the effects of dramatic climate change effects. Fifth, Vietnam is unlikely to face a serious military challenge from either Cambodia or Laos. China will likely be constrained by its growing economic and political interdependence with the Southeast Asian states, but it could be emboldened to assert its territorial claims in the South China Sea region more aggressively if Vietnam were to become preoccupied with the impact of climate change on its mainland. And finally, Vietnam is likely to react to the stress of global climate change through both self-help and multilateral mechanisms. In particular, Vietnam will work with global and regional multilateral institutions to mitigate environmental stress. And Vietnam will be willing to work in concert with the United States and other major powers if they demonstrate inclusive leadership. The historical evidence points to the durability of the Vietnamese state and the resilience of its society to cope and endure in the face of the challenges of global climate change.

Notes

1. Vietnam's ranking on the United Nations' Human Development Index (HDI) has consistently risen since 1990 when it was first compiled. Although Vietnam ranks 105th out of 177 countries and territories covered by the HDI in 2007, it ranked 56th for life expectancy and 57th for adult literacy. See United Nations Development Program, *Human Development Report 2007/2008 Fighting Climate Change—Human Solidarity in a Divided World* (New York: United Nations, 2007), 230.

2. "Nation's Legal Framework a Challenge to Saving Energy," *Viet Nam News* (Hanoi), November 29, 2007.

3. The Millennium Development Goals are a set of eight globally recognized targets for improving living conditions in developing countries by 2015.

4. In June 2008 Vietnam's National Assembly passed the Law on Atomic Energy, thus laying the legal basis for the construction of its first nuclear power plant. Construction is planned to start in 2015, and the nuclear plant is expected to become operational by 2020. Vietnam News Agency, June 4, 2008.

5. Richard Armitage and Joseph Nye, *The U.S.–Japan Alliance: Getting Asia Right through 2020* (Washington, DC: Center for Security and International Studies, 2007), 10.

6. Ashton B. Carter and William J. Perry, "China on the March," *The National Interest* 88 (2007): 21.

7. The World Conservation Union in Viet Nam, *IUCN Viet Nam Strategic Framework 2007–2010: Finding the Balance in a Changing World* (Hanoi: Viet Nam Country Office, World Conservation Union, 2007), 30–31.

8. Khai Hoan, "Vietnam's Response to Climate Change," *Nhan Dan* (Hanoi), last updated November 10, 2009, www.nhandan.com.vn/english/life/291107/life_vnr.htm.

9. Pham Manh Cuong, director of the United Nations Convention to Combat Desertification Office in Hanoi, quoted by Australian Broadcasting Corporation Radio, April 9, 2008; "From Basket Case to Rice Basket," *Economist*, April 24, 2008.

10. "Vietnam: Climate Change Threatens Economic Progress," United Nations Office for the Coordination of Humanitarian Affairs, Integrated Regional Information Network, May 22, 2008, www.irinnews.org/Report.aspx?ReportId=78353.

11. World Bank, *Global Monitoring Report 2008: MDGs and the Environment Agenda for Inclusive and Sustainable Development* (Washington, DC: World Bank, 2008), www.google.com/url?sa=t&source=web&ct=res&cd=6&ved=0CB4QFjAF&url=http%3A%2F%2Fwww.imf.org%2Fexternal%2Fpubs%2FFT%2FGMR%2F2008%2Feng%2Fgmr.pdf&ei=atltS8_0OIngtgOBhNyxDQ&usg=AFQjCNEaANTtdSqe_MHIZ4nMhjH9EzLiPw&sig2=GJPGr0IgDO0AiWRegPdl3g; "Top Development Success Vietnam Also Top Climate Change Victim," Deutsche Press-Agentur, May 20, 2008, www.topnews.in/top-development-success-vietnam-also-top-climate-change-victim-242844.

12. "Vietnam Listed among Countries Lacking Water-Report," *Vietnam News Briefs*, August 12, 2008, http://mailman.anu.edu.au/pipermail/enviro-vlc/2008-August/000540.html.

13. United Nations Development Program, *Human Development Report 2007/2008*, 100; and "Climate Change Could Hit Viet Nam Hard, Says UN," *Viet Nam News*, November 29, 2007, www.dwf.org/blog/Lists/Posts/Post.aspx?ID=7. The most pessimistic climate change scenario developed by the IPCC projects sea-level rise of 59 centimeters by 2100 compared with 1990. Susmita Dasgupta, Benoit Laplante, Craig Meisner, David Wheeler, and Jianping Yan, *The Impact of Sea-Level Rise on Developing Countries: A Comparative Analysis*, World Bank Policy Research Working Paper 4136 (Washington, DC: World Bank, 2007), 30, provides maps showing incremental inundation effects for Vietnam if the sea level were to rise from 1 to 5 meters, at www-wds.worldbank.org/external/default/WDSContentServer/IW3P/IB/2007/02/09/000016406_20070209161430/Rendered/PDF/wps4136.pdf.

14. "Climate Change Could Hit Viet Nam Hard, Says UN," "Climate Change, Sea-Level Rise to Cost Vietnam Dearly," *Thanh Nien*, 12 January 2008, www.thanhniennews.com/education/?catid=4&newsid=34955; Tran Thuc, head of the Institute of Hydrometerology and Environment, quoted in "Conference Discusses Vietnam's Vulnerability to Climate Change," Thanh Nien, February 29, 2008, www.thanhniennews.com/education/?catid=4&newsid=36173.

15. Cited by Vo Van Kiet, "Climate Change Could Submerge Mekong Delta," *Thanh Nien*, March 1, 2008, www.thanhniennews.com/others/?catid=13&newsid=36207. The Ministry of Irrigation was incorporated into the Ministry of Agriculture and Rural Development. Kiet is a former prime minister who conducted an inspection tour of the central region in 2007 after a series of storms and floods.

16. "Conference Discusses Vietnam's Vulnerability to Climate Change."

17. "Vietnam to Be Severely Affected by Rising Sea Levels," *VietNamNet Bridge*, August 7, 2008, http://english.vietnamnet.vn/tech/2008/08/797563/.

18. Dasgupta et al., *Impact of Sea-Level Rise*, 44.

19. "Climate Change, Sea-Level Rise to Cost Vietnam Dearly."

20. Ibid.

21. A 101-year flood event is a commonly accepted risk assessment standard. Over the coming decades, the unprecedented growth and development of the Asian megacities will be a key factor in driving the increase in coastal flood risk globally.

22. R. J. Nicholls, S. Hanson, N. Patmore, S. Hallegatte, J. Corfee-Morlot, J. Chateau, and R. Muir-Wood, *Ranking Port Cities with High Exposure and Vulnerability to Climate Extremes: Exposure Estimates*, Environment Working Paper 1 (Paris: Organization for Economic Cooperation and Development, 2007), 7, www .preventionweb.net/files/1434_39889422.pdf.

23. Dasgupta et al., *Impact of Sea-Level Rise*, esp. charts on 41–43.

24. Edward C. O'Dowd, *Chinese Military Strategy in the Third Indochina War: The Last Maoist War* (London: Routledge, 2007), 89–107.

25. Carlyle A. Thayer, "Vietnam's Strategic Readjustment," in *China as a Great Power: Myths, Realities and Challenges in the Asia-Pacific Region*, edited by Stuart Harris and Gary Klintworth (New York: St. Martin's Press), 187–88.

26. David G. Marr, *Vietnam 1945: The Quest for Power* (Berkeley: University of California Press, 1995), 96–107.

27. "Vietnam Must Improve Sea Defence in Climate Change Fight," Agence France-Presse, March 27, 2008, http://afp.google.com/article/ALeqM5h95IXQSjgJnPNrKGjXtBk8-UmsKA.

28. See John T. McAlister and Paul Mus, *The Vietnamese and Their Revolution* (New York: Harper & Row, 1970); and, as popularized by Frances FitzGerald, *Fire in the Lake: The Vietnamese and the Americans in Vietnam* (Boston: Little, Brown, 1972).

29. The author conducted interviews with Vietnamese government officials in Hanoi in November–December 2007 and November 2009.

30. World Bank Global Facility for Disaster Reduction and Recovery, *Climate-Resilient Cities 2008 Primer: Reducing Vulnerabilities to Climate Change Impacts and Strengthening Disaster Risk Management in East Asian Cities* (Washington, DC: World Bank, 2008), 77, http://siteresources.worldbank.org/inteapregtop urbdev/Resources/Primer_e_book.pdf.

31. In other words, transplacement rather than replacement or transformation, to use terms coined by Samuel P. Huntington in *The Third Wave: Democratization in the Late Twentieth Century* (Norman: University of Oklahoma Press, 1991).

32. Milton Osborne, *Southeast Asia: An Introductory History* (Sydney: Allen & Unwin, 2010), 36–37.

33. Carlyle A. Thayer, "Building Socialism: South Vietnam Since the Fall of Saigon," in *Vietnam since 1975: Two Views from Australia*, by Carlyle A. Thayer and David G. Marr, Centre for the Study of Australian-Asian Relations Research Paper 20 (Nathan: Griffith University, 1982), 36–38.

34. Carlyle A. Thayer, "Vietnamese Refugees: Why the Outflow Continues," in *Refugees in the Modern World*, Canberra Studies in World Affairs 25, edited by Amin Saikal (Canberra: Department of International Relations, Australian National University, 1989), 45–96.

35. For a discussion of monoorganizational socialism, see Carlyle A. Thayer, "Mono-Organizational Socialism and the State," in *Vietnam's Rural Transformation*, edited by Benedict J. Tria Kerkvliet and Doug J. Porter (Boulder, CO: Westview Press, 1995), 39–64.

36. Carlyle A. Thayer, "One-Party Rule and the Challenge of Civil Society in Vietnam," paper presented to conference on Remaking the Vietnamese State: Implications for Viet Nam and the Region, University of Hong Kong, Hong Kong, August 21–22, 2008.

37. For a study that discusses the linkages between climate change, sea-level rise, and terrorism in Indonesia, the Philippines, and Bangladesh (but omits Vietnam), see Paul J. Smith, "Climate Change, Weak States and the 'War on Terrorism' in South and Southeast Asia," *Contemporary Southeast Asia* 29, no. 2 (2007): 264–85.

38. *Climate Resilient Cities*, 101–2; "Climate Change Scenarios Will Guide Government's Planners," *Viet-NamNet Bridge*, August 24, 2009, http://english.vietnamnet.vn/reports/2009/08/865038/.

39. "Vietnam Urged to Upgrade Dike Systems, Coastal Residents Threatened," *Thanh Nien*, March 25, 2008; "Vietnam Must Improve Sea Defense in Climate Change Fight," Agence France-Presse, March 26, 2008, http:// tinquehuong.wordpress.com/2008/03/27/vietnam-must-improve-sea-defence-in-climate-change-fight/.

40. The Ministry of Agriculture and Rural Development's strategy on forestry development by 2020 was approved in 2008. See "Vietnam Manages Forests to Deflect Climate Change," *Nhan Dan*, May 27, 2008, www.nhandan.com.vn/english/life/260408/life_v.htm.

41. "Vietnam to Spend $143.6 Mln on Climate Change Program," *Thanh Nien News*, December 10, 2008, http://thanhniennews.com/print.php?catid=4&newsid=44432.

42. McAlister and Mus, *Vietnamese and Their Revolution*.

43. Jon M. Van Dike, *North Vietnam's Strategy for Survival* (Palo Alto, CA: Pacific Books, 1972), 100–110.

44. John Kleinen, *Facing the Future, Reviving the Past: A Study of Social Change in a Northern Vietnamese Village* (Singapore: Institute of Southeast Asian Studies, 1999).

45. "President Triet Warns: Get Ready for Climate Change," *VietNamNet Bridge*, May 21, 2008, http://english.vietnamnet.vn/social/2008/05/784283/.

46. Joint Global Change Research Institute and Battelle Memorial Institute, *China: The Impact of Climate Change to 2030—A Commissioned Research Report*, NIC 2009-02D (Washington, DC: National Intelligence Council, April 2009), 33–34.

47. O'Dowd, *Chinese Military Strategy in the Third Indochina War*.

48. "Climate Change Requires Concerted Response," *VietNamNet Bridge*, April 16, 2008, http://english.vietnamnet.vn/tech/2008/04/778602/.

49. "Vietnam Moves to Implement Kyoto Protocol," *Thanh Nien*, February 16, 2008, www.thanhniennews.com/education/?catid=4&newsid=35759.

50. "Gov't Adopts Plan to Combat Effects of Global Warming."

51. *Viet Nam: Climate Change, Adaptation and Poor People* (Hanoi: Oxfam, 2008), 35–45; Arief Anshory Yusuf and Hermina Francisco, *Climate Change Vulnerability Mapping for Southeast Asia* (Singapore: Economy and Environment Program for Southeast Asia, 2009), 6 and 13–14; Asian Development Bank, *The Economics of Climate Change in Southeast Asia: A Regional Review* (Manila: Asian Development Bank, 2009), 30.

4

The Philippines

Paul D. Hutchcroft

The Philippines is the second-largest archipelagic country in the world, with 7,150 islands and 36,289 kilometers of coastline.[1] Rising sea levels caused by global warming would pose an enormous risk to the country, given that more than half the population resides in coastal areas, more than half the population is urban, and most of the large cities are on the coast.[2] According to CIESIN data, the Philippines is third (after India and China) in the number of people living in the 1-meter low-elevation coastal zone: roughly 15 million persons. It is ranked fourth (after India, China, and Vietnam) in the population that resides at the 3-meter level: roughly 18 million persons.

The Philippines

Along with its archipelagic character and high-density coastal population, the Philippines also possesses other characteristics that the IPCC has associated with high levels of vulnerability to climate change, notably susceptibility to extreme weather events and an economy "closely linked with climate-sensitive resources such as agriculture."[3] Two additional factors also pose strong obstacles to the capacity of the Philippines as it would seek to address the problems of global warming: severe levels of prior environmental degradation and an increasingly dysfunctional political system. As this chapter demonstrates, the Philippines faces grave challenges as it confronts the prospect of major climate change.

This brief analysis begins by surveying both the extraordinary environmental richness of the Philippines and the enormous stress experienced by its various ecological zones over the past several decades. Second, I examine the political problems that have both contributed to environmental degradation and obstructed the diligent and persevering efforts of civil society groups to promote more responsible policies (as well as more effective policy implementation). The third section proceeds, in a prospective vein, to examine how the country might deal with the coming challenges of global warming. I analyze "fight" (conflict), "flight" (migration), and reform as plausible societal responses to climate change; variance in the anticipated response of local governments to a new law on climate change; the overall capacity of the central state to manage the consequences of global warming; and the way in which environmental challenges could affect the country's relations with key international actors. In conclusion, I argue that a concerted and well-considered program of political and bureaucratic reform, combined with effective policy change, is a critical prerequisite for the Philippines to begin to confront the mounting environmental dangers that lie ahead.

A Biodiverse "Hot Spot" of Environmental Degradation

The Philippines is one of the world's most biodiverse nations. Its shores are surrounded by 27,000 square kilometers of coral reef, and only Australia's Great Barrier Reef contains a greater diversity of coral and fish species. Similarly, Philippine waters contain among the most diverse collections of seagrass species. Roughly two-thirds of the country's flora and fauna are not found anywhere else in the world. In terms of particular types of fauna, a large proportion are unique to the Philippines: 78 percent of amphibians, 68 percent of reptiles, 64 percent of mammals, and 44 percent of birds. Although the Philippines has only 0.2 percent of the world's land area, it is home to 6 percent of the world's bird species, 5 percent of its flora species, and 4 percent of its mammals. "As a gauge of biological diversity," one study explains, "it is reasonable to think of the Philippines as the Galapagos Islands multiplied tenfold."[4]

Even in the best of times, the frequency of typhoons, floods, earthquakes, and volcanic eruptions makes the Philippines one of the most disaster-prone countries in the world.[5] From an environmental standpoint, however, recent decades have brought unprecedented and mounting levels of stress in every major ecological zone. Given other root causes—including spiraling population growth, inequalities in access to land and resources, and politico-administrative deficiencies—it is increasingly misleading to treat disasters as primarily "natural" in origin.[6] The country's population has grown from 36.7 million in 1970 to roughly 95 million today. Given the weakness of population control policies during the past quarter century—as two of the four presidential administrations between 1986 and 2010 have been particularly responsive to the desires of the Catholic Church's hierarchy—it is estimated that the country may be home to more than 140 million people by 2040.[7]

Rural populations, most of whom reside in the uplands or on the coast and rely upon natural resources for their basic livelihoods, experience the country's ecological stress most directly.[8] Because the forest cover declined from 70 percent of total land area in 1900 to 19 percent in the late 1990s, timber and wood products now constitute a negligible proportion of total exports. Deforestation, along with the conversion of land to agriculture, has contributed to major problems of soil erosion. Writing in 2003, the economists Ian Coxhead and Sisira Jayasuriya asserted that "watershed degradation as a consequence of deforestation has emerged as perhaps the most important environmental problem in the Philippines, given that impacts are felt not only in the uplands but also very widely in the lowlands."[9] They also note increased annual fluctuation in stream flow in watersheds, with associated problems of drought and flash flooding. Many of the country's major rivers and lakes are severely polluted, with highly inadequate measures to control waste from industry, mining, households, and agriculture. Sedimentation has reduced the storage capacity of the country's major reservoirs, to the detriment of power generation, water consumption, and irrigation. In the last quarter of the twentieth century, there was a 20 percent to 30 percent decline in dry-season irrigated area. As Coxhead and Jayasuriya conclude: "With the upland frontier virtually closed and emerging signs of productivity growth slowdown—or even reversal—in the 'best' lowland irrigated areas, the degradation of the agricultural land base is a source of serious concern."[10]

Environmental stress in coastal areas is equally striking. Mangroves were reduced from 450,000 hectares in 1920 to 120,000 hectares in 2001. Ninety-five percent of coral reefs have had some degree of degradation, with one-third judged to be in poor condition. Others have been damaged by pollution, sedimentation from deforestation, population growth, coastal development, tourism, and destructive fishing methods.

Philippine urban areas are also under mounting stress, with major infrastructural deficiencies in water, sewage and drainage, transportation, and pollution control. This is most dramatically exemplified by the megalopolis of Manila, with its huge concentrations of urban poor. A mere 13 percent of Metro Manila households are linked to a sewage system, and 40 percent of solid household waste is dumped illegally. Many rivers and streams in the country's capital region are biologically dead. More than a decade ago, Philippine development planners explained that "the absence of far-reaching comprehensive land use and human settlement plans has resulted in the deterioration of the country's cities as human habitats beset with interrelated problems like inadequate mass transportation and road systems; pollution and inadequate and inappropriate waste disposal systems; flooding; water shortage; deterioration and lack of basic social services; and proliferation of crime and other social evils."[11] Since this bleak assessment was written in 1998, the degradation of the urban environment has become even more pronounced.

In summary, the Philippines confronts problems of climate change after decades of sustained environmental degradation.[12] "The Philippines is both a megadiversity country and one of the world's highest-priority hot spots," proclaimed Russell Mittermeier, president of Conservation International in 1998. "Indeed, its combination of high diversity, very high endemism, and extremely high levels of threat make it the 'hottest of all hot spots' and the country that deserves maximum attention from the international community."[13] Though the Philippines has little responsibility for global greenhouse gas emissions, generating only 0.27 percent of the world's total, it has the potential to bear a substantial burden in terms of the costs of climate change.[14] According to projections cited by the Asian Development Bank, the Philippines could face a temperature increase of 1.2 to 3.9 degrees Celsius by 2080; at the middle of this range, with a 2 degree Celsius increase, rice yields would decline by 22 percent. Sea levels are projected to increase by 0.19 to 1.04 meters by 2080. At the upper end of this range, as already explained above, there are currently 15 million people living within a 1-meter low-elevation coastal zone.[15]

A Political Environment also under Stress

As it addresses the problems ahead, the Philippines can draw on its extraordinarily talented human resources combined with a favorable strategic location. As summarized by the World Bank, "the Philippines has a tremendous range of assets to draw upon for its development—a relatively educated populace with proficiency in English, plentiful managerial and entrepreneurial talent; global demand for its labor force, . . . as confirmed by a huge number of Filipino overseas workers (8 million, or 25 of active workforce); . . . a vibrant private sector: a strong positioning in several dynamic sectors (e.g., electronics, offshore [information technology], tourism), a strategic location in the fast-growing East Asia region; and an active civil society, free media, and articulate urban class."[16]

Long-standing deficiencies in political and administrative structures, however, pose major obstacles to confronting the challenges ahead, whether it be ongoing issues of environmental degradation or future projections of climate change. The Philippines has experienced substantial political instability over the past quarter century, with a particularly volatile period since 1998.[17] President Joseph Estrada, who was elected with a decisive majority in 1998, was deposed via a "people power" uprising in 2001. He faced widespread anger over corruption charges, including allegations that he was receiving regular cash payments through ties to illegal gambling syndicates. After the ascension of Gloria Macapagal-Arroyo to the presidency in January 2001, an already crisis-prone democracy faced an unusually high number of travails, including an uprising by the urban poor that nearly breached the walls of the presidential palace on May Day 2001; a botched military mutiny in July 2003; corruption scandals involving the first family; allegations of presidential involvement in fixing the 2004 elections; a failed coup attempt–cum–popular uprising in February 2006 that led to the declaration of emergency rule; concerted attacks on the press; an alarming spike in extrajudicial killings; impeachment attempts in 2005, 2006, and 2007; two major bribery scandals in late 2007, one involving the chief election officer and the other brazen cash payouts to congresspersons and governors at the palace; a November 2007 bombing at the House of Representatives that killed a notorious warlord congressman from Mindanao; an incident the same month in which junior officers barricaded themselves at a luxury Manila hotel in protest against the Arroyo administration; ongoing conflict between the government and Muslim secessionist groups in Mindanao, and also between the government and Communist rebels; and a late 2009 election-related massacre of fifty-seven persons in the Mindanao Province of Maguindanao.

Macapagal-Arroyo very effectively wielded the substantial powers of the presidency to keep herself in office, and in the process she exhibited no qualms about further undermining the country's already weak political institutions. As the Philippines suffers one political crisis after another, its long-standing democratic structures become increasingly imperiled. "All across the political spectrum," observed leading political analyst Sheila Coronel in 2007, "there is deep dissatisfaction with the country's dysfunctional democracy but no consensus on how it should be reformed."[18] The decisive May 2010 election of Benigno "Noynoy" Aquino to the presidency has raised new hopes, with opinion polls registering extraordinary levels of public trust in his leadership as he assumed the helm. In the midst of heightened expectations, however, there is no assurance that the country's political institutions will be able to respond to the pressing needs of the Philippine citizenry—particularly the poor and excluded mass of the population.[19]

In the quarter century since the fall of Ferdinand Marcos's crony-capitalist authoritarian regime in 1986, spirited hopes for democratic change have alternated with dispirited frustration over the character of the country's democracy. The problems faced by the political system combine with deeper deficiencies of the Philippine bureaucracy, which is widely regarded to be low in capacity and high in levels of corruption.[20] There is a high degree of political control over the bureaucracy, with a well-entrenched spoils system dating to the American colonial period. Within this system, both the national executive and legislators are able to intervene on a particularistic basis in matters of appointment and promotions. The formulation of national goals is undermined by the prevalence of the pork barrel, in which national legislators have a high degree of discretion in allocating funds to their localities. At the base of the system are local political forces, which sometimes rely on illegal activities (gambling, smuggling, drugs, prostitution, etc.) and coercive resources to enhance their power. The more control that local politicians can exert over their town, city, or provincial vote base, the more leverage they can gain with the center.

The territorial presence of the state throughout the archipelago is inconsistent and often very weak. As an old joke says, with obvious exaggeration, the only two truly national institutions in the Philippines are the Catholic Church and San Miguel Brewery. At the periphery, especially in parts of Muslim Mindanao, local authority is fragmented. State officials must negotiate with a range of local political forces (governors, mayors, military commanders, rebel commanders, and warlords) in order to effect basic tasks. In sum, the Philippines is a patronage-based state, with weak command and control by Manila through formal bureaucratic structures.[21] Important functions of the state are subcontracted to local politicians, some of whom use their authority for reformist purposes and some of whom are intent only on solidifying control over their bailiwick. The relative powers of national versus local politicians are continually negotiated and renegotiated, with the latter enjoying the greatest leverage at election time, when they can exchange control over votes for patronage resources from the center. As long as local allies supply the votes, national politicians look the other way when they build up coercive power and engage in illegal activities.

Historically, the Philippine state has had difficulty raising sufficient revenue, and its overall "tax effort" (total revenue as a percentage of GDP) has been low (in 2008, only 14.1 percent).[22] In a 2005 report, the World Bank observed that "tax evasion, avoidance, and corruption in revenue agencies and special tax exemptions and incentives have undermined fiscal revenue generation."[23] As part of the 1991 Local Government Code, local governments benefit from a major revenue-sharing program, whereby 40 percent of internal revenue taxes are given out to provinces, cities, and municipalities throughout the archipelago. Given the availability of this easy money from Manila, local politicians have few incentives to promote local revenue generation.[24] Under President Arroyo, new tax measures were put into place. But fiscal stability remains a long-term struggle, particularly given major infrastructural deficiencies and the need to meet the challenges posed by major environmental stress.

Early in the new century, restive elements of the military once again revealed putschist tendencies. This follows the bout of coup attempts against the administration of Corazon Aquino (1986–92), and then a hiatus in military adventurism under the presidency of former general Fidel Ramos (1992–98). The military and the police played a major role in tipping the balance against Estrada when he was deposed in 2001. There were three military actions against Macapagal-Arroyo's government, in 2003, 2006, and 2007. In each case, junior officers disaffected with her corruption and electoral manipulation led botched attempts to rally popular opposition to the government. Those who want to see the perpetuation of civilian democratic structures might find some comfort in the fact that—despite numerous attempts over the past quarter century—elements of the Philippine military have never launched a successful grab for power. Given the underlying weakness of political institutions, however, it would be a mistake to dismiss recurring possibilities of a coup. After all, the odds that such an attempt might succeed depend not only on the capability of a group of disgruntled soldiers but also upon the nature of the political institutions that are being targeted. Although the Philippines can boast the oldest democratic structures in Asia, they are prone to crisis and often lacking in legitimacy.[25]

The weakness of Philippine political parties contributes to a lack of stability in the political sphere. The country's political parties are nearly indistinguishable from one another in terms of programmatic and policy positions. Politicians have little allegiance to party labels, frequently bolting from one party to another in search of the greatest access to patronage resources. Political divides are ever shifting, uniting former rivals and dividing former allies in a continual process of alignment and realignment almost entirely divorced from coherent positions on policies or programs. The political scientist Nathan Quimpo provides perhaps the best description of contemporary Philippine political parties: "convenient vehicles of patronage that can be set up, merged with others, split, reconstituted, regurgitated, resurrected, renamed, repackaged, recycled, refurbished, buffed up or flushed down the toilet anytime."[26]

In recent years, elements of the so-called "covert netherworld" have enjoyed growing influence over the political process.[27] As Philippine elections have become increasingly costly, they have encouraged politicians to become more creative in raising funds, whether through the promise of legislative and regulatory favors, real estate scams, involvement in gambling syndicates, or links to drug lords and the underworld. In a surprisingly candid moment in late 2007, House Speaker Jose de Venecia explained the system: "It's the drug lords and the gambling lords . . . who finance the candidates. So from Day One, they become corrupt. So the whole political process is rotten."[28]

Assessing Responses to Climate Change: Society, State, and International Actors

As the above discussion emphasizes, the Philippines faces major challenges even in the absence of global warming. Environmental degradation is already well advanced throughout the country, in both rural and urban areas, and its political and administrative structures are under mounting stress. Climate change is thus proceeding in the context of both environmental and political instability, which suggests a low degree of adaptive capacity to the projected effects of climate change. The societal responses to climate change are sometimes summarized in terms of "fight or flight," and in the Philippines one should anticipate both—increased levels of societal conflict along with continuing high levels of migration, both domestic and international.

There are already strong fissures in Philippine society. The first is large socioeconomic disparities, quantified in the form of a Gini ratio of 45.8.[29] This can be attributed in part to the absence of effective measures of land reform at critical points in the country's history, notably in the 1950s. Land issues are an important factor in the continuing strength of the New People's Army of the Communist Party of the Philippines, which has for nearly forty years been engaged in Maoist armed struggle against the government. They endure, despite major internal party conflict in the late 1980s and early 1990s (over which much blood was spilled), a lack of charismatic leadership, and major strategic blunders (e.g., the boycott of the 1986 elections, a decision that put the party on the sidelines during the subsequent "people power revolution" that toppled Ferdinand Marcos and brought Corazon Aquino to power).

Yet the Philippines has long been a highly inequitable society, and perhaps the most important question is why there has not, historically, been *more* mobilization along class lines. Especially puzzling is the absence of a mainstream party able to provide consistent representation to the interests of those at the bottom end of the socioeconomic spectrum. At least four factors help to answer these questions. The first is government repression of the left, and the active exclusion of the left from mainstream politics. These patterns can be traced to early efforts at peasant organization in the 1930s, and they were found most strikingly in the 1946 expulsion of the leftist Democratic Alliance from the House of Representatives. This expulsion, in turn, encouraged the left to abandon parliamentary democracy in favor of an armed struggle that was not defeated until the early 1950s. Second, the Philippines has historically had strong patron–client ties, classically exemplified in the relationship between landlord and tenant but also present in other types of interclass relations. In many (but certainly not all) cases, these linkages provide some small measure of social protection to those who have ties to a patron of higher social status.

Third, a complex system of patronage binds together the national polity from the presidential palace downward to the provincial, city, municipal, and *barangay* (barrio) levels. Politicians derive the greatest benefit from the system, with the quantity of largesse decreasing as one moves further down the political ladder, but there are some minimal concrete benefits to those at the bottom of society. For ordinary citizens who derive such minimal benefits, explains the economist Emmanuel de Dios, "government is an abstraction, an alienated entity, whose only palpable dimension is the episodic patronage dispensed by bosses and politicians, which merely reinforces the poor's real condition of dependence."[30] Fourth and finally, the patronage-oriented nature of the political system presents major obstacles to efforts to build a strong and well-institutionalized political party able to represent the interests of the poor in the electoral arena. The specific character of Philippine electoral and representational institutions, as they have evolved over the past century, have very strongly tended to nurture a political culture that privileges particularistic benefits over policy, and pork over programs.

Class divides remain strong, and they were evidenced most dramatically in a May Day 2001 uprising, in which huge numbers of the urban poor almost broke through the gates of the presidential palace as they were protesting the arrest of former president Joseph Estrada. This type of mobilization has not been repeated, and it is difficult to predict what might precipitate it again in the future. One can easily imagine, for example, a scenario of substantial social stress over access to such basic human needs as water—in effect, a fight for resources at least partly drawn along class lines. But there are factors, historically, that have mitigated and discouraged conflict along class lines. These factors may continue to dampen the prospects of such conflict in the future.

The second major fissure to consider is religion, primarily the majority Christian population and the minority Muslim population (which makes up roughly 5 percent of the population and is geographically concentrated in Mindanao and the Sulu Archipelago). There are long-standing historical grievances among the Mindanao Muslims at their treatment by Manila and by Christian settlers, and the Muslim areas of Mindanao are among the very poorest in the country.[31] Manila faces two major Muslim secessionist movements, the Moro National Liberation Front (MNLF, formed in 1972) and the Moro Islamic

Liberation Front (MILF, initially a splinter group of the MNLF, formed in 1977). Although the government forged a peace agreement with the MNLF in 1996, peace has been elusive between the government and the MILF. A major agreement was supposed to be signed in August 2008, but, amid fervent opposition from Christian settler politicians, it was struck down by the Supreme Court. After an escalation of conflict that displaced hundreds of thousands of persons, peace negotiations resumed—albeit with little substantial progress.[32]

At the same time, there have historically been many incentives for Muslim politicians to cooperate with Manila, and to integrate themselves within patronage networks that provide resources to them and to their constituents. It must also be emphasized that major divisions exist within Muslim Mindanao, based on differences of ethnicity, region, and class. On balance, there is likely to continue to be simmering conflict in Mindanao for some time to come, involving particular groups that have grievances against Manila or disputes with neighboring Christian settler populations over land or other resources. There is no reason, however, to anticipate conflict strictly along religious lines, Muslim versus Christian.

A third possible fissure is ethnicity, which one might anticipate in a country that has a large number of distinct ethnolinguistic groups. Strikingly, however, ethnic divisions are not a major source of political contention among lowland Christian groups, which make up more than 90 percent of the population and encompass great ethnolinguistic diversity (with Tagalog, Ilocano, Kapampangan, and Bicolano populations predominating in Luzon; and Cebuano, Waray, and Ilonggo populations constituting the major groups in the Visayas). Largely because of strong intra-elite ties at the national level, and a well-developed system of distributing patronage outward from Manila, ethnolinguistic diversity is not a major political fissure among lowland Christian Filipinos.

When one shifts from ethnic identity to pan-ethnic identity, there are indeed two very important fissures in Philippine society: one between lowland populations and the peoples of the mountainous Cordillera region of Luzon, and another between lowland Christian populations and the populations of Muslim Mindanao. In each area, there is substantial resentment of Manila and lowland Christians. But this is balanced by the very substantial ethnolinguistic diversity that exists among the populations of both the Cordillera and Muslim Mindanao. The division and rivalry between the MILF and the MNLF, for example, is based in large part on ethnic differences (Maguindanao vs. Tausug, respectively).

To summarize, should we anticipate conflict to come forth from climate change? A positive prognosis is that conflict is not an inevitable result of climate change, particularly given the mitigating factors noted above. The more cautionary prognosis is that climatic shifts can be expected to exacerbate and intensify already-existing lines of conflict. This accords with a recent World Bank analysis, which asserts that "disasters do not by themselves cause conflict. . . . Adverse environmental changes are expected to act as a threat multiplier by *adding strains to already-vulnerable populations in poor and fragile societies*" (emphasis added). As the World Bank report explains in terms of cross-national variation, one can also note in terms of subnational variation in the Philippine archipelago—areas "which already suffer disproportionally from instability and violence . . . face a double security challenge through additional climate-imposed strains on human health and livelihood."[33] This is evident by examining subnational variation in vulnerability to sea-level rise. Of the six provinces of the Philippines that are most vulnerable to sea-level rise (in terms of area that is within the 1-meter low-elevation coastal zone), four are in areas of Mindanao and the Sulu Archipelago heavily affected by banditry, clan warfare, and conflict between the government and Muslim secessionists. A fifth province is on Samar Island, long a stronghold of the Communist New People's Army.[34] Sadly, then, many areas of the Philippines that are particularly prone to the ill effects of climate change already experience particularly high levels of poverty, instability, and conflict.[35]

Moving from "fight" to "flight," the most basic point to emphasize is that the Philippines is a country in which there is already a great deal of internal as well as outward migration. The late 2008 resurgence of conflict in Mindanao forced some four hundred thousand persons from their homes, and the past four decades of conflict in Mindanao has encouraged further hundreds of thousands of Filipinos to relocate to nearby Sabah in Malaysia. Each year, typhoons, floods, mudslides, and other natural disasters force tens of thousands of Filipinos from their homes. Severe climate change, it can be anticipated, would only further these long-standing trends.

Spurred by better economic opportunities abroad, there are currently at least 8 million Filipinos (25 percent of the active labor force) working overseas. They are the country's largest export, bringing in nearly $21 billion per year in remittances in 2010. The recent rapid growth of telephone call centers helps to absorb some of the growing domestic labor force, but the impulse toward external migration can be expected to continue, and probably accelerate, into the indefinite future. In sum, patterns of flight are already well established. These patterns will continue as long as people feel the need to flee dangers of many forms, not only military conflict and natural disasters but also environmental degradation, water shortages, and rising sea levels.

On a more positive note, the Philippines has important latent reserves of strength in its strong and well-organized civil society. There are many nongovernmental organizations dedicated to environmental causes, and from their base in Manila they enjoy strong linkages with local groups throughout the country. Through their efforts, important environmental reforms have been passed, including, for example, laws establishing "protected areas" and seeking to safeguard water supplies.[36] During the past twenty-five years, dedicated groups of investigative journalists have emerged and demonstrated a seemingly fearless desire to expose corruption and environmental abuses. There are also important reform elements within the central government and at the local level. This includes governors and mayor who work with citizens' groups to actively promote policies of environmental protection. In some localities, for example, citizens join Bantay Dagat (Sea Patrol) and guard against illegal fishing. The tourism industry has a vested interest in preserving the coral reefs and other environmental treasures that attract visitors to their resorts.

These are very positive trends, but they need to be backed up with consistent and effective action from the central government. Due to ineffective implementation, environmental laws do not achieve their intended goals. Nongovernmental organizations can push for important changes, but they do not have the institutional strength that would come from the emergence of stronger and more programmatic political parties. Investigative journalists have in the past exposed illegal logging, for example, but this did not halt the practice or lead to the prosecution of those who violated the law. The efforts of reformist and environmentally oriented local politicians are unfortunately counterbalanced by the host of other local politicians who do not hesitate to despoil the environment for personal gain.

One positive development at the national level is the October 2009 passage of the Climate Change Act. As the law declares, it is "the policy of the State to systematically integrate the concept of climate change in various phases of policy formulation, development plans, poverty reduction strategies and other development tools and techniques by all agencies and instrumentalities of the government." A new Climate Change Commission has been established, headed by the president and provided with an initial appropriation of 50 million pesos ($1.125 million). Disaster risk reduction efforts are to be integrated into programs and initiatives relating to climate change, and a twelve-year framework strategy on climate change was adopted in April 2010. Many national agencies are given explicit roles in promoting the act's objectives; the Department of Education is to integrate climate change issues into the curriculum of the nation's schools, the Department of the Interior and Local Government is to train local government units in climate change issues, the Department of the Environment and Natural Resources is to maintain a "climate change information management system," the Department of Foreign affairs is to continue the Philippines' long-standing involvement in international agreements on climate change, and government financial institutions are to "provide preferential financial packages for climate-change-related projects."[37]

For all its well-considered provisions, it remains to be seen if this law will have any more success in terms of implementation than previous well-considered environmental laws. "It's always easy to say that we are mainstreaming a certain issue into the mandates of [government] agencies," explained Yeb Saño, head of the World Wide Fund for Nature, "but it's another thing to be able to actually see that into implementation."[38] It is notable, for example, that the law proclaims that local governments "shall be the frontline agencies" in supporting climate change actions plans within "their respective areas" but provides no clear mechanism for ensuring their involvement. No penalties are specified for nonengagement, and one can anticipate that local politicians will follow their own instincts within highly divergent local settings. As part of well-established patterns of Philippine politics, moreover, it is unlikely that national politicians will challenge poorly performing local counterparts as long as they deliver the votes from their bailiwicks at election time.[39]

Indeed, variations in local leadership, local political dynamics, and local government capacity are likely to result in major differences in how Philippine localities are able to respond to the challenges of climate change. Some local leaders are dedicated to sound environmental management, but others blatantly abuse their position to secure personal enrichment at the expense of the environment.[40] Some localities have monopolistic control by one family, whereas others witness ongoing battles for control among rival clans, and still others have well-developed mechanisms of civil society engagement. As the economist Emmanuel S. de Dios explains: "Hobbesian situations where contending political clans vie for power with inconclusive or impermanent results will be associated with the worst development outcomes." The optimal situation, he further explains, is found when "political power becomes less based on the charisma or personal reputation of the local leaders and instead relies more on a civic constituency with a pronounced stake in the provision of local public goods."[41]

Given the variability that exists at local levels, strong central government initiatives and oversight are essential. As a 2009 Asian Development Bank report rightfully observes, "Climate change is the most serious market failure the world has ever witnessed. Like any market failure, it can only be resolved through the intervention of public policy."[42] How might the Philippine state cope with the consequences of climate change? Over the past decade, one can discern a major decline in already-weak Philippine political institutions. Further erosion should be expected, with probable increases in de facto regional autonomy. Under a worst-case scenario, political decay could benefit drug lords and underworld figures along with radical Islamists anxious to forge stronger linkages with armed groups in Mindanao. To the extent that the central state becomes even weaker, they would have the opportunity to consolidate their political influence and possibly enjoy de facto territorial gains.

At the same time, it must be emphasized that the Philippines often exhibits a surprising degree of resilience. Though the patronage-based state is highly inefficient in delivering public goods, it displays a remarkable capacity to muddle through from one crisis to the next. Patronage serves as important glue for the polity as a whole, binding politicians to the center in their search for particularistic resources and gain. In addition, informal networks among major political-economic elites have in the past been able to foster ad hoc means of resolving a national crisis; traditionally, both Catholic Church leaders and American ambassadors played an important role in brokering agreements among rival parties. Can this resilience endure, and will the country be able to continue to muddle through the mounting challenges of political and environmental stress? One can hope so, but there are certainly reasons not to be overly sanguine.

Given the weakness of the Philippine state, international actors can be expected to play a major role as the country confronts the challenges of climate change. Historically, there are many occasions in which the United States, as the former colonial power, has rescued the Philippine state from crisis. The removal of U.S. military bases from the Philippines in the early 1990s greatly reduced the strategic incentives to continue this policy. Beginning in the late 1990s, and especially after September 11, 2001, however, strategic ties between the two countries were renewed on many fronts and the Philippines came to play a prominent role in the United States–led "war on terror." This may give the United States a strategic incentive to provide sustained assistance to the Philippines as it attempts to meet the challenges of climate change. In the short term, this could come in the form of assistance in dealing with natural disasters—particularly to the extent that these disasters become more frequent and more intense. Through historical ties as well as more current trends in immigration, the Philippines will continue to have reason to maintain strong linkages to the United States.

At the same time, rivals to the United States could also find this an opportunity to build stronger ties with Manila—and reorient the country away from its strong ties to America. The most likely such rival, of course, is China. Also for strategic reasons, other regional actors (notably Japan, South Korea, Taiwan, and Singapore) will likely find it advantageous to assist the Philippines with the challenges of climate change. Deepening environmental stress could make the country ever more dependent on outsiders able to provide assistance in responding to typhoons, flooding, mudslides, drought, earthquakes, tsunami, and rising sea levels. This dependence would reduce the Philippines' scope of options as it deals with perceived security interests. In areas of disputed sovereignty, notably the Spratly Islands in the South China Sea, the Philippines would have increasing difficulties defending its claims against those being pressed by countries with stronger military power. Given the weakness of the Philippine Coast Guard, one can anticipate ongoing problems with incursions of fishing vessels from neighboring countries. As the chair of the Senate committee on defense and security declared in the mid-1990s, with only slight exaggeration: "We have an air force that can't fly and a navy that can't go out to sea. When the U.S. left, we were basically naked."[43] Plans for military modernization in the 1990s were shelved in the face of the Asian economic crisis, and there has been no substantial increase in the capacity of the Philippines to protect its more than 36,000 kilometers of shoreline.

Coping with Climate Change: The Imperatives of Political Change

Projected increases in sea levels would pose a major challenge to any country with a substantial coastline, but for the many reasons outlined above, the Philippines faces a particularly grave threat—millions of persons residing in low-elevation coastal zones, historical susceptibility to extreme weather and natural disasters, agriculture and fishing and other climate-sensitive resources as primary sources of livelihood for the country's rural poor, severe environmental degradation in recent decades, and political and administrative structures under major stress. Coping with this threat demands a very carefully considered program of reform, with the goal of helping the country prepare for the enormous challenges ahead. The institutional reform of political structures needs to be crafted with the goal of nurturing a polity more oriented toward the long-term public good and less toward the short-term goals of the patronage system. Administrative structures need to be freed from the constraints of the spoils system, with incentives for effective implementation of public policy rather than mere responsiveness to the particularistic demands of dominant elites. With a more effective bureaucracy in place, creative new policy measures must be implemented to move toward the goal of reversing the myriad causes of environmental degradation. Civil society groups have long articulated the urgency of change, but the country has lacked the political and administrative structures required to bring forth a coherent program of environmental restoration. Climate change forces the issue. If the Philippines is to begin to confront the mounting environmental dangers that lie ahead, a concerted and well-considered program of political and administrative reform is a necessary first step.

Notes

The author is grateful for comments received from participants in the December 2007 conference at the Naval Postgraduate School, and most especially for editorial suggestions from Daniel Moran and the anonymous readers. Thanks also go to Helen Mendoza, president of the Philippine Network on Climate Change, for kindly providing copies of key recent government reports; to Doracie Zoleta-Nantes of the Australian National University for citations in the field of natural hazards analysis; and to Thuy Thu Pham of the Department of Political and Social Change at the Australian National University for invaluable research assistance. All errors, of course, are the author's alone.

1. Of the Philippines' 7,150 islands, 6,700 have an area less than one square mile (2.6 square kilometers). The vast bulk of the population resides on the 11 largest islands: Luzon in the North, Mindanao in the South, Palawan and Mindoro in the West, and 7 major islands in the central portion of the archipelago: Panay, Negros, Cebu, Bohol, Masbate, Leyte, and Samar. Many smaller islands, however, have fairly substantial population (e.g., Catanduanes, Marinduque, Romblon, Guimaras, Mactan, Siquijor, Camiguin, Basilan, Jolo, and Tawi-Tawi). The Philippine coastline is exceeded only by those of Canada (202,080 kilometers), Indonesia (54,716 kilometers), Greenland (44,087 kilometers), and Russia (37,653 kilometers). The United States (19,924 kilometers) has slightly more than half the coastline of the Philippines. *CIA World Fact Book*, www .cia.gov/library/publications/the-world-factbook/fields/2060.html.

2. Four of the country's five largest urban areas are on the coast: Metro Manila (with a population of more than 12 million), metropolitan Cebu (more than 1 million), Davao (more than 1 million), and Zamboanga City (more than 500,000). Only Angeles City (also more than 500,000) is landlocked. World Bank, East Asia Infrastructure Department, "Issues and Dynamics: Urban Systems in Developing Asia," n.d., http://site resources.worldbank.org/INTEAPREGTOPURBDEV/Resources/Philippines-Urbanisation.pdf.

3. CIESIN, "Country-Level Exposure to Potential Sea-Level Rise," 2007, 6, 8; IPCC, *Climate Change 2007: Impacts*, 50. The Philippines also faces many of the problems that the IPCC report (p. 63) highlights as characteristic of "small islands" generally.

4. Lawrence R. Heaney, "Discovering Diversity," in *Vanishing Treasures of the Philippine Rain Forest*, edited by Lawrence R. Heaney and Jacinto C. Regalado Jr. (Chicago: Field Museum, 1998), 9, www.field museum.org/vanishing_treasures/.

5. "Between January 1900 and May 2006, . . . 379 disasters . . . caused economic damages worth USD 7 billion and killed more than 48,000 people." J.-C. Gaillard, C. C. Liamzon, and J. D. Villanueva, "'Natural' Disaster? A Retrospect into the Causes of the Late-2004 Typhoon Disaster in Eastern Luzon, Philippines," *Environmental Hazards* 4, no. 4 (2007): 257. Other data, examining the period 1905–2006, categorize 406 disasters as follows: typhoons, 60 percent; floods, 19 percent; and volcanic eruptions and earthquakes, each 5 percent. Asian Development Bank, *The Economics of Climate Change in Southeast Asia: A Regional Review* (Manila: Asian Development Bank, 2009), 32, www.adb.org/Documents/Books/Economics-Climate -Change-SEA/PDF/Economics-Climate-Change.pdf.

6. This point draws on Gaillard, Liamzon, and Villanueva, "'Natural' Disaster," whose analysis is based on typhoons in Quezon Province in late 2004. The Asian Development Bank's 2009 report (*Economics of Climate Change*, 31) notes a significant increase in the frequency of typhoons in the Philippines from the years 1990 to 2003. A host of recent storms are listed in Greenpeace Southeast Asia, "The Philippines: A Climate Hotspot," Quezon City, April 2007, 7; and Greenpeace Southeast Asia, "Climate Change and Water: Impacts and Vulnerabilities in the Philippines," December 21, 2009, 22, www.greenpeace.org/seasia/en/press/reports/ climate-change-water-impacts-philippines. The latter includes the disastrous floods in Metro Manila and Central Luzon that came in the wake of Typhoon Ondoy (Ketsana), September 26–28, 2009.

7. This is based on estimates of the National Statistics Office, www.census.gov.ph/data/pressrelease/2006/ pr0620tx.html.

8. World Bank Group in the Philippines, *Country Assistance Strategy for the Philippines, 2006–2008* (Washington, DC: World Bank, 2005), 19.

9. Ian Coxhead and Sisira Jayasuriya, "Environment and Natural Resources," in *The Philippine Economy: Development, Policies, and Challenges*, edited by Arsenio M. Balisacan and Hal Hill (New York: Oxford University Press, 2003), 385.

10. Ibid., 386.

11. Republic of the Philippines, *The Philippine National Development Plan: Directions for the 21st Century* (Manila: Republic of the Philippines, 1998), quoted by Coxhead and Jayasuriya, "Environment and Natural Resources," 390–91.

12. In discussing these patterns from the past, it is worth noting an important point made by one analyst: "Inter-temporal inferences may be invalid" because climate change is "a novel problem" and "the future is likely to be different from the past." See Joshua W. Busby, "Who Cares about the Weather? Climate Change and U.S. National Security," *Security Studies* 17, no. 3 (2008): 471. But if a country has already had major problems of resolving environmental problems in the past, and if these problems are likely to be greatly heightened in the process of climate change, and if the larger political system seems to be increasingly dysfunctional, there does indeed seem to be a strong basis for moving from a focus on historical patterns to some tentative "inter-temporal inferences."

13. Heaney and Regalado, *Vanishing Treasures*; quoted from the back cover.

14. Overall, according to the World Bank, developing countries are expected to "bear 75–80 percent of the costs of damage caused by changing climate." "Philippines: World Bank Report Urges 'Climate Smart' Growth Strategy," www.preventionweb.net/english/professional/news/v.php?id=12668.

15. Asian Development Bank, *Economics of Climate Change*, 27, 34, 43.

16. World Bank, "Philippines and the World Bank: Overview," www.worldbank.org.ph/WBSITE/EXTER NAL/COUNTRIES/EASTASIAPACIFICEXT/PHILIPPINESEXTN/0,,contentMDK:20203936~menuPK:33 2990~pagePK:141137~piPK:141127~theSitePK:332982,00.html.

17. This section draws on Paul D. Hutchcroft, "The Arroyo Imbroglio in the Philippines," *Journal of Democracy* 19, no. 1 (January 2008): 141–55.

18. Sheila S. Coronel, "The Philippines in 2006: Democracy and Its Discontents," *Asian Survey* 47, no. 1 (2007): 175–82.

19. Paul D. Hutchcroft and Joel Rocamora, "Patronage-Based Parties and the Democratic Deficit in the Philippines: Origins, Evolution, and the Imperatives of Reform," in *Handbook of Southeast Asian Politics*, edited by Richard Robison (London: Routledge, in press).

20. See Paul D. Hutchcroft, *Booty Capitalism: The Politics of Banking in the Philippines* (Ithaca, NY: Cornell University Press, 1998). In the World Bank's 2008 Worldwide Governance Indicators, the Philippines is given a score of 26.1 in the category of "control of corruption." This compares with 99.5 for Singapore and 1

for Myanmar. In the category of "rule of law," the Philippines scores 39.7; Singapore, 93.8; and Myanmar, 4.8. See http://info.worldbank.org/governance/wgi/mc_chart.asp.

21. See Paul D. Hutchcroft, "Dreams of Redemption: Localist Strategies of Political Reform in the Philippines," in *The Politics of Change in the Philippines*, edited by Yuko Kasuya and Nathan Quimpo (Manila: Anvil Press, 2010). Further analysis of patronage ties between the center and localities is given by Emmanuel S. de Dios, "Local Politics and Local Economy," in *The Dynamics of Regional Development: The Philippines in East Asia*, edited by Arsenio M. Balisacan and Hal Hill (Cheltenham, U.K.: Edward Elgar, 2007).

22. World Bank data from http://data.worldbank.org/indicator/GC.TAX.TOTL.GD.ZS. According to an analysis by the head of the National Tax Research Center, the Philippines ranked among the lowest in Southeast Asia in terms of tax effort in the years 2001–5. See Lina D. Isorena, "The Philippines' Continuing Quest for Financial Stability," www.adb.org/Documents/Events/2007/Seventeenth-Tax-Conference/L-Isorena-paper.pdf.

23. World Bank Group in the Philippines, *Country Assistance Strategy*, annex M, 108.

24. Rosario G. Manasan, "Decentralization and the Financing of Regional Development," in *Dynamics of Regional Development*, ed. Balisacan and Hill.

25. On the oscillations of popular satisfaction with democracy, see Hutchcroft and Rocamora, "Patronage-Based Parties."

26. Nathan Gilbert Quimpo, "The Left, Elections, and the Political Party System in the Philippines," *Critical Asian Studies* 37 (March 2005): 4–5.

27. Alfred W. McCoy, "Covert Netherworlds: Clandestine Services and Criminal Syndicates in Shaping the Philippine State," in *Government of the Shadows: Parapolitics and Criminal Sovereignty*, edited by Eric Wilson and Tim Lindsey (London: Pluto Press, 2009).

28. "De Venecia Calls on Arroyo to Set Up New Administration," *Philippine Daily Inquirer*, October 18, 2007; De Venecia's candid appraisal came amid mounting tensions between him and the president, in the course of which his influence was curbed.

29. The Gini ratio measures income inequalities, with 0 signifying perfect equality and 100 perfect inequality. This figure, based on 2006 data, is taken from the *CIA World Fact Book*, www.cia.gov/library/publications/the-world-factbook/rankorder/2172rank.html.

30. Emmanuel S. De Dios and Paul D. Hutchcroft, "Philippine Political Economy: Examining Current Challenges in Historical Perspective," in *Philippine Economy*, ed. Balisacan and Hill, 65.

31. In comparison with other provinces in the Philippines, the Muslim Mindanao provinces have higher infant mortality, lower levels of educational achievement, higher poverty rates, and lower levels of life expectancy. See Hal Hill, Arsenio M. Balisacan, and Sharon Faye A. Piza, "The Philippines and Regional Development," in *Dynamics of Regional Development*, ed. Balisacan and Hill, 22–25.

32. See International Crisis Group, "Running in Place in Mindanao," Asia Briefing 88, February 16, 2009, www.crisisgroup.org/home/index.cfm?id=5921&l=1.

33. Halvard Buhaug, Nils Petter Gleditsch, and Ole Magnus Theisen, *Implications of Climate Change for Armed Conflict* (Washington, DC: World Bank Social Development Department, 2008), 37, 40, 42, http://siteresources.worldbank.org/INTRANETSOCIALDEVELOPMENT/Resources/SDCCWorkingPaper_Conflict.pdf.

34. This provincial-level data is found in Greenpeace Southeast Asia, "Philippines," 9. The four provinces include Sulu (a major base of the MNLF) and Basilan (a stronghold of the Abu Sayyaf group of bandits), as well as two nearby provinces, Zamboanga del Sur and Zamboanga Sibugay. Of the twenty most vulnerable provinces, four are part of the Autonomous Region for Muslim Mindanao, three are part of the Zamboanga Peninsula, and several others are historical areas of New People's Army strength (including two provinces on Samar Island as well as Camarines Sur, Camarines Norte, Quezon, Bohol, Negros Occidental, and Davao).

35. As the website of the Presidential Task Force on Climate Change readily acknowledges, "the poor and the disadvantaged are the most vulnerable to the negative consequences of climate change"; www.doe.gov.ph/cc/nrp.htm.

36. For a list of major environmental laws, see Asian Development Bank, *Economics of Climate Change*, 189–90.

37. Republic Act 9729, the "Climate Change Act of 2009"; "GMA Okays 12-Year Action Plan on Climate Change," *Philippine Star*, April 29, 2010, www.philstar.com/Article.aspx?articleId=570851&publicationSubCategoryId=200.

38. Stephen de Tarczynski, "Guarded Optimism for New Climate Change Law," Interpress Service News Agency, November 10, 2009, http://ipsnews.net/news.asp?idnews=49199.

39. These challenges are indeed apparent in the very ambitious twelve-year (2010–22), 240-page strategy document recently produced under the auspices of the Department of Environment and Natural Resources with the support of the Deutsche Gesellschaft für Technische Zusammenarbeit. From the outset it is acknowledged both that government policies toward climate change need to be made more consistent and that there is "inadequacy in the ability and capacity of government agencies and communities to respond or adapt to climate change vulnerability and extreme events." At the same time, a key strategy is "creating an enabling environment for mainstreaming climate change adaptation based on a decentralized framework of good governance" (pp. v–vi). The report asserts that "leadership by the government on adaptation is essential" and notes that "there will be annual reporting of implementation results and outcomes" (pp. 89, 93), but there is seemingly no clear road map for moving from enabling environment to ensuring implementation. In other words the success of the plan seems to depend entirely on the voluntary participation of myriad decentralized and substantially autonomous actors. Department of Environment and Natural Resources, *The Philippine Strategy on Climate Change Adaptation* (Quezon City: Department of Environment and Natural Resources, 2010).

40. Positive examples of local initiatives—one from Bicol, and one from the Cordillera region—are chronicled by Asian Development Bank, *Economics of Climate Change*, 111–12, 134–35.

41. De Dios, "Local Politics and Local Economy," in *Dynamics of Regional Development*, ed. Balisacan and Hill, 154.

42. Asian Development Bank, *Economics of Climate Change*, 186. Unfortunately, there is a tendency in some of the literature on climate change to assign roles and advocate strategies for government without any clear analysis of the government's capacity, national or local. See for instance E. B. Capili, A. C. S. Ibay, and J. R. T. Villarin, "Climate Change Impacts and Adaptation on Philippine Coasts," from the Proceedings of the International Oceans Conference, Washington, September 19–23, 2005, 5–6; and Greenpeace Southeast Asia, "Philippines," 15. More generally, there is a need for literature on natural hazards and climate change to better engage literature on politics, and vice versa.

43. Quoted in Paul D. Hutchcroft, *The Philippines at the Crossroads: Sustaining Economic and Political Reforms* (New York: Asia Society, 1996), 17.

5

Indonesia

Michael S. Malley

In late November 2007, just a week before Indonesia hosted the thirteenth United Nations climate change conference, severe floods washed over the country's capital city, Jakarta. A tidal surge sent water up to a mile inland, blocked traffic to the main international airport, and covered large parts of the city in water several feet deep. At the start of that year, a torrential rainfall had wreaked even greater havoc. Unusually intense storms occurred over several days, dropping up to ten inches of rain on some parts of the city during a single day. As a result of these heavy downpours, more than four hundred thousand people were displaced from their homes, more than fifty died, and parts of the city were covered in water more than ten feet deep. Similar storms had occurred in 2002, but even though floodwaters touched the presidential palace that year, no significant steps had been taken over the next five years to prepare the capital for the impact of severe weather conditions.[1]

Climate change was hardly the most important cause of these events. Far more direct contributions were made by clogged, colonial-era canals, poorly maintained seawalls, and the widespread conversion of water catchment areas to residential and commercial use. Nevertheless, the impact of these weather-related events resembles some of the most likely effects of climate change in Southeast Asia, such as rising sea levels and more extreme rainfall. And Indonesia's response to these recent events offers an indication of how it is likely adapt to climate change. So far, the evidence is not encouraging. As this chapter shows,

Indonesia

Indonesia's newly democratic institutions have found it difficult to address environmental challenges even when they have been persistent, severe, and focused directly on the country's center of political and economic power.

To assesses Indonesia's likely reaction to climate change during the next two decades, this chapter proceeds in three steps. The first section summarizes the main scientific findings about the observed and projected effects of climate change on Indonesia. The second section describes how recent political and economic crises have constrained the country's ability to address changes in its climate. The third section examines the prospects for various kinds of responses to climate change in the country, placing this chapter firmly within the framework outlined in the introduction to this volume and making possible comparisons with the other cases discussed in its various chapters. This final section draws special attention to the potential impact of climate change on regional security and on domestic stability and prosperity. Rising sea levels alone threaten the habitability of major cities, the productivity of key rice-growing regions, and even the existence of the islands on which its international borders depend.

Throughout this chapter, two points should be borne in mind. The first is that there are very few scientifically reliable predictions about the impact of climate change on Indonesia, let alone on its specific regions. Even though the country is among the largest in the world, the islands where the majority of its 240 million people live "are not even represented as land" in the most commonly used climate models.[2] The problem is that those models lack sufficient resolution to capture the islands of Java and Bali, let alone the interaction between their diverse topographies and general climate changes. The implications for this chapter are clear: If predictions about the effects of climate change are imprecise, then so too must be our assessment of responses to it.

The second point that needs to be borne in mind is that Indonesia's vastness and diversity create additional obstacles to predicting either the impact of climate change or the country's response to it. Indonesia includes more than 17,000 islands that stretch over 5,000 kilometers along the equator between the Indian and Pacific oceans. These islands contain snow-capped peaks as well as tropical swamps and rain forests. As the largest archipelagic state in the world, Indonesia has a coastline of more than 54,000 kilometers. It is also the world's fourth most populous country. Although many people live in sparsely populated rural areas, about half of all Indonesians live in industrialized urban areas, and nearly all the major cities are located in coastal areas vulnerable to rising sea levels.

This diversity also extends to the political arena. During the past decade, the shape of the country's basic political institutions has changed tremendously. Although democracy appears increasingly well entrenched, no political party has come close to securing a majority and no dominant coalition of parties has emerged. Moreover, democratization has been accompanied by a radical decentralization policy that has dispersed power to provincial and local governments. Because of these changes, it is difficult to predict how Indonesia's public policy is likely to be made over the next two decades, especially with respect to a set of issues that is only beginning to acquire a prominent place on the national agenda.

Observed and Predicted Climate Change in Indonesia

It is not easy to describe how Indonesia's climate is changing, let alone predict precisely how it may change. The reports produced by the IPCC provide few details about the observed or predicted effects of climate on Indonesia. Even in the chapter devoted solely to Asia, there are few direct references to Indonesia.[3] Some of the reasons for this scarce information may be inferred from the IPCC report devoted to the physical science basis of climate change. This report notes that the main scientific "models' ability to simulate observed interannual rainfall variations was poorest in the Southeast Asian portion of the domain."[4] The same report notes that other researchers have found that "no generalisation could be made on the impact of global warming on rainfall" in Southeast Asia because the models they examined yielded "highly contrasting results."[5]

Similarly, the data assembled by CIESIN suggest that climate change is likely to have an unremarkable impact on Indonesia. They indicate that temperature changes will be "average" and that agricultural productivity change will be "negligible" (see appendix A). In addition, CIESIN's data show that only a small percentage of Indonesians live in the low-elevation zones likely to be affected by the most commonly predicted changes in sea levels; just 6.4 percent of the population lives in the 3-meter low-elevation coastal zone, and 1.1 percent in the 1-meter zone (see appendix B).

However, by narrowing their focus to Southeast Asia, and particularly to Indonesia, other prominent institutions have managed to construct a clearer picture of Indonesia's changing climate. Chief among these are recent surveys of the scientific literature that have been conducted by (or on behalf of) the Asian Development Bank,[6] the World Bank,[7] the World Wildlife Fund,[8] and the Indonesian Ministry of the Environment.[9] They agree quite broadly on how temperature, rainfall, and sea levels are changing; how they are likely to change; and how these changes will affect Indonesians' livelihoods.

Most studies anticipate that temperature change across Indonesia, as in most equatorial countries, will be modest in comparison with countries farther from the equator. They find that mean temperatures are likely to rise by about 0.1 to 0.3 degree Celsius per decade for the remainder of this century. However, these projections present two problems. One is that evidence of temperature change in Indonesia and elsewhere in Southeast Asia indicates that the region may have warmed *more* than the global mean over the past century. The IPCC reports that global mean surface temperatures rose by 0.74 degree Celsius during the period 1906–2005.[10] But according to data compiled by the Asian Development Bank, temperatures in Indonesia and neighboring countries rose by more than 1 degree Celsius during the twentieth century.[11] At the very least, this difference indicates that Indonesia's equatorial location alone cannot protect it from experiencing temperature increases that exceed the global mean. A second problem arises from the lack of data on temperatures across Indonesia. Most studies appear to rely on data drawn from about thirty stations located mainly urban areas. And because the country has experienced rapid urbanization in the past few decades, it is possible that data gathered from these stations have been affected by changes in local conditions rather than broader changes in the climate.[12]

If temperatures change only modestly, severe consequences are unlikely to occur for several decades. But if the climate warms more quickly, or if substantially warmer temperatures occur more frequently, at least two important effects could be felt within the next few decades. One is a decline in the production of food crops, including rice, which is Indonesia's staple food. A commonly cited study indicates that rice yields may decline by 5 percent for every 1 degree rise in nighttime temperatures.[13] The outputs of other important crops, such as maize, are also likely to suffer from the stress caused by rising temperatures. Second, there is a broad consensus that rising sea temperatures pose a threat to fisheries, especially the extensive aquaculture that has developed in coastal regions. Higher sea temperatures are also associated with the process of "coral bleaching," which threatens coral reefs' function in sustaining fisheries and protecting coastlines.

Indonesia has already experienced significant changes in mean sea levels, but these vary sharply from one part of the country to another. According to data gathered by the

Indonesian Environment Ministry on various periods since the 1980s, the mean sea level has risen by more than 9 millimeters per year in Semarang, a major city in Central Java, and by nearly 8 millimeters per year in Medan, the country's fourth-most-populous city and the largest one in Sumatra. However, other ports in Java, including Jakarta, reported rises of only about 1 millimeter per year.[14] Predictions about changes in sea levels are subject to great uncertainty, and efforts to estimate future sea levels in Southeast Asia remain rare. However, nearly every study expects sea levels to continue rising, and dramatically so in some areas. For instance, the Asian Development Bank cites an Indonesian study that indicates "the mean sea level in Jakarta Bay will increase as high as .57 cm per year."[15]

Even at more modest levels, such as the ones Indonesia has recently experienced, the effects of rising sea levels are likely to be especially severe in its coastal cities. As a result of rapid urbanization since the 1980s, groundwater has been depleted to the point that the ground is sinking even faster than sea levels are rising. The combination of these trends presents serious challenges to the sustainability of major urban centers, especially along the densely populated north coast of Java. For instance, some parts of Jakarta have been falling at rates of 6 centimeters per year.[16] And in Semarang, farther east along Java's north coast, the rate of subsidence has generally ranged from 2 to 10 centimeters per year, but in some places has reached as much as 16 centimeters.[17]

The effects of these processes are far-reaching. In the first place, they make urban coastal areas more vulnerable to flooding, whether due to tidal surges or heavy rainfall. In addition, they increase the rate of salt water intrusion into underground aquifers, which reduces the supply of freshwater to households. It is already common for residents in large cities to purchase water in bulk from vendors who circulate through neighborhoods near the coast. The effects will also be felt beyond urban areas. In particular, as salt water intrusion spreads, it is also likely to hurt the aquaculture industry, which farms shrimp and other seafood in brackish coastal waters. This, in turn, may raise the cost but reduce the supply of protein.

Like the IPCC's own study, those by other institutions anticipate that the most significant impact of climate change will arise from changing patterns of rainfall. Since the middle of the twentieth century, rainfall appears to have declined over most parts of Indonesia, which is consistent with observations for Southeast Asia as a whole.[18] One pattern that emerges from the rainfall data since the 1960s suggests a difference between the country's northern and southern parts: In the northern regions, rainfall declined during the rainy and dry seasons; in the southern regions, rainfall rose during the rainy season and fell during the dry season.[19] A second important pattern concerns the shift in seasons: In Java and most of Sumatra, the wet season now begins ten to twenty days later than in previous decades, and the dry season begins ten to sixty days earlier.[20] In practice, this means that droughts have become more likely in the dry season, while flooding, landslides, and other effects of short, intense rainfall have become more likely. Little is known about conditions in regions outside Java and Sumatra.

Indonesian rainfall projections remain "limited."[21] There is a general presumption that total annual rainfall will increase across all of Indonesia during the twenty-first century but will increase most in the east-central region that stretches from Maluku in the north to the Nusa Tenggara Islands in the south.[22] Although different climate models produce widely varying predictions of future rainfall, their results are consistent in other important respects. In particular, they indicate that in the southern parts of the country, recent trends are likely to persist; dry seasons will lengthen, and rainy seasons will be shorter but rainfall will be higher. Because the data on the northern regions are much more limited, predictions are far more uncertain. It seems possible, however, that their experience will be the opposite of the southern areas.[23]

The flooding of major urban areas is likely to increase in severity and frequency. Even though only a small percentage of Indonesian live in the lowest-lying coastal areas, eight of the country's ten most populous cities are located in those zones. As a result, the political and economic consequences of severe and increasingly frequent floods are likely to be quite large.

Agricultural output, especially of rice, is likely to suffer. Though densely populated, Java continues to produce more than half of Indonesia's rice. Yet many climate models suggest that it will experience longer, drier dry seasons, which will make it increasingly difficult for farmers there to achieve multiple harvests each year, as they have traditionally done. The major rice-producing regions outside Java, located in southern Sumatra and southern Sulawesi, fall within the same climatic region as Java.[24] Thus they can be expected to experience similar changes in rainfall and declines in output, further diminishing the country's food supply. Although the more northerly parts of Indonesia are less populated and may receive more rainfall, they cannot be expected to compensate for climate-change-induced declines in rice production in the south. The reason is that most of those regions simply lack appropriate soils for that purpose.

Much less is known about the potential impact of changing rainfall patterns on the tree crops produced outside Java, or even on its rain forests. In most discussions of climate change, forests are understood to play an important mitigating role. By contrast, the challenge of "linking adaptation and tropical forests is a new frontier for science and policy."[25] And in Indonesia, it will be an important one. The size of its rain forests, and the biodiversity they contain, rival those of the Amazon and Congo River basins, but little is known about how such complex ecosystems are likely to respond to climate change. Even less is known about their potential impact on plantation crops. As rain forests have been cut down, vast swaths of Sumatra and Kalimantan have been planted in oil palm. Although this process has also occurred elsewhere in the world, it has been far more dramatic in Indonesia; the amount of land planted in oil palm throughout the world has tripled since the 1980s, but in Indonesia it has grown by *more than 2,100 percent*. Today, nearly 5 million hectares, a third of the world's total, are planted in oil palm in Indonesia, and the government is encouraging further expansion into Papua.[26] Thus, Indonesia will be uniquely vulnerable to climate changes that affect a single crop whose susceptibility to those changes is not well understood.

Finally, the lengthening of the dry season in much of the country presents its own challenges. Two of these are especially prominent and have already been experienced during the protracted dry seasons associated with El Niño years. The first is declining water levels in the reservoirs used to generate electricity. Because power generation capacity frequently falls short of demand, resulting in blackouts in major cities, any decline in supply has immediate effects on consumers. The other is the outbreak of forest fires, which are nearly impossible for authorities to control. The main problem is that extraordinarily deep peat deposits lay below many Indonesian forests, and during long dry spells the peat becomes dry. When forest fires occur, the flames spread underground and smolder until sufficient rain arrives to extinguish them. The results can be devastating not only for Indonesia but also for the global environment.[27]

Crisis-Ridden Indonesia's Limited Response to Climate Change

Evidence of global climate change has mounted over the past decade, but a series of crises during this time has prevented Indonesians and their government from preparing for its impact. In the late 1990s, Indonesia experienced its worst economic crisis since the mid-1960s. This downturn shrank the economy by 15 percent and produced widespread social discontent, which led to the downfall of the military-dominated authoritarian regime that had ruled the country for more than thirty years.[28]

Today, Indonesia's economy has largely recovered, and its political system has been dramatically transformed. Poverty rates have fallen to their precrisis levels, and economic growth rates have risen to more than 6 percent annually. More surprisingly, democracy has replaced authoritarianism. In fact, Indonesia has become the most politically open country in Southeast Asia and the most democratic country in the Muslim world. Nearly 90 percent of Indonesians profess Islam, which means that more Muslims live in Indonesia than in any other country.

However, Indonesia's political and economic transitions were so troubled that its own leaders, as well as many foreign observers, worried that it would become another failed state.[29] As official authority weakened and the economy contracted, armed separatist movements intensified their efforts in three far-flung provinces while communal conflicts erupted in several others, often along Muslim–Christian lines. By the early 2000s, these conflicts had killed several thousand people and displaced more than a million from their homes. Then, in 2002, the largest terrorist attack since the September 2001 strikes against the United States occurred in Indonesia. More than two hundred people, mainly foreign tourists, died. These attacks revealed a widespread network of terrorists, some linked to al-Qaeda, which had deep roots in Indonesia and elsewhere in Southeast Asia.

In the short run, the net impact of these changes on Indonesia's capacity to respond to climate change has undoubtedly been negative. Throughout the country's protracted political transition, leaders have devoted their energy to gaining and maintaining control over the state, not to long-term planning and investment in the sort of public infrastructure that could mitigate rising sea levels, warmer temperatures, and heavier rainfall (broadly speaking, the major changes that are anticipated in years to come). Moreover, the protracted political transition that began at the end of the last century has sharply diminished what would otherwise be significant capacity for long-term planning and investment, which the previous regime had developed, and which its successors have not yet achieved.

The long-run impact of Indonesia's difficult but far-reaching political and economic transitions is not yet clear. Most observers now regard the country as a "normal developing democracy."[30] This means that the political system is stable—indeed, more stable than it has been since the 1990s—but the state's capacity to make and implement policies to address complex long-term issues like climate change remains limited. As discussed in the previous section, the government has been slow even to take action against the severe floods that regularly occur in the country's capital. Still, awareness is growing of the need to address chronic problems such as corruption, underinvestment in infrastructure, and the challenges posed by global climate change. Translating awareness into action will require political will, which so far remains in very short supply.

The public generally shares politicians' focus on short-term challenges, which of course strongly reinforces such an outlook. But the extreme weather events that have occurred in recent years have made the public increasingly aware of the major environmental changes taking place around them. Changes in seasonal rainfall patterns have been gradual but palpable. Many Indonesians believe what climate scientists know: The rainy season now starts later than it used to, and it has become less predictable. This awareness is reflected in recent polls that show about 80 percent of Indonesians consider global warming to be a serious problem. However, the public holds another belief that suggests they are unlikely to support government actions to make the expensive investments required to adapt to climate change: Only one-third say they would be willing to pay higher prices to address climate change.[31]

In recent years, Indonesia has begun establishing a framework to address the challenges of climate change. In late 2007 the government released a "national action plan," which describes the problems the country is likely to face and outlines steps that can be taken to mitigate climate change and adapt to its impact.[32] A year later the president established the National Climate Change Council and charged it with implementing the action plan and overseeing all aspects of climate policy. Initially, these steps appeared to show that substantial political will existed to tackle the problem. This optimism was bolstered by an announcement that the government would allocate five times as much money to fighting climate change as it typically spends on all the activities of the Environment Ministry. However, critics complained that little was being done to address basic problems, such as rehabilitating irrigation infrastructure, and the government began to scale back its goals as it realized that foreign donors were unwilling to foot the bill for developing countries' adaptation policies.[33]

Climate Change and Social Change

On its own, climate change is unlikely to spur disruptive sociopolitical change in Indonesia. Its most significant short-term manifestation—more intense and redistributed rainfall, and somewhat higher temperatures and sea levels—are, however, likely to exacerbate existing problems that stem from a combination of environmental abuse and inadequate infrastructure.

In urban areas, it is quite possible that increasingly frequent and severe floods may lead to stronger public discontent with official policies and create pressure for the government to respond more effectively. Such pressure is likely to be intensified by government concerns about floods' negative impact on the economy. In extreme flooding conditions, the government already must make decisions about which sluice gates to open and which to close in Jakarta, with clear implications for richer and poorer neighborhoods. Such concrete policy trade-offs could exacerbate social conflict along class lines.

There is evidence that government officials are beginning to respond to this sort of pressure. In early 2008, floods similar to those described in the introduction to this chapter again struck the capital city. As in the past, economic activity was badly disrupted. But this time, the president himself was forced to abandon his vehicle en route to a cabinet meeting and switch to one of his guards' jeeps.[34] Under this sort of pressure, the cabinet set targets for protecting strategic infrastructure; in the future, the ministers agreed, flooding of the airport toll road must be overcome within five hours.[35] The likely increase in frequency of events of this kind is probably the most direct means by which climate change will put pressure on the government and its institutions.

In rural areas, the impact of climate change is likely to result in demands for more public investment in irrigation. This will be essential to contain flooding and provide a reliable source of water throughout the year in order to sustain intensive forms of rice cultivation in what will become progressively drier parts of Java. To the extent that these pressures are concentrated on the densely populated island of Java, where social and political organization is greater and mobilization easier, the potential for disruptive sociopolitical change will be increased. However, the potential for mobilization elsewhere should not be dismissed. In recent years, organized protests have disrupted activities of mines, plantations, and shrimp farms outside Java.[36]

Given the uncertain impact of climate change on the highly varied forestry, plantation, and aquaculture sectors outside Java, it is difficult to predict either higher or lower levels of social conflict. The outbreaks of communal and separatist conflict that accompanied Indonesia's transition to democracy during the late 1990s and early 2000s often stemmed partly from competition for access to and control over natural resources.[37] However, this was not the principal cause of these conflicts, which suggests that intensified competition for land and other resources alone will not produce disruptive sociopolitical change.

Fight or Flight?

In Indonesia emigration (or internal migration) would appear to be a far more likely response to climate change than increased levels of social conflict and violence. During the past century millions of Indonesians have moved in search of better economic opportunities, and the rate of migration increased during the last two decades of the twentieth century as it industrialized and urbanized. However, the impact of climate change is likely to upset well-established migration patterns. In the past, most people moved from rural, inland regions to urban, coastal ones. Under most scenarios climate change will make coastal area less habitable, due to frequent floods and shortages of freshwater; but at the same time, food production in rural areas is also likely to decline. As a result, people will have new reasons to leave rural areas but less reason to move to the main cities. And people in the main cities may wish to move, but not to rural villages.

Under these conditions, three possibilities arise. The first is that traditional patterns will persist, despite the negative effects of climate change on the quality of life in coastal cities. The second is that inland cities may experience rapid growth by attracting migrants from inland agricultural areas as well as the large coastal cities. The third is that people may move from Java, where most large coastal cities are located, to islands elsewhere in the archipelago where environmental stresses are, by comparison, lower. Such changes could occur quite rapidly. Because of a serious economic crisis in the late 1990s, which particularly hurt urban industrial workers, the traditional rural-to-urban flow was reversed. Labor appears to have flowed out of Java's cities into agricultural jobs, and even out of Java to other islands.[38]

Emigration may also become more attractive. For people in some parts of Indonesia, especially Java, this is a well-established option. There are already about 4 to 5 million Indonesians working abroad, mainly in nearby Malaysia and in Saudi Arabia and other Middle Eastern countries.[39] Whether these destinations remain attractive to Indonesians, or whether others become more attractive, depends at least partly on the impact of climate change elsewhere in the world. However, because Malaysia falls into the same climate zone as most of northern Indonesia, it is likely to become more attractive to Indonesians from Java and the nation's other southern islands, which also are the most heavily populated. But it cannot be assumed that this option will remain available in the future. The presence of hundreds of thousands of Indonesian workers, mainly unskilled and many undocumented, has been a frequent source of tension in relations between Jakarta and Kuala Lumpur since the late 1990s.[40]

Conversely, Indonesia, which is currently a net exporter of labor, may become a relatively desirable destination for migrants from parts of the world where the social consequences of climate change prove more immediate and severe. It is estimated, for instance, that a 40-centimeter rise in the average sea level would displace 55 million people in South Asia, and 21 million in Southeast Asia.[41] The arrival of any significant fraction of these people on Indonesian shores would certainly have far-reaching effects, because Indonesia, like other developing countries, is poorly equipped either to absorb or to fend off such a flood. Though this possibility would amount to a sharp reversal of recent trends, it should not be dismissed lightly. The intensification of conflict in far-off Afghanistan, Iraq, and Sri Lanka during the 2000s has produced a steady flow of illegal migrants seeking to reach Australia. For the most part, their arrival is facilitated by organized human trafficking groups, but the numbers have been large enough to become a source of tension in Australia's relations with Indonesia.[42]

State Capacity

State failure appears improbable in the next twenty years. During the past decade, Indonesia survived major economic, political, and environmental challenges, and it seems clear that modest and gradual climate changes would not place greater pressure on the state than the ones it has recently endured. However, state failure results from the cumulative weakening of the state, not just dramatic shocks that push weak and failing regimes into complete collapse. And in this sense, the stress of climate change certainly may push the Indonesian state in the direction of failure. The more severe such climate changes are, the faster the state will move in that direction. The reason is that adaptation to climate change requires expensive investment in the sort of public infrastructure that can protect the public from floods, drought, rising sea levels, and threats to the supply of reasonably priced food and water, among other things. It is precisely these sorts of investments that the state has become less able to provide since the political and economic crises that began in the late 1990s.

As a consequence, Indonesia is not especially well prepared to win the battle with climate change, despite the seemingly modest effects that are anticipated there in the short to intermediate terms. Nevertheless, there are groups within the country that are likely to demand action to achieve at least a partial victory. One of these is the tourism industry, which attracts several million foreign visitors each year. In general, tourists come to enjoy the beaches and other outdoor activities, many of which would be threatened by rising sea levels, though not necessarily by rising temperatures or increased rainfall. This industry has a powerful incentive to protect itself against climate change, but many Indonesians perceive its interests as parochial and essentially "local," because only a few regions receive the bulk of foreign tourists.

Another group that can be expected to press for adaptive or mitigating measures is made up of government actors and nongovernmental organizations that perceive the international desire to protect tropical forests as an opportunity to obtain payments in exchange for agreeing to limits on logging.[43] The salience of this issue is heightened by the fact that the preservation of rain forests itself is commonly regarded as a form of climate change mitigation, and their progressive destruction, if it were to continue as a consequence of Indonesia's drive for economic growth, or simply through recklessness, is likely to bring international pressure to bear on the government.

Other groups perceive climate change as an opportunity to achieve policies victories of their own. Since 2004, Indonesian officials have sought to revive plans for a nuclear power industry, for instance, which they justify partly on the basis that nuclear energy does not produce greenhouse gases and therefore does not contribute to global climate change.[44] Their agenda must compete with policies that promote the use of coal to generate electricity, and rising public concern about climate change may favor those who advocate the nuclear option.

Similarly, there are agribusiness interests that promote Indonesian's oil palm industry—one of the largest in the world, as noted above—as one capable of mitigating the impact of global climate change in two ways. In the first place, replanting deforested areas with fast-growing tree crops will help absorb carbon dioxide from the atmosphere. In addition, such tree crops can be used to produce biofuels, thereby reducing the world's reliance on fossil fuels. Another potential winner may be the agricultural producers in Indonesia's more sparsely populated northern regions, which are expected to experience rising amounts of rain. Depending on the precise combination of precipitation and soil type, some of these areas may enjoy rising levels of agricultural productivity, perhaps based upon new crop types that are not currently economical but may become so under changing conditions.

International Relations and Foreign Policy

It is extremely unlikely that Indonesia will become the target of foreign aggression, at least in the conventional sense. Since achieving independence in 1949, it engaged in limited military skirmishes with its neighbors only in the mid-1960s. After bringing those conflicts to an end, it helped found the Association of Southeast Asian Nations, which has contributed to peaceful relations among its members since the late 1960s.

Even as the threat of terrorism and worries about state failure in Indonesia have mounted over the past ten years, concerned governments have sought Indonesia's cooperation and have not taken unilateral action to root out terrorist networks there. Moreover, the United States and other nations have repeatedly emphasized their support for Indonesia's territorial integrity in order to reassure Indonesian leaders that foreign intervention in East Timor was an exception that proves the rule, rather than the start of a new trend.

Nevertheless, there are two persistent sources of tension between Indonesia and its neighbors that may be exacerbated by climate change and that, if severe enough, may provide grounds for limited intervention in Indonesia. One is the problem of migrants leaving the country to seek work abroad. As noted above, hundreds of thousands of Indonesians already work in Malaysia, and tensions between the two countries have been high in recent years as Malaysians have blamed rising crime on illegal Indonesian workers. The authorities in Kuala Lumpur have already forced many illegal residents to return to Indonesia, a policy that might well spur considerable public resentment if it were to become more adamant in the face of increasing migratory pressure.

The other source of tension arises from disputes over the country's maritime borders, especially in areas thought to be rich in natural resources or in areas of strategic significance, such as the Malacca Strait. Rising sea levels promise to swamp some low-lying islands on which the government bases its territorial claims. In particular, the government believes that a 1-meter rise in sea levels would submerge 2,000 of its 17,000 islands. Although most of these are small and located away from its borders with other countries, "at least 8 out of 92 outmost small islands that serve as a baseline" for Indonesian borders are considered "very vulnerable" to rising sea levels.[45] For this reason, researchers have urged the government to accelerate the completion of its negotiations with neighboring countries over unresolved borders.[46]

Concern over the loss of territory has been growing since 2002, when the International Court of Justice awarded two disputed islands to Malaysia. The court based its decision chiefly on Malaysia's long record of conducting activities on those islands. Since then, the Indonesian government has faced public pressure to identify and protect territories that are at risk. Since 2004, the government has undertaken an ambitious effort to identify and defend its claim to its outermost islands. The most prominent of these is Nipah Island, not far from Singapore.[47] By the early 2000s, this island had nearly disappeared as a result of erosion and sand quarrying. In response the government took action on three fronts. Diplomatically, it entered into negotiations with Singapore to fix its border as though Nipah would continue to exist and form part of the baseline from which Indonesia's border was determined. The countries reached an agreement in 2009. In addition, Indonesia undertook concerted efforts to reclaim the island. At a cost of more than $30 million, the island now stretches more than 60 hectares at low tide and up to 1.5 hectares at high tide. Finally, the Indonesian military established a permanent presence there, staffed by eighty personnel.

Since Indonesia achieved independence in 1949, its governments of all stripes have defined the country's foreign policy as "independent and active." Although this generally means avoiding alliances and other forms of alignment with great powers, it has accommodated both the left-leaning foreign policy of Sukarno (1945–66) and the anticommunism of Suharto (1966–98) about equally well, and thus it is unlikely to change in the near term.[48]

Control over foreign policy is effectively centralized in the professional bureaucracy of the Foreign Ministry, which generally adopts a pragmatic and nonideological approach to problems as they arise. Nevertheless, democratically elected politicians have already sought to influence foreign policy in ways that appeal to the public's sense of nationalism, a practice that is not likely to cease. For instance, since 2004 many elected leaders have declared support for Iran's nuclear program and have opposed their own government's official policy of demanding that Iran comply with requests from the International Atomic Energy Agency. This experience illustrates a common tension in Indonesian foreign policy, between the Foreign Ministry, which seeks to cultivate the country's image as a good global citizen by upholding international commitments, and its elected leaders, who seek to improve their own image in the eyes of a public that believes international institutions are controlled by the great powers.[49] Climate change is a recent addition to Indonesia's domestic and foreign policy agendas, however, and in light of the newness of Indonesian democracy, it is unclear how climate change will influence the tone and outlook of its foreign policy over the next three decades.

Indonesia is a party to the Kyoto Protocol of the UNFCCC, and in December 2007 it hosted the parties' thirteenth meeting, in Bali. Like most developing countries, however, its commitment to the UNFCCC process is predicated on the UN's recognition that "economic and social development and poverty eradication [remain] global priorities."[50] Bali nevertheless committed poorer countries for the first time to propose "mitigation actions" at the next meeting of the parties, which was held in Copenhagen in December 2009.

Indonesia's "first national communication," submitted pursuant to its accession to the Kyoto Protocol, presented an extensive survey of the nation's potential environmental risks while remaining remarkably reticent about what it intended to do about them.[51] One way or another, this stance can be expected to change. Even if the perils that climate change poses for Indonesia are, for the time being, relatively modest compared with those for the rest of Southeast Asia, it cannot escape the consequence of environmentally induced stress upon its neighbors. At the moment, moreover, it is disproportionately "part of the problem" as far as greenhouse gas emissions are concerned, a position that seems likely to expose it to increasing international pressure with the passage of time. Although it is unlikely that the resulting stresses and strains will pose a direct security risk in themselves, they will present an important test of the relationship between Indonesia's new democracy and the larger world.

Notes

1. These floods are described by Mark Caljouw, Peter J. M. Nas, and Pratiwo, "Flooding in Jakarta: Towards a Blue City with Improved Water Management," *Bijdragen tot de Taal-, Land- en Volkenkunde* 161, no. 4 (2005): 454; and Pauline Texier, "Floods in Jakarta: When the Extreme Reveals Daily Structural Constraints and Mismanagement," *Disaster Prevention and Management* 17, no. 3 (2008): 358.

2. Rosamond L. Naylor, David S. Battisti, Daniel J. Vimont, Walter P. Falcon, and Marshall B. Burke, "Assessing Risks of Climate Variability and Climate Change for Indonesian Rice Agriculture," *Proceedings of the National Academy of Sciences* 104, no. 19 (May 8, 2007): 7754. For a similar view, see E. Aldrian, L. Dümenil Gates, and F. H. Widodo, "Seasonal Variability of Indonesian Rainfall in ECHAM4 Simulations and in the Reanalyses: The Role of ENSO," *Theoretical and Applied Climatology* 87, nos. 1–4 (January 2007): 42.

3. See chapter 10, "Asia," in IPCC, *Climate Change 2007: Impacts*, 469–506.

4. See chapter 11, "Regional Climate Projections," in IPCC, *Climate Change 2007: Science*, 881.

5. Ibid., 885.

6. Asian Development Bank, *The Economics of Climate Change in Southeast Asia: A Regional Review* (Manila: Asian Development Bank, 2009), 21–61.

7. Pelangi Energi Abadi Citra Enviro, *Indonesia and Climate Change: Current Status and Policies* (Jakarta: PEACE, 2007); Josef Leitmann et al., *Investing in a More Sustainable Indonesia: Country Environmental Analysis* (Washington, DC: World Bank, 2009), 45–73, www.reliefweb.int/rw/rwb.nsf/db900SID/ MYAI-7Y47SY?OpenDocument. Note that the study by the consultant Pelangi Energi was also supported by Britain's Department for International Development.

8. Michael Case, Fitrian Ardiansyah, and Emily Spector, *Climate Change in Indonesia: Implications for Humans and Nature* (Jakarta: World Wildlife Fund, 2007), www.worldwildlife.org/climate/Publications/ WWFBinaryitem7664.pdf.

9. Indonesian Ministry of the Environment, *Indonesia Country Report: Climate Variability and Climate Changes and Their Implications* (Jakarta: Ministry of the Environment, 2007), www.undp.or.id/pubs/docs/ Final%20Country%20Report%20-%20Climate%20Change.pdf.

10. IPCC, *Climate Change 2007: Science*, 241.

11. Asian Development Bank, *Economics of Climate Change*, 23.

12. Indonesian Ministry of the Environment, *Indonesia Country Report*, 17.

13. Shaobing Peng, Jianliang Huang, John E. Sheehy, Rebecca C. Laza, Romeo M. Visperas, Xuhua Zhong, Grace S. Centeno, et al., "Rice Yields Decline with Higher Night Temperature from Global Warming," *Proceedings of the National Academy of Sciences* 101, no. 27 (July 2004): 9971, www.pnas.org/content/101/27/9971. full. They find the impact of maximum daytime temperatures on rice insignificant.

14. Indonesian Ministry of the Environment, *Indonesia Country Report*, 27.

15. Asian Development Bank, *Economics of Climate Change*, 50.

16. Caljouw, Nas, and Pratiwo, "Flooding in Jakarta," 465.

17. M. Aris Marfai and Lorenz King, "Tidal Inundation Mapping under Enhanced Land Subsidence in Semarang, Central Java Indonesia," *Natural Hazards* 44, no. 1 (January 2008): 99.

18. Asian Development Bank, *Economics of Climate Change*, 25.

19. Indonesian Ministry of the Environment, *Indonesia Country Report*, 19

20. Ibid., 20.

21. Asian Development Bank, *Economics of Climate Change*, 26.

22. Leitmann et al., *Investing in a More Sustainable Indonesia*, 49.

23. Indonesian Ministry of the Environment, *Indonesia Country Report*, 33–34.

24. Edvin Aldrian and R. Dwi Susanto, "Identification of Three Dominant Rainfall Regions within Indonesia and Their Relationship to Sea Surface Temperature," *International Journal of Climatology* 23, no. 12 (October 2003): 1435.

25. Bruno Locatelli, Markku Kanninen, Maria Brockhaus, Carol J. Pierce Colfer, Daniel Murdiyarso, and Heru Santoso, et al., *Facing an Uncertain Future: How Forests and People Can Adapt to Climate Change* (Bogor, Indonesia: Center for International Forestry Research, 2008), 2.

26. Douglas Sheil, Anne Casson, Erik Meijaard, Meine van Noordwijk, Joanne Gaskell, Jacqui Sunderland-Groves, Karah Wertz, and Markku Kanninen, et al., *The Impacts and Opportunities of Oil Palm*

in *Southeast Asia: What Do We Know and What Do We Need to Know?* (Bogor, Indonesia: Center for International Forestry Research, 2009), 1.

27. Mark E. Harrison, Susan E. Page, and Suwido H. Limin, "The Global Impact of Indonesian Forest Fires," *Biologist* 56, no. 3 (August 2009): 156.

28. Michael S. Malley, "Indonesia: The Erosion of State Capacity," in *State Failure and State Weakness in a Time of Terror*, edited by Robert I. Rotberg (Washington, DC: Brookings Institution Press, 2003), 183–218.

29. Donald K. Emmerson, "Will Indonesia Survive?" *Foreign Affairs* 79, no. 3 (May–June 2000): 95–106.

30. Douglas E. Ramage, "Indonesia: Democracy First, Good Governance Later," in *Southeast Asian Affairs 2007* (Singapore: Institute of Southeast Asian Studies, 2007), 135.

31. Pew Research Center, *25-Nation Pew Global Attitudes Survey* (Washington, DC: Pew Global Attitudes Project, 2009), 175–76.

32. Indonesia, *Rencana Aksi Nasional Dalam Menghadapi Perubahan Iklim* [National Climate Change Action Plan] (Jakarta: Ministry of the Environment, 2007).

33. "Perubahan Iklim: Program Adaptasi Sulit Dioperasionalkan" [Climate Change: Adaptation Program Difficult to Put into Practice], *Kompas* (Jakarta), February 13, 2009, http://cetak.kompas.com/read/xml/2009/02/13/00461021/program.adaptasi.sulit.dioperasionalkan; "Fokus Pendanaan Iklim dari Dalam Negeri" [Climate Funding Should Be Domestic], *Kompas*, January 28, 2009, http://cetak.kompas.com/read/xml/2009/01/28/00330673/fokus.pendanaan.iklim.dari.dalam.negeri.

34. Desy Nurhayati, "SBY Trapped in Flood," *Jakarta Post*, February 2, 2008, www.thejakartapost.com/news/2008/02/02/sby-trapped-flood.html.

35. "Banjir Hancurkan Infrastruktur," [Floods Destroy Infrastructure], *Kompas* (Jakarta), February 6, 2008, www.kompas.com/kompascetak/read.php?cnt=.kompascetak.xml.2008.02.06.02195828&channel=2&mn=2&idx=2.

36. See, e.g., Elizabeth Fuller Collins, *Indonesia Betrayed: How Development Fails* (Honolulu: University of Hawaii Press, 2007); and Nancy Lee Peluso, "Violence, Decentralization, and Resource Access in Indonesia," *Peace Review* 19, no. 1 (January–March 2007): 23–32.

37. Gerry van Klinken, *Communal Violence and Democratization in Indonesia: Small Town Wars* (London: Routledge, 2007); Jacques Bertrand, *Nationalism and Ethnic Conflict in Indonesia* (Cambridge: Cambridge University Press, 2004).

38. Graeme J. Hugo, "The Impact of the Crisis on Internal Population Movement in Indonesia," *Bulletin of Indonesian Economic Studies* 36, no. 2 (August 2000): 115.

39. Graeme J. Hugo, "International Migration in Indonesia and Its Impacts on Regional Development," in *Global Migration and Development*, edited by Ton van Naersse and Ernst Spaan (London: Routledge, 2008), 49.

40. Alexander R. Arifianto, "The Securitization of Transnational Labor Migration: The Case of Malaysia and Indonesia," *Asian Politics & Policy* 1, no. 4 (October–December 2009): 613–30; Joseph Liow, "Malaysia's Illegal Indonesian Migrant Labour Problem: In Search of Solutions," *Contemporary Southeast Asia* 25, no. 1 (April 2003): http://findarticles.com/p/articles/mi_hb6479/is_1_25/ai_n29007898/.

41. Herminia A. Francisco, "Adaptation to Climate Change: Needs and Opportunities in Southeast Asia," *ASEAN Economic Bulletin* 25, no. 1 (2008): 7.

42. Lindsay Murdoch, Tom Allard, and Dan Oakes, "Yudhoyono Will Visit to Discuss Boat Arrivals," *Sydney Morning Herald*, March 8, 2010, www.smh.com.au/national/yudhoyono-will-visit-to-discuss-boat-arrivals-20100307-pqlv.html.

43. Adianto P. Simamora, "Carbon-Absorbing Tropical Forests a Potential Gold Mine," *Jakarta Post*, February 25, 2009, www.thejakartapost.com/news/2009/02/25/carbon-absorbing-tropical-forests-a-potential-gold-mine.html-0.

44. Andrew Symon, "Southeast Asia's Nuclear Power Thrust: Putting ASEAN's Effectiveness to the Test?" *Contemporary Southeast Asia* 30, no. 1 (April 2008): 118–39.

45. Indonesian Ministry of the Environment, *Indonesia Country Report*, 40–41.

46. Adianto P. Simamora, "Outlying Islands May Disappear," *Jakarta Post*, February 17, 2009, www.thejakartapost.com/news/2009/02/17/outlying-islands-may-disappear.html.

47. Ferry Santoso, "Simbol Pertahanan Negara Kepulauan" [Symbol of the Defense of an Archipelagic State], *Kompas* (Jakarta), August 12, 2009, http://cetak.kompas.com/read/xml/2009/08/12/03175578/simbol

.pertahanan.negara.kepulauan; "Batas Laut Teritorial Disepakati" [Maritime Borders Settled], *Kompas*, February 3, 2009, http://cetak.kompas.com/read/xml/2009/02/03/0014570/batas.teritorial.laut.disepakati; "Satu Kompi Marinir Dikirim ke Nipah" [One Company of Marines Sent to Nipah], *Jawa Pos News Network*, January 17, 2010, www.jpnn.com/index.php?mib=berita.detail&id=56583; "Pemerintah Prioritas Penyelamatan 92 Pulau Terluar" [Government Prioritizes Preservation of 92 Outermost Islands], *Kompas*, May 5, 2009, http://sains.kompas.com/read/2009/05/05/00552723/pemerintah.prioritas.penyelamatan.92.pulau.terluar.

48. Paige Johnson Tan, "Navigating a Turbulent Ocean: Indonesia's Worldview and Foreign Policy," *Asian Perspective* 31, no. 3 (2007): 147–81.

49. Michael S. Malley, "Bypassing Regionalism? Domestic Politics and Nuclear Energy Security," in *Hard Choices: Security, Democracy, and Regionalism in Southeast Asia*, edited by Donald K. Emmerson (Stanford, CA: Walter H. Shorenstein Asia-Pacific Research Center, 2008), 257-261.

50. "The Bali Action Plan," Decision 1 of the Report of the Conference of the Parties [to the UNFCCC] on its thirteenth session, Bali, December 3–15, 2007, 3, http://unfccc.int/meetings/cop_13/items/4049.php.

51. Government of Indonesia, *First National Communication on the [UNFCCC] Climate Change Convention* (1994), http://unfccc.int/resource/docs/natc/indoncl.pdf. One attentive reader of the communication has noted that, of its 116 pages, only three refer to measures of adaptation. See Ancha Srinivasan, "Mainstreaming Adaptation Concerns into Agriculture and Water Sectors: Progress and Challenges," paper presented at Seventeenth Asia-Pacific Seminar on Climate Change, Bangkok, July 31–August 3, 2007, http://unfccc.int/files/adaptation/adverse_effects_and_response_measures_art_48/application/pdf/ancha_water.pdf.

6

India

T. V. Paul

Global climate change has been occurring at a rapid pace in recent years and is predicted to accelerate further in the coming decades. Public awareness of its risks is, if anything, increasing even more rapidly. With rising average temperatures have come melting glaciers, rising sea levels, and unpredictable weather patterns, such as severe droughts in some places and unexpected rains and storms in others. Developing countries are especially vulnerable to the economic, social and political dislocations, and periodic natural disasters that are anticipated to accompany climate change in years to come. The security aspects of climate change, broadly defined to include the security of individual lives and futures, along with that of the state and the international system generally, have become a major source of global concern. The capacity of the state is a crucial determinant of how severely climate-induced changes will affect a country and its inhabitants. State capacity is defined here as the ability of the political elite to effectively carry out its responsibilities fairly and legitimately without being hindered by societal forces.[1] States with a low institutional capacity are especially vulnerable in facing the myriad challenges that climate change may bring to their political structures and social fabric in general and their economies in particular. Large states with a multitude of preexisting schisms, whether social or economic, are particular sources of concern in the security arena.

This chapter deals with the likely impact of climate change on a broad range of issues pertaining to the security of India, a country where more than one-sixth of humanity lives

India

(1.3 billion people in 2008) but that occupies only 2.4 percent of the globe's geographical area. The central argument I make here is that climate changes could accentuate preexisting conflict patterns. By itself, climate change may not be sufficient to cause large-scale security problems for the South Asian nation. However, because India is already exposed to a wide range of potential internal and external conflicts, climate change will likely heighten these preexisting conflict patterns. Further, climate change is linked to a host of developmental challenges, including food production and water distribution, in which there is limited tolerance for failure—people cannot go for long without food and water. Protracted deficiencies in these areas can easily generate internal and interstate conflicts if proper state management is missing.[2]

India's challenge will be to improve its state capacity to deal with climate-change-induced conflicts, and also to help develop a regional cooperation process for confronting issues that affect the entire South Asia region. Longer-term development strategies are necessary in this regard. India currently stands at the early stages of accelerated economic development. As such, it could offer a major platform for establishing the energy-efficient and low-carbon industrial development options promised by new and emerging technologies. If it is able to do this, it may mitigate some of the worst consequences of climate change while avoiding some of the pitfalls revealed by the historical development of older industrial economies.

India, as a large multiethnic state, already experiences considerable internal conflicts, arising from historically rooted socioeconomic disparities, ethnic and linguistic issues, and caste-based allegiances. Although parts of India are rapidly developing as a result of its economic liberalization and increasing integration into the global economy since 1991 (especially in the information technology area), nearly 64 percent of the population still relies on agriculture and subsistence farming. In addition, estimates for 2004–5 suggest that nearly 26 percent of Indians lived below the poverty line, nearly 35 percent of the population was without food security, and 53 percent of children were malnourished.[3] It is estimated that India has emerged as the world's fourth-largest emitter of carbon dioxide and that these emissions are increasing, although admittedly from a low base in per capita terms; per capita energy consumption in India in 2008 was less than half that of China, and only 18 percent of the U.S. rate. This is largely because nearly 500 million Indians have little or no access to electricity.[4]

With its large reserves of coal—nearly 10 percent of the world's supply—and with constraints in obtaining other sources of power such as nuclear energy and natural gas, India is likely to use more coal for electricity production. The increased use of coal and other biofuels by India is likely to aggravate the problem of greenhouse gases globally, unless radical alternate energy systems are developed and used widely. Like that of China, India's future economic success inevitably includes some risk that it will be perceived as a major contributor to climate change. Such perceptions may compromise its relations with allies and trading partners that believe more should be done to limit the environmental consequences of what, from an Indian perspective, may well be regarded as no more than the arrival of a long-overdue prosperity.

The extent of negative security outcomes will depend on how severe climatic changes will be, and how rapidly they unfold. Many questions remain unanswered: Will there be a sudden increase in the average temperatures of South Asian countries? What subregions will be most exposed to this increase? What will be the impact on weather patterns, regional rainfall, and carbon dioxide concentration? What is the likely impact of climate change on different crops in different regions and over what period of time can we expect the sea levels to rise? Given the somewhat speculative nature of climate projections, it is difficult to make hard predictions for a specific period. We know that temperature in India has been increasing at a rate of 0.57 degrees Celsius per 100 years. But this increase is expected to accelerate according to various studies, and therein lies the major challenge.[5]

Environmental degradation could result in further competition and conflict over depleting resources, and demands for their allocation by the state at both national and local levels in India. Climate changes can add to the mix of problems that India faces, especially by slowing down the economic and political development of large sections of Indian society, despite the fact that GDP growth, mainly driven by private investment in the services sector, may continue at a high rate. This would mean a very lopsided economic growth pattern, in which the fruits of development may not reach traditionally underprivileged segments of Indian society.

There is an important qualification to be made here: India's economy is growing at an annual rate of 8 to 9 percent, and if this can be sustained India will also increase the resources required to confront environmental challenges as they emerge. Whatever options may emerge for developing states as they confront climate change, the option to "stay poor" is not going to be one of them. By 2030 more than half the Indian population will have joined the ranks of the middle class, and a large number of poor people will have emerged out of abject poverty. Even then, more than one-fourth of the population will still remain in extreme poverty. India needs major improvements in state capacity and distributional efficiency at the national, provincial, and local levels, areas where it has been lacking. It will be crucial for the state to gain the capacity to deliver developmental assistance in the locations where climate change can have the most adverse impact.

Climate change is expected to have different subregional effects within India. Water shortages will affect the country's agricultural production, especially in the already-arid regions of Rajasthan, Uttar Pradesh, Bihar, and the southern states of Andhra Pradesh, Karnataka, and Tamil Nadu. It could curtail agricultural production even in India's granary states of Punjab and Haryana. A depletion of internal food supplies would mean inflation in food prices, which would affect the poorer sections of India disproportionately. Added to this is the increasing global demand for food, shortages of supply in the global market in general, and the price increases that would follow from the increasing need to import food. The melting of snow from the Himalayan glaciers means that India's major rivers—especially the Ganges, its tributaries, and the Brahmaputra—could alternate between abnormally low flows in the early summer and winter months and extraordinarily high flows during the monsoon, posing the double risk of drought followed by flood. The glaciers on both sides of the Tibetan Plateau and its Himalayan rim, which are the source of water for the Indus, the Ganges, and Brahmaputra, along with several river systems in China and Southeast Asia, have been melting rapidly, although the extent of this melting is a matter of controversy.[6] India's vulnerability in this respect is amplified by the fact that more than 60 percent of its population still depends on the vagaries of the monsoon for subsistence farming. The impact of climate change would therefore depend on whether the annual Indian monsoon, which is the single biggest influence on agriculture, will remain stable and cover its normal area during the June–August period of the year.[7]

If monsoonal rains become increasingly erratic as a consequence of global warming, there will likely be serious food shortages in the regions that depend on them, unless new farming technologies are implemented quickly. The agricultural sector, on which a large majority of the population depends for basic sustenance, has been growing at the lowest rate of any of India's economic sectors in recent years. The economic reforms initiated since 1991 have primarily affected services and industry. Increased food imports are a very constrained solution, because of the likelihood that major food shortages in India will be echoed elsewhere around the world, and because the cost of imported food exceeds what those Indians most likely to be affected will be able to pay. Food prices in India and throughout the developing world shot up in 2008 due to increased demand (itself a reflection of economic success), high energy costs, and the conversion of arable land cultivation in North America to corn cultivation for ethanol production. If agriculture is widely affected by climatic changes, the rich/poor and urban/rural gaps in India could widen further. A large section of the Indian poor could continue to be ensnared by the "poverty trap."[8] Further, substantial numbers of landless agricultural laborers could migrate to India's already-overburdened cities and generate instability and violence, especially if they demand permanent status for slums and shantytowns, as happened in Mumbai in 2004–5.[9] Indian cities often find it difficult to provide services to long-term inhabitants. If landless migrants move from the countryside to the cities in large numbers, the capacity of those cities will be strained even further. And human security issues could worsen in the process.[10]

The potential hardship that poor people might face is reflected in the dramatic increase in the suicide rate among farmers in India. Something on the order of 150,000 Indian farmers are believed to have killed themselves between 1997 and 2007.[11] Often, poor farmers commit suicide out of an inability to pay back high-interest loans from local lenders or banks. This has reached epidemic proportions in some of the farming villages in Andhra Pradesh and Karnataka.[12] More and more incidents of suicide can be expected if climatic changes produce crop failures for poor framers.

Another major issue facing India in the climate area would be a rise in sea levels. India has a coastline of 12,700 kilometers. According to one study, a 1-meter rise could cost the economy approximately 2 trillion rupees due to the impact on fisheries, shipping, and port facilities in the three main cities of Chennai, Mumbai, and Calcutta, while displacing 7.1 million Indians.[13] This will also have an impact on agriculture due to the loss, by flooding, of low-lying arable land along the coastline. Fisheries would also suffer, and the effect would be devastating for those who rely on small-scale traditional fishing.[14]

Climate Change and Internal Conflicts

The environmental stress from climate change may accentuate both internal and external security challenges for India. It is unlikely that climate change alone will be the proximate cause of internal conflict, except perhaps as a result of exceptional weather events that create conditions with which the authorities cannot cope. More generally, however, climate change could exacerbate the problems associated with state weakness in multiple areas, including the preservation of social and political order. It is unlikely that Indians will migrate to neighboring countries in large numbers, because nearby states do not offer anything better in terms of living conditions or economic opportunities. The result, instead, would be internal displacements, particularly, as has been suggested above, from the countryside to the cities.

The displacement caused by environmental stress could also encourage political extremism, again threatening stability within the country. Currently, a large swath of India, from the Nepal border to Andhra Pradesh—often called the tribal belt—is shot through with Maoist guerillas, called Naxalites, who have been challenging India's central and state governments and are drawing support from poor landless tribal people, especially in Andhra Pradesh, Chhattisgarh, Jharkhand, and Bihar.[15] It is estimated that thirteen of

India's twenty-nine states have been affected by Naxalism in one form or other, and that the insurgency killed 698 people in 2007 alone, an increase from 678 in 2006.[16] The prime minister, Manmohan Singh, has characterized the Naxalite movement as the number one internal security threat that India faces. The Naxalites use guerilla tactics and, by offering certain services to deprived people, gain the sympathy of the local population. The virtual absence of state services, severe unemployment among rural poor people, the absence of land reforms, and ineffective law and order in many parts of India help their cause.

The attraction of the Naxalites to the poorest segments of Indian society is heightened by their opposition to ongoing caste-based repression in rural areas.[17] The Naxalite movement's upsurge is a reflection of the highly lopsided economic development in India, which is concealed by its impressive aggregate economic growth rate since the early 1990s. Although the number of people in the middle class has increased substantially, more than 300 million people still live on the equivalent of less than $1 a day.[18] Poor people's increased energies are mostly channeled through electoral politics, but some have resorted to violent responses, such as the Gujjars in Rajasthan, who want to be treated as a community with reserved quotas in jobs and educational institutions. Various conflicts over demands for quotas, especially in employment and education, will intensify as dislocations caused by climate-induced changes to agriculture and traditional economic structures become more prevalent. The Naxalites will likely attract more adherents as the consequences of environmental stress mount.

Uneven economic development exacerbated by climate change could also increase the insurgency movements in India's Northeast region, where the states of Nagaland, Mizoram, Manipur, Tripura, and Assam have seen bloody conflicts during the past forty years between rebel movements and India's security forces. At present there are more than a hundred insurgency movements in these states, and some are demanding outright separation from India while others are fighting for more autonomy or recognition by the central government.[19] New Delhi has treated these conflicts largely as a law-and-order problem and has employed harshly coercive tactics that have alienated the local population even further. The main root of the problem lies in the lack of development of the region and the inattention of successive central governments in New Delhi toward the region's problems. Climate-induced struggles would almost certainly increase the unrest in the Northeast, unless remedial measures are taken to alleviate the palpable alienation in this region. The increased migration from Bangladesh is part of the milieu of the conflict in the Northeast, a theme that I discuss below.[20]

Caste- and religion-based conflict could also occur in regions where much economic dislocation takes place. The upper castes in many Indian states still refuse to give equal rights to low-caste Hindus, and this has caused periodic violence. There has been a resurgence of lower-caste movements, and the parties supported by them have gained political power in states such as Uttar Pradesh and Bihar. Perhaps the most alarming security outcome would be increased terrorist strikes within India, which is already one of the most terrorism-stricken nations globally. In fact, in 2007 it registered the largest number of terrorism-inflicted casualties (2,300) of any country in the world.[21] Among the causes is the continuing insurgency in Kashmir, which has created a fertile ground for increased terrorist activity within India. Some of the terrorist activity is generated by foreign-trained individuals who, in concert with local sympathizers, engage in violent activities in India's crowded cities. India's security forces face severe constraints in coping with this scourge. Climate-induced changes could accentuate the terrorist threats to India as disgruntled youth could be attracted to dramatic shows of defiance. Further, the relative economic status of Muslims in India—who now make up nearly 14 percent of the total population—has not been improving significantly, even with the country's rapid economic growth rates in recent years. This has produced an underclass, and there are links between ultraradical Islamic groups within India and neighboring countries that, together, have regularly engaged in terrorist violence during roughly the past decade. The response of the Hindu nationalist party often has produced violent outcomes (as in Gujarat and the Babri Masjid Mosque demolition), causing further alienation of the Muslim community.

One other issue that is likely to be exacerbated by anticipated climate change is the ongoing internal conflicts between Indian states over water sharing and damming. The row over the sharing of Cauvery River water between Karnataka and Tamil Nadu has probably been the most high-profile and long-winded of these disputes. The issue has been of such importance to the states that the eventual award of more water to Tamil Nadu by an arbitrator in 2006 led to major violent protests in Karnataka. Similarly, Tamil Nadu and Kerala have been in a bitter conflict over the reliability of the British-era dam in Mullperiyar in the eastern Ghats. Currently, a number of Indian states have disputes over water sharing, which results in periodic small-scale violence. The proposal to link the Ganges and Cauvery rivers had to be shelved due to the potential for disputes and violence within and among the states concerned.[22] The construction of Sardar Sarovar Dam in Gujarat caused internal displacement and environmental damage and a large-scale agitation during the 1990s that attracted international attention.

Will India Fail as a State?

South Asia has already witnessed the breaking-up of a nation-state, at least partly as a consequence of environmental calamity. An inadequate response from West Pakistan to the Bay of Bengal typhoon in 1970, which killed more than three hundred thousand people, resulted in the strengthening of the separatist movement in erstwhile East Pakistan (supported by India), which would subsequently lead to the emergence of Bangladesh.[23] Although a comparable fragmentation of India would be less likely to occur due to the coercive power of the state, and the difficulties in obtaining international recognition, groups disenchanted with the status quo often have a temptation to seek independence or more autonomy. The natural tendency of the Indian state has been suppression, rather than integration, wherever the latter is impossible to accomplish.

Although dislocations from climate change could cause internal instabilities, they are likely to be centered in pockets or subregions rather than the whole of India. India is a fairly strong state in some dimensions but is weak in others. Among all the South Asian states, India has the most institutional capacity due to its fairly strong democratic and federal structures. However, internal conflicts still act as a drag on India, and these schisms are likely to be exacerbated by climate change. The Indian state has survived many crises since independence, and it is unlikely to end up in complete failure. It has considerable coercive power to suppress secessionist movements, even if they linger for decades. By 2030 it will also have substantial resources at its command, because India's economy is expected to grow and double in size every decade or so.

There could be state failure in specific regions of the country, however, such as the Maoist-controlled tribal areas and the economically weaker regions of the North and Northeast. Some of India's states exhibit chronic weaknesses, similar to Sub-Saharan Africa. These states, unless they address their problems on a war footing, could become the epicenters of conflict caused by climate change and agricultural dislocations.

Although India has yet to incorporate environmental security as part of its national security outlook, a determined effort to do so might still allow it to significantly ameliorate the challenges posed by climate changes. India's adaptive capacity lies primarily in its institutions of democracy, secularism, and a free press, which have stood it in good stead in the face of famines, invasions, caste and communal divisions, and disease throughout its history. A study by Amartya Sen showed that, unlike China in the 1960s, India averted large-scale famines due to its free press, which reported impending food shortages and forced the bureaucracy to act in time.[24] In the last two decades or so, India has experienced a number of environmental crises—for example, the Gujarat earthquake in 2001, several floods, especially in Orissa and Andhra Pradesh, and the Southeast Asian tsunami in 2004. During all these, India largely depended on its own resources, and the state was able to mobilize help to the affected, although there were reports of slow implementation and poor management.

India also has a reasonably strong administrative service, and an army and police force that can be drawn in to help people affected by natural disasters. Where India is weak is in its ability to create the sense of urgency needed to tackle problems that are not immediately visible or have not reached crisis proportions. The country is often reactive and not proactive in dealing with crises. Its policy responses are not prompted by a concern for crisis prevention. Its bloated bureaucracy can be insensitive to the plight of the victims and rather slow acting in bringing relief. Though its armed forces are increasingly playing the disaster-relief role, they could end up being overstretched because there will also be continuing external security threats. India certainly requires a better disaster relief and emergency system, especially in dealing with floods and droughts along with the resettlement of the affected. With more long-term effects in mind, the country also needs to develop mechanisms capable of anticipating, preventing, and managing future environmental changes, rather than persisting with its crisis-driven approach.[25]

Climate Changes in Neighboring States and Their Implications for India

India is the largest geographical component of the South Asian region. Its policies and actions necessarily influence conditions in neighboring states, and it is in turn influenced by theirs. Conflict with Bangladesh and Nepal would become more likely if environmental changes induce large-scale migration from Bangladesh to India's northeastern states, or from Nepal to India's North. Some 20 million Bangladeshis have already migrated illegally to the Indian states of Bihar, West Bengal, and Assam.[26] Many of these people are in fact environmental migrants who seek better living conditions, but they have not been integrated into Indian society despite having lived in the country for decades. As a result, they live in legal limbo and have generated conflicts in several Indian states. This trend will increase if, as is widely expected, heavily populated and low-lying Bangladesh suffers some of the worst effects of global climate change.[27] A 45-centimeter increase in sea level would reduce Bangladesh's territory by 10.6 percent and displace 5.5 million people.[28] Similarly, a 2 degree Celsius increase in global temperature could increase annual peak discharge of water in Bangladesh by 25 percent. Warns the IPCC study: "With a 1-meter rise in sea level, 2,500 square kilometers of mangroves in Asia are likely to be lost; Bangladesh would be worst affected by the sea-level rise in terms of loss of land. Approximately 1,000 square kilometers of cultivated land and sea product culturing area is likely to become salt marsh."[29] A large-scale migration of Bangladeshis to India could produce major conflicts or reignite preexisting ones. During the past three decades, Assam especially has witnessed intense conflict because of the Bangladeshi refugees who have already migrated to the state. In-deed, the Assam autonomy movement has become very violent over the years on the issue of refugee repatriation, and it is likely to become further aggravated if additional refugees from Bangladesh arrive in the state.

India's fencing of the 2,500-mile-long border with Bangladesh has generated a hostile response from the Bangladeshi side. There have been incidents of Bangladeshi border guards killing Indian military personnel, and some counterviolence by Indian forces. This hostility is likely to increase if there is large-scale migration from Bangladesh and India attempts to stem such a flow. India also argues that Bangladesh has become a major source of al-Qaeda-led terrorism in India, which may provide a further reason for a forceful anti-migration response on India's part.[30]

India's relations with Pakistan are likely to be further complicated by disputes over water. The Indus Waters Treaty of 1960 between the two countries, regulating the flow of water to Pakistan, has held steady even during major crises and wars between the two states. How-ever, during the 1999 and 2002–3 India/Pakistan crises, there were calls in India for uni-laterally abrogating this treaty. India also has started preliminary work on building several dams in the catchments areas of the Indus waters in Kashmir, and Pakistan has been object-ing.[31] If there are major water shortages in both countries due to droughts, water could be-come a further source of conflict alongside the many other issues that divide the two states.

India's relations with China could also suffer if Beijing attempts to build dams in Tibet and reduce the water flows to the Brahmaputra and its tributaries that receive water from the Chinese side. Recent reports of China building a dam to reduce the flow have caused a considerable uproar in India and some disquiet in Bangladesh. Given the recurring claims that China makes on India's Arunachal Pradesh State and the lingering suspicions over each others' intentions in Tibet, the relationship between the two has not improved much, despite a substantial growth in bilateral trade. India also has disputes over water with Nepal and Bangladesh. The Farakka barrage dispute between India and Bangladesh led to considerable tension in the 1980s.[32] The Ganga Water Treaty of 1996 helped reduce the tensions somewhat, but the dispute continues over other issues of water sharing. These water disputes, although never leading to open war, have the potential to emerge as major sources of tension, independent of other traditional sources of conflict.

Dealing with environmental concerns will require cooperation at the regional level, which would require extensive interaction between the various states of the region. This dynamic might generate some momentum toward the development of a more cooperative approach in relations between India and its neighbors, which would provide the potential for a resolution of further matters at issue among the states.[33] Despite the ongoing dis-putes, water sharing is an area where India has achieved some level of regional cooperation with its neighbors, in particular Pakistan and Bangladesh, as highlighted by the above-noted durability of the 1960 Indus Waters Treaty and the ameliorative impact of the Ganga Water-Sharing Agreement with Bangladesh.[34] India also has several hydroelectric projects in collaboration with Bhutan that have benefited that small landlocked country. India and Nepal are also in the early stages of jointly building a dam on the Sapta-Koshi River, which flows from Nepal to India.

Foreign Policy Implications

There are fears that, as India becomes one of the world's leading carbon emitters, its ef-forts to deal with the problem will be compromised by its concern not to disrupt famil-iar patterns of industrialization and economic growth. There is already a dispute between industrialized and industrializing countries over deadlines for controlling carbon emissions. India, like China, may not agree to caps or may not be able to fulfill the caps to which it agrees. This could result in conflict not only between these two countries and the industrial West but also with other developing states. In addition, both India and China are now competing globally for oil and gas and new sources of energy, especially in Africa. As their economies continue to grow, they will need additional resources, and this could become a source of contention if other countries have accepted plans for curtailing the production of hydrocarbons in the near term.

India is likely to seek multilateral engagement with the world's leading powers, including the United States, in addressing its environmental problems, although it is also likely to insist that the developed world bear a share of responsibility for environmental remediation that takes account of historical, and not merely contemporary, patterns of consumption. This was evident in the Indian position at the December 2009 Copenhagen meeting on the environment.[35] Although India has signed the Kyoto Protocol, it is not keen on the inclusion of large developing countries under the treaty's reduction targets, and it would prefer to postpone increasing its obligations in this area until at least 2020.

Although the principle of "common but differentiated responsibility" exempts developing countries from the reduction of emission targets, the credibility of the multilateral climate change regime hinges on these states' active participation in it. It has been argued that India, along with other key developing countries, should make an effort to achieve a "transition to a low-carbon growth path at a rate consistent with their capabilities."[36] In June 2008, India produced a draft National Action Plan on Climate Change that emphasizes the "avoidance of emissions," the fast development of alternate energy sources instead of "caps," and the setting up of "eight missions for 'multi-pronged, long-term and integrated strategies' for achieving goals on climate change in areas that include solar, enhanced energy conservation, sustainable habitats, agriculture, water and sustaining Himalayan ecosystems, [and the] Green India project."[37]

India is more likely to be persuaded to act if China comes around to also agreeing to greenhouse gas emissions targets. The Indian claim that, on a per capita basis, India is emitting much lower levels of carbon has some merit, but ultimately it is a self-defeating argument. India's cities are already uncomfortably polluted, and it is in India's own interest to reduce its pollution and hydrocarbon emissions, and to follow the path of sustainable development from an early stage of its accelerated growth.[38] Even if it would prefer to avert its eyes from the global consequences of high-carbon emissions, the local consequences of pollution caused by the burning of carbon-based fuels will be formidable.

India is likely to engage its partners in the South Asian Association for Regional Cooperation (SAARC) in finding solutions to some of these problems. However, SAARC remains a feeble institution, due largely to the India/Pakistan conflict and the weaknesses of its smaller members. This fact has not resulted in any concrete steps by India thus far, but India may resort to more engagement if the environmental crisis becomes imminent. Aggressive pressures are unlikely to produce favorable results for India or its neighbors, given that these newly emerging states are very sensitive about their sovereignty. One area of particular concern is how India can manage its relations with Bangladesh, which, as discussed above, will become a major source of immigrants if environmental problems worsen.

Because the twenty-first century will see the rise of India and China as significant global powers, it is important that climate change not become a source of contention in bilateral relations between them, or in their relations with the United States. Instead, climate change might be an area where opportunities for cooperation can emerge, given that the ill effects of climate change will affect all nations, particularly the rising powers. Thus, the potential for military conflict involving the existing and rising powers could be reduced if climate change is accepted as an arena for collective action. As India is seeking major power status and improving relations with the United States, it is likely that New Delhi will be at least be open to Washington's advice and assistance in solving some of the problems caused by climate change. The nuclear deal signed by the two countries in 2007 is expected to allow India to establish nuclear power plants producing 20 gigawatts of electricity by 2015. It is estimated that "the annual savings from the India deal could be nearly as large as the entire commitment of the twenty-five EU nations to reducing emissions under the Kyoto Protocol," assuming that, otherwise, India would be generating the same electricity using coal-fired power plants.[39] Because the industrial countries are still emitting far more carbon dioxide per person, an aggressive stance by Washington or other Western capitals with respect to India's greenhouse gas emissions will not be effective. Multilateral engagement and manageable solutions will be needed for India to be brought into an emerging global climate control regime. This may also require a willingness on the part of the industrial world to offer pollution-limiting technologies to both China and India in return for emissions control commitments.

Conclusions

This chapter has identified several areas where climate change can affect India's security, broadly conceived. This change has the potential to disrupt internal life in India, exacerbate existing social conflicts, and generate additional ones. Although India is on a high growth trajectory, the country has a lopsided and highly unequal economic system, which has left a large section of its society vulnerable. This vulnerability is likely to be worsened by climate change. In terms of external security, India's relations with China, Pakistan, Bangladesh, and Nepal could deteriorate as a result of climate-induced changes, especially in the area of water sharing. For the larger global system, India's rise as a major power is occurring against the backdrop of a global climate crisis, and this provides opportunities as well as pitfalls for engagement between India and the West, especially the United States.

Notes

1. On state capacity, see Joel S. Migdal, *Strong Societies and Weak States: State-Society Relations and State Capabilities in the Third World* (Princeton, NJ: Princeton University Press, 1988), 4–5; and Peter Evans, *Embedded Autonomy* (Princeton, NJ: Princeton University Press, 1995). I have developed this in the context of South Asia. See T. V. Paul, "State Capacity and South Asia's Perennial Insecurity Problems," in *South Asia's Weak States: Understanding the Regional Insecurity Predicament*, edited by T. V. Paul (Stanford, CA: Stanford University Press, 2010), 3–27.

2. This conclusion is somewhat consistent with other studies on this subject. See, e.g., Jon Barnett, "Security and Climate Change," *Global Environmental Change* 13, no. 1 (April 2003): 7–17; and Jon Barnett and W. Neil Adger, "Climate Change, Human Security and Violent Conflict," *Political Geography* 26, no. 6 (August 2007): 639–55.

3. Government of India, *Poverty Estimates for 2004–05* (New Delhi: Government of India, 2007), www.planningcommission.gov.in/news/prmar07.pdf; and World Food Program, *India 2007*, www.wfp.org/country_brief/indexcountry.asp?country=356.

4. United Nations Development Program, *Human Development Report 2007–2008* (New York: United Nations Development Program, 2007), 152, http://hdr.undp.org/en/reports/global/hdr2007-2008/. It was reported in June 2008 that China had emerged as the leading emitter of carbon dioxide. However, in terms of per capita emissions, the top ranks go to the United States, with 19.4 tons; Russia, 11.8 tons; Western Europe, 8.6 tons; China, 5.1 tons; and India, 1.8 tons. Elisabeth Rosenthal, "China Clearly Overtakes U.S. as Leading

Emitter of Climate-Warming Gases," *New York Times*, June 13, 2008, www.nytimes.com/2008/06/13/business/worldbusiness/13iht-emit.4.13702154.html.

5. R. K. Mall, Ranjeet Singh, Akhilesh Gupta, G. Srinivasan, and L. S. Rathore, "Impact of Climate Change on Indian Agriculture: A Review," *Climate Change* 78 (Spring 2006): 450, 453, www.springerlink.com/content/?Author=R.+K.+Mall.

6. "Melting Asia," *Economist*, June 7, 2008, 29. The extent of this melting is a matter of controversy. The IPCC report warned of complete melting by 2035, a prospect that has been questioned. See "World Misled over Himalayan Glacier Meltdown," *Sunday Times* (London), January 17, 2010, www.timesonline.co.uk/tol/news/environment/article6991177.ece.

7. The poor level of rainfall in 2007, combined with unusually intense rains in May of 2008 in some parts of the country, suggests that the melting of the Himalayan glaciers might already be affecting rain patterns in India. See "Melting Asia."

8. A "poverty trap" is a structural or systemic condition that makes it difficult or impossible for a nation (or an individual) to advance economically. At the national level the "poverty trap" most commonly means that net capital accumulation is insufficient to keep up with net population increase. See Jeffrey D. Sachs, *The End of Poverty* (New York: Penguin Books, 2006), 244.

9. This argument is tempered by other analyses. An exhaustive literature review in one study predicts that the net effect of climate change on Indian agriculture will be positive, or at worst, have a negligible negative impact up to 2050. The higher-risk categories are the winter crops in Central and Southern India. Droughts, floods, and other sudden environmental disasters will pose a greater threat to some of India's summer crops in the coming years. See Mall et al., "Impact of Climate Change," 471; and appendix A.

10. Barnett and Adger, "Climate Change, Human Security and Violent Conflict."

11. Sonya Fatah, "Suicide Rate Growing as Debt Cripples India's Farms," *Star* (Toronto), March 24, 2008, www.thestar.com/News/World/article/350058.

12. Somini Sengupta, "On India's Farms, a Plague of Suicide," *New York Times*, September 19, 2006, www.nytimes.com/2006/09/19/world/asia/19india.html.

13. Joyeeta Gupta, *The Climate Convention and Developing Countries: From Conflict to Consensus?* (New York: Springer, 1997), 46–47. See also A. S. Unnikrishnan and D. Shankar, "Are Sea-Level-Rise Trends along the Coasts of the North Indian Ocean Consistent with Global Estimates?" *Global and Planetary Change*, June 2007, 301–7; A. S. Unnikrishnan, K. Rupa Kumar, Sharon E. Fernandes, G. S. Michaell and S. K. Patwardhan, "Sea-Level Changes along the Indian Coast: Observations and Projections," *Current Science* 90, no. 3 (2006): 362–68.

14. Roger Harrabin, "How Climate Change Hits India's Poor," BBC, February 1, 2007, http://news.bbc.co.uk/2/hi/south_asia/6319921.stm; Jyoti K. Parikh and Kirit Parikh, "Climate Change: India's Perceptions, Positions, Policies, and Possibilities," Organization for Economic Cooperation and Development, 2002, www.oecd.org/dataoecd/22/16/1934784.pdf; and Jayant Sathaye, P. R. Shukla, and N. H. Ravindranath, "Climate Change, Sustainable Development and India: Global and National Concerns," *Current Science* 90, no. 3 (February 10, 2006): 314–25, www.iisc.ernet.in/currsci/feb102006/314.pdf.

15. "India's Naxalites: A Spectre Haunting India," *Economist*, August 17, 2006.

16. Milan Ridge, "India: Discontent, Poverty Fueling Naxalite Rebels," *Christian Science Monitor*, April 29, 2008.

17. On the possible causes by which climate change can cause conflict, see Barnett, "Security and Climate Change."

18. On the question whether or not India's liberalization has helped the poor, see Baldev Raj Nayar, *The Myth of the Shrinking State: Globalization and the State in India* (Oxford: Oxford University Press, 2009).

19. On the prevalence of conflict in the subregion, see Sanjib Baruah, *Postfrontier Blues: Toward a New Policy Framework for Northeast India*, East-West Center Policy Study 33 (Washington, DC: East-West Center, 2007), www.eastwestcenter.org/fileadmin/stored/pdfs/PS033.pdf.

20. On possible pathways by which climate change could induce internal violence and extremism, see Clionadh Raleigh and Henrik Urdal, "Climate Change, Environmental Degradation and Armed Conflict," *Political Geography* 26, no. 6 (August 2007): 674–94; and Ragnhild Nordås and Nils Petter Gleditsch, "Climate Change and Conflict," *Political Geography* 26, no. 6 (August 2007): 627–38.

21. U.S. Department of State, *Country Reports on Terrorism 2007*, www.state.gov/documents/organiza tion/105904.pdf.

22. Neil Pelkey, "The Cauvery Water War," Division of Environmental Studies, University of California, Davis, n.d., www.des.ucdavis.edu/staff/pelkey/cauvery.htm; and "Bangalore Tense at River Dispute," BBC, February 6, 2007, http://news.bbc.co.uk/2/hi/south_asia/6333907.stm.

23. Schubert, H., J. Schellnhuber, N. Buchmann, A. Epiney, R. Grießhammer, M. Kulessa, D. Messner, S. Rahmstorf, and J. Schmid, *World in Transition: Climate Change as a Security Risk*, Report from German Advisory Council on Climate Change (Earthscan: London, 2008), 31, www.wbgu.de/wbgu_jg2007_engl.html.

24. Amartya Sen, *Poverty and Famines* (Oxford: Oxford University Press, 1983).

25. See Ministry of Environment and Forests, *India: Addressing Energy Security and Climate Change* (New Delhi: Government of India, 2007), http://envfor.nic.in/divisions/ccd/Addressing_CC_09-10-07.pdf.

26. Sreeram Chaulia, "Bangladeshi Immigrants Stoke Terror in India," *Asia Sentinel*, May 15, 2008, www .asiasentinel.com/index.php?option=com_content&task=view&id=1200&Itemid=35.

27. Schubert et al., *World in Transition*, 123.

28. Barnett, "Security and Climate Change"; appendix B.

29. IPCC, *Climate Change 2007: Impacts*, 481.

30. Roland Buerk, "Villagers Left in Limbo by Border Fence," BBC, January 28, 2006, http://news.bbc.co .uk/2/hi/programmes/from_our_own_correspondent/4653810.stm.

31. See John Briscoe, "War or Peace on the Indus?" *SouthAsianIdea Weblog*, http://thesouthasianidea .wordpress.com/2010/04/03/war-or-peace-on-the-indus/.

32. See Ashok Swain, *The Environmental Trap: The Ganges River Diversion, Bangladeshi Migration and Conflicts in India* (Uppsala: Uppsala University Press, 1996).

33. For further information, see Toufiq A. Siddiqi, "India–Pakistan Cooperation on Energy and Environment: To Enhance Security," *Asian Survey* 35, no. 3 (March 1995): 280–90.

34. Reeta Tremblay and Julian Schofield, "Institutional Causes of the India–Pakistan Rivalry," in *The India-Pakistan Conflict: An Enduring Rivalry*, edited by T. V. Paul (Cambridge: Cambridge University Press, 2005), 225–48.

35. Anjana Pasricha "India Satisfied with Copenhagen Summit," VOAnews.com, December 22, 2009, www1.voanews.com/english/news/India-Satisfied-with-Copenhagen-Climate-Summit—79888187.html.

36. United Nations Development Program, *Human Development Report 2007–2008*, 13.

37. Sonu Jain, "India's Climate Change Action-Plan Takes the Safe Way: No to Caps, Yes to Efficiency," *Indian Express* (New Delhi), June 4, 2008, www.indianexpress.com/story/318373.html.

38. Despite this, India has made some progress. It has a high level of recycling, it is the world's fourth-largest producer of wind power, and "its solar yield is also bigger than any country except America"; "Melting Asia," 31.

39. David G. Victor, "The Indian Nuclear Deal: Implications for Global Climate Change," testimony before the U.S. Senate Committee on Energy and Natural Resources, July 18, 2006, www.cfr.org/publication/ 11123/india_nuclear_deal.html.

7

Pakistan

Daniel Markey

During the next twenty years, climate change will stress the Pakistani state and exacerbate its current fragility. However, the threats posed by climate change will almost certainly be dwarfed by other political, economic, and military factors in determining the fate of this rapidly expanding South Asian nation. By 2030 Pakistan's population is expected to top 240 million, which would make it the fifth-largest country in the world.[1] Strategically located and nuclear armed, Pakistan will remain an extremely high priority for U.S. and international policymakers.

At present Pakistan's state institutions are relatively weak and unlikely to serve as bulwarks against the challenges of climate change. The country's feudal and tribal social structures are not typically viewed as reserves of social resilience or ingenuity, but rather of repression and limited human capital. That said, during the past few years its civil society and media have shown growth potential. Over time they could begin to fill the void between public institutions and traditional social structures, informing the public and creating informal mechanisms (e.g., nongovernmental organizations) to respond to threats posed by climate change.

Although the near-term risk of climate-induced state failure in Pakistan is low, during the next several decades it could become an even more fragile state for reasons unrelated to climate change, and the impact of climate change could serve as the proverbial straw that breaks the camel's back. The present-day stresses of fighting militancy in the country's

Pakistan

tribal areas are likely to continue for at least another five to ten years, further weakening its army and the federal government's capacity to respond to secessionist movements in Sindh and Baluchistan. It is also conceivable, if unlikely, that heightened tensions with India could ruin Pakistan by 2030.

Islamabad's success in managing domestic political extremism and militancy will shape Pakistan's receptiveness to Western, and particularly American, influence. Greater success is likely to bring Pakistan closer to the West, especially the United States. Conversely, frustration or outright failure will drive a wedge between Islamabad and Washington. To the extent that climate change constitutes an added drag on the Pakistani economy, a possible trigger for conflict with India, or a destabilizing influence in national politics, it will render the Pakistani government's effort to fight militancy—and, by extension, to engage with the Western world—even more difficult.

Alternatively, constructive foreign interventions by international organizations or friendly states to help Pakistan husband its water resources, produce and deliver its electricity more efficiently, and respond to natural disasters are likely to make Pakistanis receptive to expanded engagement. In this respect, issues related to climate change provide an opening for the United States and other members of the international community to build mutually beneficial, long-term partnerships with Pakistan. U.S. policymakers charged with providing assistance in Pakistan would do well to seize this initiative for both strategic and humanitarian reasons.

The Science: What We Know about Climate Change in Pakistan

Recent studies indicate that climate change in the 2030 time frame is likely to threaten Pakistan in two fundamental ways: by limiting freshwater resources and by damaging the coastline. During the next twenty years, these threats are likely to impose significant, if not devastating, political, economic, and humanitarian costs. Projecting current trends into the decades beyond 2030, climate change might well decimate Pakistan's agricultural base, financial system, and commercial hub, threatening the livelihoods of tens of millions of people.

Pakistan's Limited Water Resources

During the next twenty years, global climate change is likely to stress Pakistan's already limited freshwater supply, with serious negative implications for the nation's agricultural and hydropower sectors. Pakistan already faces a water shortage, with an average annual rainfall of only 250 millimeters.[2] Practically all of Pakistan's available surface water and groundwater is currently used, and some studies project a 30 percent shortfall by 2030 for agricultural, domestic, and industrial needs.[3] Water availability has decreased from 5,000 cubic meters per capita in 1951 to about 1,100 this decade, and by 2025 it will dip below 700.[4] Rainfall projections suggest that the dry winter months will become drier, while the rest of the year will be marked by increased, if unreliable, precipitation.[5]

Pakistan's primary freshwater sources are rivers fed by the melting snow and ice of the region's glaciers.[6] Rising temperatures are shrinking these glaciers, which appear to be retreating more rapidly than any others in the world.[7] The Siachen Glacier, famously contested and still occupied by the Indian and Pakistani militaries, has retreated 1.7 kilometers in the last seventeen years.[8] By 2035, the region's glacier fields may be only one-fifth their current size, transforming regional rivers into seasonal ones.[9] In the interim period, as glaciers melt more rapidly, river levels are expected to become more variable before they fall and, eventually, run dry.[10] During the next twenty years, this variability of water levels, along with a growing absolute water scarcity, will constitute a major challenge for Pakistan's economy and infrastructure.

The best way to regulate water flow is through the construction of large dams and increased storage capacity. Pakistan's existing dam and irrigation system is vast, feeding water to farmlands that cover about 65,640 square miles.[11] But in recent decades Pakistan has severely underinvested in these areas.[12] Today, Pakistan's existing irrigation system is technically inefficient, and water distribution is plagued by corruption.[13] The necessary megadam projects run into the tens of billions of dollars in associated costs and would take years to become operational.[14]

The negative implications of water shortages and inconsistent supply are most clearly relevant to Pakistan's agriculture sector. By one estimate, climate change could reduce South Asia's cereal crop productivity up to 30 percent by 2050.[15] Pakistan's coast and plains have experienced a drop-off in precipitation over the past century, but it is important to note that most of Pakistan's agriculture currently depends upon river-fed irrigation.[16] In fact, the country's ratio of irrigated to rain-fed land is the highest in the world.[17] And because 70 percent of Pakistan's land is arid, it would be vulnerable to desertification if irrigation canals ran dry.[18] In recent years reduced river flows into the Indus River Delta region have raised salinity levels and rendered land unsuitable for crop cultivation.[19] Nationwide, a quarter of Pakistan's irrigated cropland already suffers from heightened salinity. As of 2006 it is estimated that the total costs imposed by high salinity levels amounted to nearly 1 percent of Pakistan's GDP.[20]

Aside from crop cultivation, Pakistan's rangeland and mangrove forests are also vulnerable to climate change. Pakistan's rangeland, and the herding communities that depend upon it, will also suffer from increased temperatures and reduced rainfall.[21] The Indus Delta's mangroves are largely dependent upon glacier-fed freshwater and have rapidly diminished in recent decades, a decline that is attributed to the intrusion of seawater.[22]

Agriculture occupies a disproportionately large share of Pakistan's national economy. According to World Bank estimates, it accounts for 25 percent of GDP, two-thirds of employment, and 80 percent of exports.[23] From 2007 to 2008, water shortages contributed to reduced crop yields, which tangibly demonstrates the implications of future water scarcity.[24] The Indus River System Authority projected up to a 40 percent water shortage during the 2008 fall harvest season.[25]

Pakistan also depends heavily on its rivers for power generation and drinking water. Roughly one-third of the nation's power capacity is provided by hydroelectric plants.[26] Recent water shortfalls have translated into less electricity for Pakistan's burgeoning population and (until recently) expanding economy.[27] In early 2008, Pakistan's government ordered the closure of steel-melting units and hundreds of textile mills in order to cope with electricity shortages.[28] For Pakistan, low water levels already have clear economic costs.

Increasing salinity in coastal water sources and diminishing access to clean groundwater already threaten most of Pakistan's urban population, which increased from 20 million to 70 million between 1980 and 2000.[29] A total of 40 million rural Pakistanis already lack access to safe drinking water.[30] The costs associated with water scarcity permeate the economy, imposing a direct or indirect drag on industries ranging from food processing to mining and manufacturing.[31]

Pakistan's Threatened Coastline

Climate change is also likely to threaten Pakistan's coastline, with rising sea levels and increasingly frequent and violent cyclones representing the greatest challenges to livelihoods and economic interests. Rising sea levels will affect Pakistanis living along the southern coast, where 700,000 people currently live in low-elevation areas (see appendix B). In general, coastal villages and urban slums are the most vulnerable to flooding. Heightened sea levels also contribute to the broader problem of salinity in Pakistan's drinking water and freshwater-dependent mangrove forests.[32]

Pakistan's megacity of Karachi, at about 8 meters elevation, is the largest population center likely to feel the effects of rising sea levels.[33] The city's impoverished residents who live in low-lying areas will bear a disproportionate share of the pain. But for these vulnerable populations, violent storm surges and cyclones are in all probability even more dangerous than steadily but slowly rising sea levels.[34] In June 2007 a series of storms killed two hundred people in Karachi, demonstrating the inability of Karachi's infrastructure to withstand even existing risks.[35] The intensity of these storms appears to be increasing, possibly driven by higher water temperatures in the Arabian Sea.[36] In the future, cyclones and storm surges could also threaten Pakistan's new port of Gwadar, potential gas pipelines running through Baluchistan, and agriculture in parts of southern Punjab.

Because Karachi is Pakistan's single largest financial, trade, and manufacturing center, the city's vulnerability to the effects of climate change has critical implications for the national economy. The city is home to 30 percent of Pakistan's manufacturing, 90 percent of its business head offices, and by far its largest stock exchange. Approximately two-thirds of its income tax, 58 percent of its sales tax, and one-fifth of its total GDP are generated in Karachi.[37] The possible losses associated with a traumatic climate-induced event in the city are suggested by the fact that in the aftermath of Benazir Bhutto's 2007 assassination, rioting shut down the city and cost an estimated $161.8 million in a single day.[38] Together, Karachi and nearby Port Qasim account for 95 percent of Pakistan's international trade. Karachi alone handles 60 percent of the nation's cargo.[39]

Dangerous Implications of Climate Change

Climate change in Pakistan will have social, political, and economic implications, which will be determined in large part by existing structural conditions and long-term trends. Pakistan already faces challenges associated with weak governing institutions, a fragile national economy, persistent social tensions, extreme demographic pressures, and long-standing regional conflicts, all of which could be aggravated by water scarcity and the coastline's vulnerability to storms and rising sea levels.

Weak Government Institutions and a Fragile National Economy

The primary determinant of whether the challenges associated with climate change will threaten national stability, or merely impose manageable economic and humanitarian costs, is the relative strength of Pakistan's governing institutions. At present, those institutions are not up to the task of mitigating climate change risks. By most accounts, corruption and inefficiency are inextricable features of the Pakistani bureaucracy.[40] As is the case in many other developing states, the policy reforms required to address threats compounded by climate change—including population relocation, energy conservation, and technological change—would be politically unpopular and costly in the near term, even if they could be vital for long-term national stability.[41]

Pakistan's investments in national infrastructure have consistently fallen below the levels needed to maintain its aging irrigation canals, dams, and power generation and distribution networks. As of 2006, for instance, the state spent only 0.25 percent of GDP on water supply and sanitation.[42] Since 1999 Pakistan's total installed power generation capacity has

Table 7.1 Pakistan's Total Installed Electricity Generation Capacity

Period	Capacity in Megawatts
1999–2000	16,764
2000–2001	17,772
2001–2	17,697
2002–3	17,728
2003–4	19,254
2004–5	19,389
2005–6	19,439
2006–7	19,440
2007–8	19,566
2008–9	19,754

Sources: Compiled from data in the following Pakistani government reports: 1999–2000 figure from *Economic Survey of Pakistan 1999-2000*, table 16, www.paksearch.com/Government/STATISTICS/Survey00/CH_16a.htm; 2000–2001 figure from *Economic Survey of Pakistan 2003–2004*, 201, www.accountancy.com.pk/docs/Economic_Survey_2003-04.pdf; 2001–2 and 2002–3 figures from *Economic Survey of Pakistan 2002–2003*, 217, www.accountancy.com.pk/docs/Economic_Survey_2002-03.pdf; 2003–4 and 2004–5 figures from *Economic Survey of Pakistan 2004–2005, Part III*, 197, www.accountancy.com.pk/docs/Economic-Survey-2004-05-Part-III.pdf; 2005–6 and 2006–7 figures from *Pakistan Economic Survey 2006–2007*, 236, www.finance.gov.pk/admin/images/survey/chapters/15-Energy.pdf; 2007–8 figure from *Pakistan Economic Survey 2007–2008*, 257, www.finance.gov.pk/admin/images/survey/chapters/15-Energy08.pdf; 2008–9 figure from *Pakistan Economic Survey 2008–2009*, 234, www.finance.gov.pk/admin/images/survey/chapters/15-Energy09.pdf.

increased by only 2,990 megawatts, or 17.8 percent (table 7.1). For its estimated population of 174.6 million, its total power generation capacity thus amounts to approximately 0.0001 megawatt per person.[43] By comparison, Americans enjoy approximately 0.003 megawatt per person.[44] A July 2009 Gallup poll indicated that 53 percent of Pakistanis live without electricity for more than 8 hours a day.[45] And about 40 percent of Pakistani households are not even connected to the electrical grid.[46] The tremendous inefficiencies of Pakistan's water and electricity distribution networks will become ever more salient as absolute water supplies dry up.

Pakistan's civil service has been diminished by decades of politicization and lengthy periods of military rule, during which time an undue proportion of national resources and attention have been directed toward the defense sector rather than socioeconomic priorities. Moreover, frequent national crises and periodic leadership purges have fostered a short-term, reactive political culture.

The national elections in February 2008 returned civilian leaders to power in Islamabad. If civilian governance endures for the next two decades, it is possible that Pakistan's state institutions, from the National Assembly down to the local police, could mature, expand, and become more capable of addressing large-scale shocks, whether climate induced or otherwise. But if history is any guide, the institutions of civilian politics are more likely to remain incapable of doing so, owing to inadequate investment, human capital, and leadership. And to make matters even worse, by 2030 many of Pakistan's national institutions could further atrophy, having fallen victim to internal conflict and disruptions completely unrelated to climate change. Under this worst-case scenario, the scale of human suffering from food and water shortages, flooding, and simple economic deprivation will be immense and nearly impossible for external actors to mitigate.

Not only will Pakistan's weak institutions make it harder for the state to protect its people from the negative consequences of climate change, but climate change is also likely to weaken state institutions still further by damaging the national economy and constricting tax revenues. This sort of vicious cycle is most easily envisioned in Karachi, simultaneously Pakistan's greatest generator of government revenue and a coastal megacity vulnerable to rising sea levels, storm surges, and freshwater scarcity. A Katrina-like storm in Karachi during the next twenty years would raise doubts about Pakistan's economic and political survival. Pakistan's devastating earthquake of October 2005 stressed civilian and military institutions to their limits, and it is hard to imagine how they would have been able to cope if the earthquake's epicenter had been closer to major urban centers. Although less dramatic, the added costs associated with doing business in Karachi (as opposed to cities less threatened by the effects of climate change) will deter business investment and slow economic growth.

Exacerbating Ethnic, Regional, and Class Tensions

Resource scarcity accelerated by climate change could drive dangerous wedges into the existing fault lines that run through Pakistani society. One would anticipate that conflict is most likely where ethnic and regional distinctions coincide with divisions between haves and have-nots, and where governing institutions are not strong enough to deter violence and provide equitable redistribution mechanisms.[47] Comparative studies of intrastate conflict driven by water scarcity indicate that conflict tends to erupt where supply is restricted suddenly in the context of existing social and economic disparities and a history of conflict.[48]

Pakistan meets the criteria that put it at risk for conflict driven by water scarcity.[49] In Pakistan's provincial politics, differences over water rights typically pit the wealthier and politically dominant province of Punjab against disadvantaged, less populous Sindh, Balochistan, and North-West Frontier Province (NWFP). The fact that before Pakistan's independence ethnic Punjabis alone were granted access to canal-irrigated land in the provinces of Punjab and Sindh even today contributes to ethnic tensions.[50] The allocation of water among Pakistan's four provinces is presently governed by the terms of the 1991 Indus Water Accord, but recent provincial assembly protests suggest that the issue is far from settled.[51] In one recent instance, the Punjab provincial government rejected a request by the national Indus River System Authority to reduce Punjab's withdrawal of water from the Tarbela Dam, which exceeded its agreed-upon intake for the season.[52]

During the past decade, interprovincial conflict has been most apparent in the context of large dam-building projects. The Kalabagh Dam, proposed for construction along the Indus River in northern Pakistan, is designed to support agriculture and power production, but it is firmly opposed in the NWFP and Sindh.[53] Residents of the NWFP would bear an inordinate share of the dam's environmental impact and, perhaps with good reason, they also tend to fear inadequate compensation from neighboring Punjab, which would benefit far more from the dam's capacity to regulate water levels and generate hydropower. Downstream, Sindh's residents contest the "theft" of water and the impact on the salinity of the Indus near its mouth at the Arabian Sea, which would hurt coastal mangroves and local agriculture. In general, the trade-off between using scarce water resources for irrigation or for power generation pits farmers against industrialists and urbanites, a dynamic that is already sparking political debates.[54]

Under the Pervez Musharraf regime, these political pressures effectively gridlocked the Kalabagh Dam project.[55] But during the next two decades, population growth and economic expansion will compel future Pakistani leaders to consider a variety of new dam construction projects, inevitably sparking subsequent bouts of provincially rooted dissent. In this sense, climate change would fuel disruptive regional tensions. But because large dam projects are not built overnight, they are unlikely to create the sort of sudden restriction on water resources that in and of itself would spark violence.

Pakistan's ethnic and political cleavages are not limited to its provincial boundaries but also extend into a number of the country's major cities. Karachi is especially plagued by competition between ethnic Sindhis, Pashtuns, and Mohajirs (Urdu-speaking Pakistanis who trace their ancestry to modern India). Each group is represented by a different political party, and each party maintains its own armed gangs. Street violence has long been an all-too-frequent feature of life in Karachi, so it is not hard to imagine that climate-change-driven resource scarcity could, at the margins, hasten future deterioration of law and order in the city. Flooding or other sorts of localized environmental degradation could spur even more intense competition over real estate, already one of the city's most significant sources of internal conflict.

At present Karachi's financial and economic primacy is not reflected in national politics, which remains a Punjab-dominated sport. Long-standing regional tensions fueled by Karachi's perceptions of inequity might boil over into sustained and intensive street protests and violence if the national government in Islamabad were to fail, for instance, to respond adequately to a climate-driven disaster along the coastline. By a similar logic, the national government is likely to be acutely sensitive to political unrest in the nation's Punjabi heartland. Sharp productivity losses in Punjab's agricultural sector (precipitated by water scarcity or higher temperatures) would threaten the political stability of any government in Islamabad.

In almost any future scenario, Pakistan's poorest farmers and fishermen are likely to suffer most acutely from climate change, and they are also the least capable of protecting themselves from the predations of corrupt officials and more powerful agricultural and industrial interests.[56] Pakistan's impoverished masses are also the least well represented in the nation's politics, which even during periods of elected civilian rule tend to be dominated by a near-feudal oligarchic elite. With few avenues for redressing their economic grievances, poor people already resort to demonstrations and violence, which could worsen over the next twenty years. Socioeconomic inequalities and poor education also contribute to widespread alienation and create a permissive environment for extreme ideologies, including those of militant and terrorist organizations.[57]

Amplifying Demographic Trends: Urbanization and Migration

During the next two decades, Pakistan's changing climate will play into existing trends of dramatic population growth and movement. The physical, political, and social pressures associated with overpopulation and migration may be exacerbated by the added stresses imposed by water scarcity, reduced agricultural productivity, and the destruction left by flooding and periodic storms.

Pakistan's population has been growing rapidly for decades; it more than doubled between 1975 and 2005. The United Nations estimates that Pakistan will grow by another 54 million during the next fifteen years (assuming no major systemic shocks) and will surpass 300 million by the middle of this century. By that time Pakistan will be the fourth-largest country in the world, after only China, India, and the United States.[58]

Pakistan is already the most urbanized country in South Asia, with more than one third of Pakistanis living in towns and cities.[59] Pakistan's shift from rural to urban has come quickly. In 1980 there were 20 million city dwellers. By 2000 that number had rocketed to 70 million.[60] Karachi alone grew at an average annual rate of 4.5 percent from 1981 to 1998. At this rate, the megacity is projected to reach a population of 26.4 million by 2020.[61]

Pakistan's population boom has already strained its physical infrastructure—roads, canals, hospitals, schools, and so on—and there is little doubt that public-sector investment has failed to keep pace with basic needs. In this context, the future implications of climate change are likely to resemble (and exacerbate) today's challenges posed by environmental degradation and shortages of food and electricity. These challenges are already immense. At present, UNICEF reports that waterborne diseases such as cholera, dysentery, and hepatitis cause roughly one-third of all deaths in Pakistan, imposing a cost of 1.8 percent of GDP and tying up 20 percent to 40 percent of the nation's hospital beds at any given time.[62] In 2008, global and national supply shortfalls sent food prices soaring, which caused especially severe hardships for tens of millions of Pakistani urbanites. The global loss in agricultural productivity associated with climate change is projected to send cereal prices still higher in the coming decades.[63]

Unless Pakistan can find a way to achieve consistently higher growth rates, translate private wealth into state revenues, and reinvest resources in infrastructure, public institutions, and basic services, these sorts of population pressures will grow worse. And if they are compounded by climate change, these pressures might well pose insurmountable obstacles to the nation's political leadership and undermine the state's legitimacy.

Although no single actor other than the state could conceivably cope with policy challenges of this magnitude, localized solutions offered or imposed by private actors are likely to fill discrete, near-term governance voids. This pattern is already apparent in pockets of the NWFP, where the breakdown of judicial institutions has encouraged militant groups to offer informal adjudication services and thereby assert their authority over that of the provincial bureaucracy.[64] Similarly, after Pakistan's October 2005 earthquake, charitable wings of prominent militant organizations emerged as some of the most effective distributors of humanitarian assistance, winning the gratitude, and quite possibly the allegiance, of the citizens who suffered major losses. Over time, the combined activities of even relatively benign nongovernmental organizations can hollow out the legitimacy, capacity, and authority of a weak state, rendering it even more fragile in the face of subsequent, more concerted threats. This dynamic raises the likelihood of civil conflict and state breakdown.

In addition to spurring greater competition for scarce resources within Pakistan, demographic pressures have led millions of Pakistanis to seek opportunities overseas. Nearly 4 million Pakistanis are registered as overseas residents, mainly in the Persian Gulf states (1.1 million in Saudi Arabia and 500,000 in the United Arab Emirates), the United Kingdom (800,000), and the United States (600,000).[65] The total number of Pakistanis living overseas, including illegals, stayovers, and students, may run as high as 7 million, or roughly 4 percent of the total population.[66] And these numbers have been increasing over time; approximately 50,000 more Pakistanis left in 2004 than in 1995.[67] Pakistanis working overseas make substantial contributions to the national economy, to the tune of roughly $6 billion in 2007–8.[68] Remittances are believed to account for at least 4 percent of GDP.[69]

As might be expected, studies of Pakistan's migration patterns indicate that regions with lower agricultural productivity tend to produce greater numbers of migrants.[70] That said, initial travel costs make the very poorest Pakistanis less likely to migrate than their low- to middle-income peers.[71]

The combined implications of climate change may accelerate Pakistan's migration trends. As agricultural productivity falls (and the population continues to increase), more and more Pakistani farmers will be unable to sustain families and will emigrate for work. If current patterns hold, these unskilled, uneducated men are most likely to seek opportunities in the energy-rich, labor-poor Gulf states. By implication, Pakistan's national economy will become more dependent upon regional stability and, indirectly, upon the global market for fossil fuels. Political or economic shocks to the Gulf region will therefore be increasingly significant to Pakistan's own stability.

Pakistan is a net exporter of labor, but over the past several decades it has also served as a destination for millions of migrants, primarily refugees from Afghanistan. Many of these refugees have returned to their home country, and many more are slated to return in the next several years. But if Afghanistan again fails to become a viable, stable state, Pakistan remains the obvious potential host. To the extent that Afghanistan's own environmental degradation (partly driven by climate change) impedes agriculture, herding, and water-intensive industries, more Afghans may choose to stay in Pakistan. Most Pakistani observers believe that the sprawling Afghan refugee camps in Pakistan have created permissive environments for smuggling, criminality, and militancy, all of which threaten the stability of the Pakistani state and the security of its citizens.

Inciting Regional Conflict

Water has long been a point of serious contention between Islamabad and New Delhi. The 1960 World Bank–brokered Indus Waters Treaty preserves for Pakistan the use of the western rivers of the Indus system—the Chenab, Indus, and Jhelum. The political dispute over the former princely state of Kashmir has been exacerbated by Pakistan's concerns over Indian dam construction projects, including most recently the Baglihar Dam over the Chenab River, which flows from Kashmir into Pakistan proper.[72] In September 2008 Islamabad filed a complaint with India, claiming that the Chenab River's water flow into Pakistan had dropped below the minimum level guaranteed by World Bank arbitration and blaming India's filling of the Baglihar Dam as the cause.[73] India attributed the reduced water flow into Pakistan to a "lean" year overall in water accumulation.[74]

In the 2030 time frame, a blatant violation of the treaty by either India or Pakistan, perhaps driven by a desperate need for water or hydropower, could trigger disruptive change, even war, between these historic rivals. India might seize Pakistani water resources by building an unauthorized dam or otherwise siphoning water from the Indian source waters of Pakistan's rivers. Pakistan could be expected to respond by escalating the dispute diplomatically or militarily.

There are, however, reasons to expect that India will refrain from aggressive actions of this sort, even as India's own water needs grow. In particular, India's increasing stake in the international system, including multilateral structures like the World Bank, will continue to sensitize New Delhi to the costs of violating treaty obligations. More generally, statistical studies suggest that although states with shared rivers do engage in low-level conflict more commonly than other states, they also tend to cooperate more.[75] Though water disputes could plausibly trigger wider Indian/Pakistani conflict, they are unlikely to set off a war in the absence of other even more serious causes.

Beyond its relationship with India, the detrimental effects of climate change could also render Pakistan more vulnerable to regional influence and competition. As it suffers from reduced agricultural productivity and is increasingly constrained hydropower resources, Pakistan will be at the mercy of regional and/or global energy and food producers and in stiff competition with other desperate consumers. At present, few regional coordination mechanisms exist that might, for instance, bring greater efficiency to trade and transit from Central Asia to South Asia, which would in turn reduce consumers' energy and food costs.

Potential Sources of Strength

Layered on top of Pakistan's other daunting challenges, climate change will contribute to domestic and regional competition, conflict, and hardship during the next twenty years. That said, several notable and positive trends could conceivably offer greater strength to Pakistan's state and society, serving as a counterweight to some of the worst implications of climate change. The gradual expansion of a vibrant civil society and media, the potential for greater transparency and accountability from a democratic leadership, and the resources provided by steady macroeconomic growth are the three most plausible stabilizing influences in the next twenty years.

Civil Society and the Media

Despite the many authoritarian tendencies of the Pakistani government under President Pervez Musharraf, his regime brought a noteworthy expansion of media freedom and activity throughout the country. Electronic media, particularly satellite television, experienced the greatest growth. The government has licensed more than forty private satellite television stations.[76] Cable television has penetrated at least 70 percent of the urban population and 50 percent of rural areas.[77] Though many viewers prefer sports (cricket) or entertainment programming, many are also tuning in for news coverage and current affairs talk shows. Print journalism also experienced a relative resurgence under the Musharraf regime, which by most accounts (and for most of its time in office) practiced less heavy-handed censorship than its predecessors.

Only a strong and independent Pakistani media is capable of investigating issues related to climate change, raising public awareness, and consistently pressuring the government to respond. That said, the long-term nature of the problems generated by climate change, especially sea-level rise and water scarcity, makes them less amenable to quick fixes or management by crisis response, and also less appealing topics for the fast-paced sensationalism of today's media. Accordingly, it will not be sufficient to rely upon the profit-motivated media alone as a mechanism to spur change.

However, the past decade has also witnessed an increasingly assertive Pakistani civil society, a phenomenon that was manifest most dramatically in 2007–9, when lawyers who objected to the dismissal of Pakistan's Supreme Court chief justice took to the streets in protest. In this context, the massive coordinating efforts of regional bar associations, independent human rights activists, and, eventually, political parties demonstrated the powerful potential of nongovernmental organizations to mobilize grassroots support for a cause (enabled, in large part, by the media and improved communications technologies).

Although it is possible that grassroots organizations concerned about climate change could parlay media exposure into greater pressure on the government, it is important to recognize that a significant environmentalist movement is still a long way off in Pakistan. However, during the next twenty years a civil society increasingly attuned to the negative impact climate change has on quality of life could coalesce into a movement to promote widespread water and energy conservation, improved information sharing and a more efficient use of the existing infrastructure, and more effective coordination of local responses to humanitarian crises produced by climate change, such as flooding and cyclone damage.[78] Moreover, given the fact that Pakistan's business and agricultural sectors have strong economic incentives to see the state ensure steady access to water and energy, influential and popular private interests could converge in support of infrastructure development and greater conservation and energy efficiency. Amplified by the megaphone of independent media, these interests could exert significant pressure on future governments in both Islamabad and the provincial capitals.

Democratic Accountability

The positive influence of the media and civil society groups will most likely translate into proactive government policies if Pakistan can maintain and strengthen its commitment to democracy. In a well-functioning democracy, accountability to the electorate helps to check the worst abuses of state power and creates incentives for leaders to anticipate and respond to the needs of the people. The failure of Musharraf's military/technocratic government to invest in significant new power generation or electricity distribution capacity is but one example of the deficiencies associated with unaccountable leadership.[79]

Of course, the reality of democratic institutions is far from the ideal. At the time of this writing, Pakistan's civilian government is fragile and the short-term preoccupations of maintaining power far outweigh serious consideration of long-term projects, including infrastructure development. The country's civilian government may well remain mired in tactical political skirmishes despite the electorate's dissatisfaction over basic economic issues, such as food shortages and inflation.[80] And public institutions, such as the nation's Water Resources and Power Development Authority, have been weakened by decades of neglect that long preceded the Musharraf regime.[81]

There are, however, encouraging signs that the civilian government in Islamabad may be more attuned to the problems of climate change than its military predecessor, having become motivated primarily by devastating electricity shortages. Stopgap energy conservation measures have included a plan designed to reduce usage, especially during peak hours.[82] Longer-term plans include ambitious dam projects, more power generation by independent producers, and greater public investment in water management.[83] Whether these plans will in fact be implemented remains to be seen, but progress along these lines would certainly help to mitigate the worst consequences of climate change.

Economic Growth

In addition to political consensus and effective leadership, the Pakistani state will require significant financial resources in order to address climate change. The country's macroeconomic performance from 2004 to 2007 was quite healthy, with annual growth of 6 to 8 percent.[84] Sustained prosperity of this sort translates into higher tax revenues, which in turn enable the state to invest in dam building, irrigation systems, and more efficient electricity grids.

Unfortunately, in 2008 soaring global energy prices and political instability sent Pakistan's economy into a tailspin, severely constraining public-sector spending. As long as the economy remains weak, infrastructure and institutional investment will be placed on hold or financed through other means. Outside sources of support, including multilateral institutions (e.g., the Asian Development Bank) and foreign governments (the United States, Japan, and the EU), are the most likely option. As part of its enhanced nonmilitary assistance in Pakistan, the Barack Obama administration has signaled its intention to devote significant resources to large infrastructure projects.[85] Pakistan also has other tools at its disposal, such as offering tax holidays to private corporations that invest in dam projects, which might represent plausible alternatives.[86]

The Worst Case: State Failure?

Cumulatively, the dangers posed to Pakistan as a result of climate change are significant in the 2030 time frame. Yet, given Pakistan's many other daunting social, economic, and political challenges, it is reasonable to consider whether state failure is a realistic outcome within the next two decades.

Pakistan's medium- to long-term future is more difficult to predict than that of most countries. This difficulty can be explained by uncertainties inherent in the nation's turbulent, personality-driven politics along with the seemingly endemic lack of transparency within those few institutions (particularly the military and intelligence agencies) that serve as the state's central pillars.[87] In other words, the robustness of the Pakistani state is hard to assess in the present tense, and even harder to project into the future. Any impulse toward optimism must be tempered by the fact that there are already parts of Pakistan where the government's writ does not run, or barely so.

Nevertheless, the Pakistani state has survived tremendous internal and external pressures in the past, including devastating wars and natural disasters, without outright failure. It would appear that the sinews of the Pakistani nation and society have held together more effectively than its formal institutions. Even so, when played out over decades, the mounting threats posed by a combination of antistate militancy, terrorism, popular alienation, poverty, ethnic and social conflict, and regional animosities turn state failure into a plausible proposition.

In this context, it is best to characterize the potential impact of climate change as a drag on Pakistan's political economy, which is likely to exacerbate other more significant threats but unlikely to become the proximate cause of conflict or state failure. If the Pakistani state collapses before 2030, it will not be because of climate change alone—or, indeed, from any other single cause. At the same time, if Pakistan achieves stability and prosperity during the next several decades, they will have come at a greater cost because of the foreseeable challenges imposed by climate change.

Looking Ahead: Opportunities for Outside Engagement

Among the many problems Pakistan faces, climate change offers some of the best opportunities for constructive engagement by outside powers, including the United States. Assistance in the areas of water management or power generation, for instance, could also pay significant dividends in policy areas that go well beyond climate change.

A number of external institutions already contribute to Pakistan's megadam projects, which are anticipated to cost tens of billions of dollars. For example, the World Bank, China Development Bank, and Asian Development Bank have pledged assistance for Pakistan's Basha Dam project.[88] International aid could similarly strengthen Pakistan's aging irrigation network and expand its undersized water storage capacity. Assistance programs of this sort bore much fruit in the years shortly after Pakistan's independence. The World Bank and other donors helped build an interlocking system of canals that fundamentally reconfigured the flow of water from colonial India's major western rivers, thereby enabling agricultural cultivation throughout much of what is now Pakistani Punjab.[89]

In today's geostrategic environment, Pakistan's stability is not merely a humanitarian concern. The weakness of the Pakistani state has contributed to its status as a safe haven for Islamist militants and terrorists. The terrorist networks that operate from Pakistan's territory, including al-Qaeda, pose a clear threat to regional and international security. These extremists exploit a vacuum of governance and feed upon popular alienation. To confront this threat, the international community has a vested interest in shoring up Pakistan's institutions and winning the support of its people.

At present, many Pakistanis are more inclined to see the United States and its "global war on terror" as more threatening than the terrorists or militants themselves.[90] But recent history also suggests that external assistance can dramatically shape Pakistani public opinion. After Pakistan's devastating earthquake in October 2005, American Chinook helicopters flew relief supplies to survivors, U.S. military and civilian officials contributed medical and engineering support, and Washington extended more than $500 million in aid from both public and private contributions to help rebuild the hardest-hit regions.[91] Subsequent opinion polls showed massive, if temporary, shifts in Pakistani sentiment. The number of Pakistanis who viewed the United States favorably doubled from 23 to 46 percent. Eighty-one percent of respondents said that American earthquake relief had informed their overall impression of the United States.[92]

Even if major contributions to Pakistan's infrastructure do not win immediate and popular acclaim in quite the same way as post-earthquake relief, they could still tangibly improve millions of Pakistani lives during the next two decades and beyond. Because a more prosperous and stable Pakistan will be more capable of withstanding a wide variety of threats, including those posed by climate change, the United States and Pakistan's other international partners would be well advised to make these investments sooner rather than later, for the strategic and humanitarian costs will only increase over time.

Notes

1. Population Division, Department of Economic and Social Affairs, United Nations Secretariat, *World Population Prospects: The 2006 Revision*, http://esa.un.org/unpp.

2. World Bank, *Pakistan Strategic Country Environmental Assessment*, 2 vols., World Bank South Asia Environment and Social Development Unit Report 36946-PK, August 21, 2006), vol. 1, 9, http://web.worldbank.org/WBSITE/EXTERNAL/COUNTRIES/SOUTHASIAEXT/0,,contentMDK:21459418~pagePK:146736~piPK:146830~theSitePK:223547,00.html.

3. John Briscoe and Usman Qamar, *Pakistan's Water Economy Running Dry*, World Bank Working Paper 44375, November 8, 2005, 29, www-wds.worldbank.org/external/default/WDSContentServer/WDSP/IB/2008/06/25/000333037_20080625013311/Rendered/INDEX/443750PUB0PK0W1Box0327398B01PUBLIC1.txt.

4. World Bank, *Pakistan Strategic Country Environmental Assessment*, vol. 1, 50.

5. Schubert, H., J. Schellnhuber, N. Buchmann, A. Epiney, R. Grießhammer, M. Kulessa, D. Messner, S. Rahmstorf, and J. Schmid, *World in Transition: Climate Change as a Security Risk*, Report from German Advisory Council on Climate Change (Earthscan: London, 2008), 144, www.wbgu.de/wbgu_jg2007_engl.html; "Melting Asia," *Economist*, June 7, 2008, 29.

6. "Almost 90 percent of the water in the upper portion of the river basin comes from glaciers located in the Himalaya, Karakoram, and Hindu Kush mountain ranges." World Conservation Union, *Indus Delta, Pakistan: Economic Costs of Reduction in Freshwater Flows*, Case Studies in Wetland Valuation 5 (May 2003): 2, www.cbd.int/doc/external/countries/pakistan-wetland-cs-2003-en.pdf.

7. IPCC, *Climate Change 2007: Impacts*, 493.

8. Ghulam Rasul and Bashir Ahmad, "Climate Change in Pakistan," 26, www2.restec.or.jp/geoss_web/pdf/0414/11.pdf.

9. Cruz et al., "Asia," 493.

10. Briscoe and Qamar, *Pakistan's Water Economy Running Dry*, xvii. The timeline for this transition is contested, as some scientists predict more rapid melt than others. For instance, Professor Syed Iqbal Hasnain estimates only a twenty- to thirty-year period until the Himalayan glaciers melt; "Melting Asia," *Economist*, 29. The IPCC's willingness to accept this general conclusion, based upon a single study by the World

Wildlife Federation, has been sharply challenged, however. See "UN Climate Panel Admits 'Mistake' over Himalayan Glacier Melting," *Telegraph* (London), January 20, 2010, www.telegraph.co.uk/earth/environ ment/climatechange/7031403/UN-climate-panel-admits-mistake-over-Himalayan-glacier-melting.html.

11. Brian Handwerk, "Some Glaciers Growing Due to Climate Change, Study Suggests," *National Geographic News*, September 11, 2006, http://news.nationalgeographic.com/news/2006/09/060911-growing -glaciers.html. See also World Conservation Union, *Indus Delta*, 1.

12. Briscoe and Qamar, *Pakistan's Water Economy Running Dry*, xviii, notes that "whereas the United States and Australia have over 5,000 cubic meters of storage capacity per inhabitant, and China has 2,200 cubic meters, Pakistan has only 150 cubic meters of storage capacity per capita."

13. World Bank, *Pakistan Strategic Country Environmental Assessment*, 15; also see Jean-Daniel Rinaudo, "Power, Bribery, and Fairness in Pakistan's Canal Irrigation Systems," in *Transparency International Global Corruption Report 2008: Corruption in the Water Sector*, edited by Dieter Zinnbauer and Rebecca Dobson (Cambridge: Cambridge University Press, 2008), 77–79.

14. For more on recent dam building plans, see Ihtasham ul Haque, "WB to Raise $9 Billion for Water Projects," *Dawn* (Karachi), April 26, 2008, www.dawn.com/2008/04/26/top11.htm; "Senate Question Hour: 'Govt Working on Generating 2,200 MW on Emergency Basis,'" *Daily Times* (Lahore), May 1, 2008, www .dailytimes.com.pk/default.asp?page=2008\05\01\story_1-5-2008_pg7_19; Zafar Bhutta, "Ministry Seeks Rs 60bn for Dams," *Daily Times*, May 29, 2008, www.dailytimes.com.pk/default.asp?page=2008\05\29\ story_29-5-2008_pg7_1; and "Bhasha Dam's Design and Tender Papers Ready, Says Wapda Chief," *Dawn*, July 1, 2008, www.dawn.com/2008/07/01/top9.htm.

15. Cruz et al., "Asia," 481. Other estimates are somewhat more conservative. See, e.g., William R. Cline, *Global Warming and Agriculture: Impact Estimates by Country* (Washington, DC: Peterson Institute for International Economics, 2007), 70. Cline estimates that agricultural productivity will decline 20 percent by 2080 (assuming carbon fertilization) or 30 percent (without carbon fertilization).

16. Eighty percent of Pakistan's farmland is irrigated by the Indus watershed; World Conservation Union, *Indus Delta*, 3.

17. Ibid., 1.

18. World Bank, *Pakistan Strategic Country Environmental Assessment*, vol. 1, 14–15.

19. World Conservation Union, *Indus Delta*, 3.

20. World Bank, *Pakistan Strategic Country Environmental Assessment*, vol. 1, 16.

21. Ibid., 18.

22. World Conservation Union, *Indus Delta*, 3. In 1990 mangrove cover was estimated at 160,000 hectares. By 2003 coverage had shrunk to 106,000 hectares. World Bank, *Pakistan Strategic Country Environmental Assessment*, vol. 1, 7.

23. Schubert et al., *World in Transition*, 145.

24. See Zafar Bhutta, "Sindh, Punjab Provinces Face Water Shortage," *Daily Times* (Lahore), May 14, 2008, www.dailytimes.com.pk/default.asp?page=2008\05\14\story_14-5-2008_pg7_1; and Sher Baz Khan, "Shortage of Wheat in All Provinces except Punjab," *Dawn* (Karachi), June 9, 2008, www.dawn.com/2008/ 06/09/top3.htm.

25. Ahmad Fraz Khan, "Severe Water Crisis Feared," *Dawn* (Karachi), September 11, 2008, www.dawn .com/2008/09/11/top8.htm.

26. Pakistan claims an installed capacity of 19,845 megawatts. In addition to hydropower, Pakistan depends on thermal stations, fuelled primarily by gas and oil. See "Pakistan Puts Clocks Forward, Hopes to Save Power," Reuters, June 1, 2008, http://uk.reuters.com/article/idUKISL7392020080601; and "Only 0.65pc Added to Power Generation Capacity in 9 Months of 2007–08," *Dawn*, June 11, 2008, www.dawn.com/2008/ 06/11/top11.htm.

27. On recent shortfalls, see Sajid Chaudhry, "Crisis as Water Reservoirs Fall to 10-Year Low," *Daily Times* (Lahore), January 4, 2008, www.dailytimes.com.pk/default.asp?page=2008\01\04\story_4-1-2008 _pg1_10; Zafar Bhutta, "PEPCO Increases Load Shedding Duration," *Daily Times*, March 13, 2008, www.daily times.com.pk/default.asp?page=2008\03\13\story_13-3-2008_pg7_10; Dilshad Azeem, "Power Generation Dropped as Water Situation Deteriorates," *News International*, May 7, 2008, www.thenews.com.pk/print3 .asp?id=14547; and Bhutta, "Sindh, Punjab Provinces Face Water Shortage."

28. "Pakistan Puts Clocks Forward."

29. Briscoe and Qamar, *Pakistan's Water Economy Running Dry*, 24.

30. World Bank, *Pakistan Strategic Country Environmental Assessment*, vol. 1, 50, 55.

31. For more, see Piet Klop and Fred Wellington, *Watching Water: A Guide to Evaluating Corporate Risks in a Thirsty World*, World Resources Institute, March 31, 2008, 10–12, www.wri.org/publication/watching-water.

32. IPCC, *Climate Change 2007: Impacts*, 35.

33. For Karachi's position relative to sea level, see Karachi Water Foundation, "Situation Analysis: Karachi Water and Sewerage Problems and Challenges Faced by the City," April 2007, http://karachiwaterpartnership.org/pdf/Situation_Analysis.pdf.

34. As of 2002, the sea level near Karachi was estimated to rise at a rate of 1.1 millimeters per year. See T. M. A. Khan, D. A. Razzaq, and Q. Chaudhry, "Sea-Level Variations and Geomorphological Changes in the Coastal Belt of Pakistan," *Marine Geodesy* 25, nos. 1–2 (February 2002): 159–74.

35. "Storms in Karachi Kill 200 People," BBC, June 24, 2007, http://news.bbc.co.uk/2/hi/south_asia/6233868.stm.

36. Cruz et al., "Asia," 476. See also *National Security and the Threat of Climate Change*, CNA Corporation, 2007, 24, http://securityandclimate.cna.org/.

37. See Asian Development Bank, *Karachi Mega Cities Preparation Report*, 2 vols., Report TA 4578-Pakistan, August 2005, vol. 1, 3–4, www.adb.org/Documents/Produced-Under-TA/38405/38405-PAK-DPTA.pdf; and Arif Hasan and Masooma Mohib, "The Case of Karachi, Pakistan," in *Understanding Slums: Case Studies for the Global Report 2003*, UN-Habitat, 2003, 8, www.ucl.ac.uk/dpu-projects/Global_Report/pdfs/Karachi.pdf.

38. Syed Fazl-e-Haider, "Pakistan's Economy Takes a Hit," *Asia Times Online*, January 3, 2008, www.atimes.com/atimes/South_Asia/JA03Df06.html.

39. Government of Pakistan, *Economic Survey of Pakistan, 2007–08*, Part 2, June 10, 2008, 223, www.accountancy.com.pk/frameit.asp?link=docs/economic-survey-pakistan-2007-08-02.pdf; Bilal Qureshi, "Pakistan's Economy Powerhouse," National Public Radio, June 1, 2008, www.npr.org/templates/story/story.php?storyId=91007621.

40. For the World Bank's governance indicators for Pakistan (in the areas of voice and accountability, political stability, government effectiveness, regulatory quality, rule of law, and control of corruption), see World Bank, *Governance Matters 2009*, http://info.worldbank.org/governance/wgi/sc_chart.asp.

41. Nils Petter Gleditsch, Ragnhild Nordås, and Idean Salehyan, "Climate Change and Conflict: The Migration Link," International Peace Academy, Coping with Crisis Working Paper Series, May 2007, 3, www.ipinst.org/media/pdf/publications/cwc_working_paper_climate_change.pdf.

42. World Bank, *Pakistan Strategic Country Environmental Assessment*, vol. 1, 56.

43. Population estimate based on latest figure in *CIA World Fact Book*, July 2009, www.cia.gov/library/publications/the-world-factbook/geos/pk.html.

44. According to the U.S. Department of Energy, U.S. power generation capacity is 1,008 gigawatts. See U.S. Energy Information Administration, *Annual Energy Outlook 2010*, December 2009, 19, www.eia.doe.gov/neic/speeches/newell121409.pdf. This is based on an estimated U.S. population of 307.2 million; see *CIA World Fact Book*, www.cia.gov/library/publications/the-world-factbook/geos/us.html.

45. Gilani Research Foundation/Gallup Pakistan, "Half of the Nation Is Deprived of Electricity for More than Eight Hours a Day," July 21, 2009, www.gallup.com.pk/Polls/21-7-09.pdf.

46. See Mukhtar Ahmed, "Meeting Pakistan's Energy Needs," in *Fueling the Future: Meeting Pakistan's Energy Needs in the 21st Century*, edited by Robert Hathaway, Bhumika Muchhala, and Michael Kugelman (Washington, DC: Woodrow Wilson International Center for Scholars, 2007), 17.

47. For a general discussion of the connections between resource scarcity and conflict, see Gleditsch, Nordås, and Salehyan, "Climate Change and Conflict," 5.

48. Schubert et al., *World in Transition*, 86; see also Alexander Carius, Geoffrey Dabelko, and Aaron Wolf, *Water, Conflict, and Cooperation*, Environmental Change and Security Project Report 10 (Washington, DC: Woodrow Wilson International Center for Scholars, 2004), 60–66, http://wilsoncenter.org/topics/pubs/ecspr10_unf-caribelko.pdf.

49. World Conservation Union, *Indus Delta*, 5.

50. Briscoe and Qamar, *Pakistan's Water Economy Running Dry*, 17.

51. Recent protests have been lodged by the NWFP assembly. See Zakir Hassnain, "NWFP PA Demands Water Royalty," *Daily Times* (Lahore), June 22, 2008, www.dailytimes.com.pk/default.asp?page=2008\06\22\story_22-6-2008_pg7_1.

52. Khaleeq Kiani, "IRSA and Punjab at Odds over Cut in Water Flow," *Dawn* (Karachi), September 13, 2008, www.dawn.com/2008/09/13/top7.htm.

53. The dam, had it been constructed, was intended to provide 11,750 kilowatt hours of electricity and irrigate 2.4 million acres; see "Goodbye, Kalabagh Dam!" *Daily Times* (Lahore), May 28, 2008, www.daily times.com.pk/default.asp?page=2008\05\28\story_28-5-2008_pg3_1). On provincial protests lodged against the Kalabagh Dam, see Ahmad Fraz Khan, "Kalabagh Shelved for Good: Minister," *Dawn* (Karachi), May 27, 2008, www.dawn.com/2008/05/27/top6.htm.

54. On the general point, see Schubert et al., *World in Transition*, 86. On the Pakistani case of water versus power, see Azeem, "Power Generation Dropped as Water Situation Deteriorates"; and Khaleeq Kiani, "Water Release Irks Punjab, Sindh," *Dawn* (Karachi), January 5, 2008, www.dawn.com/2008/01/05/top12 .htm.

55. "Goodbye, Kalabagh Dam!"

56. See Frank R. Rijsberman, "Water for Food: Corruption in Irrigation Systems," in *Transparency International Global Corruption Report 2008*, ed. Zinnbauer and Dobson, 75. Also see IPCC, *Climate Change 2007: Impacts,* 39.

57. These causal connections should not be drawn too tightly, but at least one effort has been made to link climate change with political extremism. See Johann Hari, "Bangladesh Is Set to Disappear under the Waves by the End of the Century," *Independent* (London), June 20, 2008, www.independent.co.uk/news/world/asia/bangladesh-is-set-to-disappear-under-the-waves-by-the-end-of-the-century—a-special-re port-by-johann-hari-850938.html.

58. Elizabeth Leahy, "The Shape of Things to Come: Why Age Structure Matters to a Safer, More Equitable World," *Population Action International*, April 11, 2007), 29, 31, www.populationaction.org/Publica tions/Fact_Sheets/FS34/Summary.shtml.

59. World Bank, *Pakistan Strategic Country Environmental Assessment*, vol. 1, 1.

60. Briscoe and Qamar, *Pakistan's Water Economy Running Dry*, 24.

61. Asian Development Bank, *Karachi Mega Cities Preparation Report*, vol. 1, 4.

62. World Bank, *Pakistan Strategic Country Environmental Assessment*, vol. 1, 51.

63. On rising global food prices, see Cruz et al., "Asia," 482.

64. For one instance, see "Swat Taliban Set Up Court in Piochar," *Daily Times* (Lahore), June 2, 2008, www.dailytimes.com.pk/default.asp?page=2008\06\02\story_2-6-2008_pg7_2.

65. Government of Pakistan, Ministry of Labour, Manpower, and Overseas Pakistanis (Overseas Pakistanis Division), *2004–05 Yearbook*, 26–30, www.opf.org.pk/opd/yearbk/YEARBK.pdf.

66. Ibid., 26.

67. Pakistan sent 174,864 migrants in 2004, compared with 122,620 in 1995. See Farooq Azam, "Public Policies to Support Migrant Workers in Pakistan and the Philippines," in *Assets, Livelihoods, and Social Policy*, edited by Caroline Moser and Anis A. Dani (Washington, DC: World Bank, 2008), 129.

68. The figure for the first eleven months of 2007–8, see "Pakistan Country Report," *Economist* Intelligence Unit, July 2008, 15. For more on remittances, see Government of Pakistan, *Economic Survey of Pakistan, 2007–08*, 146, www.finance.gov.pk/admin/images/survey/chapters/08-Trade%20and%20Payments08 .pdf; and Haris Gazdar, "A Review of Migration Issues in Pakistan," paper presented at Regional Conference on Migration, Development, and Pro-Poor Policy Choices in Asia, Dhaka, June 22–24, 2003, 10.

69. World Bank, *Migration and Remittances Factbook 2008* (Washington, DC: World Bank, 2008), http://siteresources.worldbank.org/INTPROSPECTS/Resources/334934-1199807908806/Pakistan.pdf.

70. Gazdar, "Review of Migration Issues in Pakistan," 10.

71. Azam, "Public Policies to Support Migrant Workers in Pakistan and the Philippines," 132.

72. "Pakistan–India Commission on Indus Holds Meeting," *Dawn* (Karachi), June 1, 2008, www.dawn .com/2008/06/01/top7.htm.

73. Khaleeq Kiani, "Protest Lodged with India over Reduced Water Flow," *Dawn* (Karachi), September 16, 2008, www.dawn.com/2008/09/16/top1.htm.

74. "Pakistan May Move World Bank against India over Chenab Water," *Times of India*, September 17, 2008, http://timesofindia.indiatimes.com/news/world/pakistan/Pakistan-may-move-World-Bank-against-India-over-Chenab-water/articleshow/3494468.cms.

75. Gleditsch, Nordås and Salehyan, "Climate Change and Conflict," 3.

76. "Country Profile: Pakistan," BBC, August 12, 2008, http://news.bbc.co.uk/2/hi/south_asia/country_profiles/1157960.stm#media.

77. Umair Zulfiqar Ali, "Media in Pakistan Booms," Televisionpoint.com, December 6, 2005, www.televisionpoint.com/news/newsfullstory.php?id=1133916559.

78. One example of effective civil society activism in Pakistan is the Edhi Foundation, devoted to free emergency, medical, and other services; see www.edhi.org.

79. "Pakistan Battles Power Shortages," BBC, May 15, 2008, http://news.bbc.co.uk/2/hi/south_asia/7403039.stm.

80. "Pakistan Puts Clocks Forward."

81. Briscoe and Qamar, *Pakistan's Water Economy Running Dry*, 18.

82. Zafar Bhutta, "Markets to Close at 9 PM under Energy-Saving Plan," *Daily Times*, May 15, 2008, www.dailytimes.com.pk/default.asp?page=2008\05\15\story_15-5-2008_pg1_4.

83. "Kalabagh Shelved for Good."

84. See "Pakistan," in *CIA World Fact Book*, www.cia.gov/library/publications/the-world-factbook/geos/pk.html.

85. U.S. Department of State, "Pakistan Assistance Strategy Report: Sec. 301(a) of the Enhanced Partnership with Pakistan Act of 2009," December 14, 2009, www.state.gov/documents/organization/134114.pdf.

86. "Ministry Seeks Rs 60bn for Dams." For other creative, lower-cost policy options, see James Dailey, "Climate Change and Pakistan's Priorities," *Grist Blog*, November 7, 2007, http://gristmill.grist.org/story/2007/11/6/164856/567.

87. For a primer on the Pakistani state, see Stephen Philip Cohen, *The Idea of Pakistan* (Washington, DC: Brookings Institution, 2004).

88. "WB to Raise $9 Billion for Water Projects"; "Ministry Seeks Rs 60bn for Dams."

89. For more extensive discussion of this history, see Briscoe and Qamar, *Pakistan's Water Economy Running Dry*, 8–13.

90. A TerrorFree Tomorrow survey (June 2008, p. 4) shows that 44 percent of Pakistanis believe the United States poses the greatest threat to their personal safety; www.terrorfreetomorrow.org/upimagestft/PakistanPollReportJune08.pdf.

91. U.S. Agency for International Development, *South Asia Earthquake February 2007 Update*, February 23, 2007), 1–2, www.usaid.gov/locations/asia_near_east/documents/south_asia_quake/SAEarthquakeWeeklyUpdate2007-02-23.pdf.

92. For complete polling data, see www.terrorfreetomorrow.org/upimagestft/Pakistan%20Poll%20Report—updated.pdf.

8
Bangladesh

Ali Riaz

Bangladesh, the seventh-most-populous country of the world, is located between India and Myanmar in the Ganges River Basin, an area of low-lying land that is the world's largest floodplain. Three major river systems—the Ganges, Brahmaputra, and Meghna, originating in the Himalayas—serve as the country's lifeline, and it is estimated that 230 rivers and their tributaries intersect the entire country.[1] The country comprises 144,000 square kilometers, of which about 7 percent is water. The country's total population, recorded in the latest census in 2001, was 124.35 million.[2] Its estimated population in 2007 was 150.44 million.[3]

Bangladesh's topography renders it extraordinarily vulnerable to the impacts of climate change, particularly rising sea levels. Its mean elevations range from less than 1 meter on tidal floodplains, to between 1 and 3 meters on the main river and estuarine floodplains, and up to 6 meters in the Sylhet Basin in the northeast.[4] Only in the extreme northwest are there level land elevations greater than 30 meters above sea level. The country's northeast and southeast portions are hilly, with some tertiary hills more than 1,000 meters.[5]

Bangladesh's location, combined with the dominance of floodplains and low elevation from the sea, has made the country exceptionally vulnerable to ongoing climate change. Its population density and high level of poverty have made its situation worse in general, particularly for the poorer segments of society. Its coastal areas, especially the 710-kilometer-long fringe zone of the Bay of Bengal, are considered the most at risk region in both the

Bangladesh

medium and the long terms. Indeed, threats arising from global climate change are not theoretical possibilities for Bangladesh but a matter of daily existence for its people. The country is already on the front lines when it comes to managing the social and political consequences of environmental challenges. According to the World Bank, the country is one of twelve with the highest risk of being adversely affected by climate change. The *Global Climate Risk Index 2010* noted that natural disasters have caused greater loss of life in Bangladesh during the past decade than in any other country of the world.[6]

According to the United Nations Development Program (UNDP) and the United Kingdom's Department for International Development, early signs of global warming are already observable in Bangladesh: "Sudden, severe, and catastrophic floods have intensified and [are] taking place more frequently owing to the increased rainfall in the monsoon. Over the last ten years, Bangladesh has been ravaged by floods of catastrophic proportion in 1998, 2004, and 2007."[7] Bangladesh heads the list of countries, prepared by the World Bank in mid-2009, most at risk of flooding.[8] It is worth noting that flooding has become more frequent throughout South Asia. As noted by experts, "the frequency of major floods in this region has increased since the mid–nineteenth century, occurring in 1955, 1974, 1988, 1998, and 2004, the interregnum being successively 19 years, 14 years, 10 years, and 6 years."[9]

Against this background, this chapter examines the security implications of the ongoing adverse climate changes for Bangladesh. As the point of departure, the chapter draws on the country-level data summarized in the appendixes to this volume and the scenarios hypothesized by the IPCC's Working Group II. The overall scenario with regard to South Asia presented in the fourth IPCC report serves as a key reference point throughout the discussion. The impact of climate change on Bangladesh, in brief, is as follows: "A one-metre rise in sea level would submerge one-fifth of the country by 2050–2075. Cyclones would be creeping deeper in the delta because of saline intrusion. Cyclone velocity would increase, and storms would be increasingly more intense. Besides, floods would be more frequent; irregular rainfall would make it difficult for farming; and the North-West would become drier increasing the chances of greater food insecurity."[10]

The trend of climate changes is discussed in different studies with different time frames in mind. Projected impacts vary accordingly, though all point to similar conclusions. As discussed in the introduction to this volume, 2030 may represent a plausible outer limit for the purposes of political and strategic analysis, but it is no more than the very near term in relation to anticipated climate change. Nevertheless, the nature and direction of climate impacts on Bangladesh will certainly be more pressing then than they are now, as will public and official consciousness of the trend they represent. As has already been suggested, some essential features of these looming threats are already familiar features of Bangladeshi life.

Climate Change as a Disruptive Factor

A causal relationship between ongoing climate change and potential development in the sociopolitical environment is difficult to establish in Bangladesh, as it would be anywhere else. The defining characteristic of the Bangladeshi political landscape is the dominance of two mainstream political parties—the Awami League and the Bangladesh Nationalist Party—along with a growing proliferation of Islamist forces, both legitimate and clandestine. The weak institutionalization of democracy, the acrimonious relationship between the two mainstream parties and endemic corruption have contributed to political instability since 1991, when the nation embarked on a new democratic era. This new era began after fifteen years of military rule, which was marked by street agitation and political violence and ended unceremoniously in January 2007 in similar circumstances. This was followed by the two-year rule of a military-backed caretaker regime that attempted to bring about structural reform and changes in the political culture but achieved little success. The country returned to elected civilian rule in January 2009, but the possibility of sociopolitical instability similar to that in the 1991–2006 era remains high.[11]

Bangladesh has maintained a reasonable rate of economic growth since the adoption of a liberalization policy in the late 1980s, but these policies have also adversely affected its poor and rural population. Thus, the number of discontented and marginalized citizens remains significant. The high rate of urbanization has brought millions of people to cities who are ill equipped to deal with the growing population and to provide services to these migrants, many of whom are poor. Slum dwellers have increased. Although the majority of the population still lives in rural areas, Bangladeshi politics is dominated by the urban population. It is not only that the ruling elites are urban; the disenfranchised poor urban population has always also been at the forefront of political agitation, mostly as foot soldiers in political parties.

Climate change has the potential to affect all these factors, and thus contribute to (and perhaps exacerbate) the nation's political instability. Its hypothesized impacts include a decline of agricultural productivity, repeated flooding, and periodic coastal storm surges, to name but a few. All these will increase in-migration to the major cities (as will be considered later in this chapter), creating a vast pool of the newly urbanized with little access to services and therefore little stake in the system. An increasing frequency of natural calamities will also require the mobilization of funds from the already-meager resources available for development (e.g., infrastructure) and social services (e.g., education and health). The post–Hurricane Sidr (2007) situation illustrates this dramatically.[12] Preventive and preparatory measures are costing the nation dearly. The country's foreign minister made this point at the Climate Vulnerable Forum held in the Maldives in November 2009: "In Bangladesh, we have been facing erratic pattern of floods, cyclones, and droughts. These climate-induced disasters have offset Bangladesh's development programmes with budgeted resources diverted to humanitarian support and disaster management. The interruptions caused by climate change have eroded our development gains made in previous decades; slowed down the attainment of MDGs [Millennium Development Goals]; threatens our food security by affecting sustainable agricultural production; and challenges climate sensitive programmes in areas such as water resources, agriculture, health, energy, urban planning, tourism, and disaster risk reduction."[13]

Needless to say, the strain on limited resources is not going to be equally shared by all segments of society. Therefore, certain sections of society will bear a disproportionate share of the costs.

The experience of recent decades indicates that in the event of resource constraints the state will rely on nongovernmental organizations (NGOs) to deliver social services. In several instances and with regard to many services, the NGOs' delivery mechanism has proved more effective; but the most debilitating effect is that the state loses its legitimacy. The absence of state agencies creates an opportunity for the growth of parallel authority. The rise of vigilante groups, particularly Islamist militants, in parts of the northwestern and western regions, testifies to this fact. Equally important is the fact that Islamists have adopted a two-pronged strategy: On the one hand they are critical of secular NGOs, labeling them "Western agents," while on the other hand they found NGOs with an Islamist agenda. As part of their criticism of secular NGOs, the Islamists have tried to raise the specter of "Western colonialism in a new guise" and have appealed to nationalist feeling blended with religious rhetoric as a means to reach out to the rural poor. The economic costs of climate change and the population's geographical displacement to politically volatile urban areas are the two most important issues likely to affect domestic sociopolitical arenas.

Externally, climate change has the potential to add more strain to the already-difficult relationship between Bangladesh and India. Notwithstanding the undocumented migration issue (discussed below), the issue of water sharing has once again come to the fore, especially due to the decision of the Indian federal government to construct the Tipaimukh Hydroelectric Project near the confluence of the Barak and Tuivai rivers in the Indian state of Manipur, within 65 miles of the Bangladesh border.[14] Bangladesh and India share fifty-four common rivers, and they have a long history of dispute over water sharing. In 1975 India built the Farakka Dam 11 miles from its border with Bangladesh, and it diverted water from the Ganges to India's Hugli River to supply Calcutta. India's water diversions are blamed for the environmental damage that spurred rural–urban migration within Bangladesh. In December 1996 India and Bangladesh reached a new thirty-year agreement on how to divide the water in the Ganges (Padma in Bangladesh) River. Many Bangladeshis allege that the country is deprived of its agreed-on share.[15]

In addition to the Tipaimukh project, India's ambitious multi-billion-dollar project to connect major rivers to augment the water supply in its southern states has become a source of contention between the two countries.[16] The project has provoked ire by threatening to alter Bangladeshi ecosystems. Water diversions by India are likely to have a serious impact on Bangladesh's access to freshwater. A lack of freshwater in the future will make it more important for Bangladesh to insist that India takes its concerns seriously. The India–Bangladesh relationship has a domestic political dimension as well. Anti-Indian sentiment runs high in certain quarters of the country, and Islamists have not been slow to take advantage of the situation.

Responses to Environmental Stress

Usually, the environmental stress brought on by climate change produces two kinds of negative responses: fight (civil conflict or external aggression) and flight (emigration). In Bangladesh, the most likely effect of this change would be emigration rather than violence. The emigration would take place at two levels: rural–urban migration at the domestic level; and migration out of the country, particularly to neighboring India. Thus, the country's finance minister predicted in December 2009, ahead of the Copenhagen Summit, that "up to 20 million Bangladeshis may be forced to leave the country in the next 40 years because of climate change."[17] Lisa Friedman, a reporter for *Environment and Energy*, has aptly titled a series of reports "Bangladesh: Where the Climax Exodus Begins."[18] These two trends—domestic and international migration—have already been under way for decades, but the process is expected to accentuate as more natural calamities affect a larger number of people.

In addition to the typical rural–urban migration pattern, environmentally induced migration from rural areas to urban centers is a regular seasonal occurrence in Bangladesh, particularly during the monsoon season. This is also due to the absence of nonfarm activities in the rural areas. Given the anticipated substantial decline in agricultural productivity throughout South Asia, the process is bound to have serious implications for

Bangladesh. The most important implications will be the decline, and perhaps demise, of food security.[19]

Although there has been significant diversification of the economy in recent years, Bangladesh still depends heavily on agriculture. About 22.7 percent of its GDP is dependent on agriculture. The hypothesized 2030 scenario with regard to the impact of temperature vulnerability and freshwater availability on agricultural productivity is described as "moderate" (see appendix B). But with a growing population (at the present annual growth rate of 1.88 percent, the population will be 229.7 million in 2030) and a lack of substantive diversification of the economy, the reliance on agriculture will increase, and so too will climate vulnerability, over both the medium and long terms. This point should be juxtaposed with the fact that 56 percent of current food production (30 million tons per year) comes from irrigated agriculture.[20] Water scarcity will have an immediate and direct impact on irrigation, and therefore on agricultural productivity.

As experts have indicated, the Ganges-Brahmaputra Delta in general is exposed to a great deal of environmental stress.[21] Within this megadelta, the inhabitants of southern coastal areas are the most vulnerable, because they will have to bear the brunt of both declining freshwater availability and rising sea level. As hypothesized, water scarcity below the notional acceptable minimum of 1,000 cubic meters per year per capita will affect an additional 2.6 percent of the population by 2030. The number will not be evenly distributed throughout the country, however. Inhabitants of the drought-prone northwestern districts and the flood-prone southern coastal areas are likely to be the most affected. Pockets in the northwestern districts already experience a famine-like crisis almost every year, primarily due to crop failures. Government intervention has succeeded in averting large-scale crisis; but if water scarcity combined with the arsenic crisis continues, the effect will be greater in the northern districts.[22]

The southern coastal zone will also be affected by any substantial rise in sea level. The country has a 710-kilometer-long coastal zone. According to the government's coastal zone policy, 19 districts out of 64 are in the coastal zone, covering a total of 147 *upazillas* (subdistricts) of the country. A total of 12 of the 19 districts border the sea or the lower tidal estuary. Forty-eight *upazillas* are officially classified as "exposed coastal zone," and a total of 99 as "interior coast." All these 147 *upazillas* will be affected by sea-level rise, while the exposed zone will be the most immediately and severely affected.

The inhabitants of the coastal zones in the south amount to 1 percent of the Bangladeshi population (1.37 million people). These would all be affected by a sea-level rise on the order of 1 meter. An additional 4 million people inhabit the 3-meter low-elevation coastal zone, and could be expected to suffer the impacts of storm surges and other extreme weather events associated with climate change. Fisheries remain the primary profession of a large number of people in the coastal zone, so any decline in fish productivity associated with changes in ocean temperature or currents will accelerate the hardship suffered by them.

Although a number of northwestern districts and a wide area of southern coastal zones are more vulnerable to climate change, particularly temperature variations and sea-level rise, other parts of the country will also see significant impacts. River bank erosion is a longstanding problem associated with extreme weather events, which changes in global temperature patterns are expected to increase, with frequent flooding, which will be exacerbated by the melting of the Himalayan glaciers. River bank erosion already displaces hundreds and thousands of people every year. In a survey conducted in the mid-1980s, 64 percent of households reported having been displaced by erosion at least once—the mean number of displacements per household was seven. The relocation usually takes place within a short distance, and displaced people tend to plan to return to their previous locations. But with a significant number of people having been on the move in recent decades, the situation is more complex than at any other time. The extent and frequency of the displacement can be understood from the experience of Abdul Majid, who has been forced to move twenty-two times in as many years. Majid lives on an island named Batikamari on the Jamuna River, 300 kilometers north of Dhaka, the capital. The island is one of those that emerge when water levels drop during the summer.[23] The total displaced population on an annual basis from a combination of these factors is estimated at 1 million. An Oxfam official in Dhaka notes that "the environmental refugee situation will turn into a dangerous problem in the future and the Bangladeshi government may find it difficult to face the challenge."[24]

The factors described above show that two regions of the country—the southern coastal zone and a part of the southwestern region—are especially vulnerable and are already being affected. Although no disaggregated data are available on the regional origins of people who move to cities, it is generally observed, and anecdotal evidence abounds, that a disproportionate number come from these two regions.

Emigration to urban areas will likely prove to be the most important secondary or social impact of climate change. Such migration will not be confined within the country. A growing number of Bangladeshis are crossing the border to India. This is a politically charged issue. Indian sources claim that as many as 15 million people from Bangladesh have moved to India, while the Bangladeshi authorities have insisted that no migration takes place. The number may be contentious, but the phenomenon cannot be denied altogether. A recent background paper on Asia prepared for the secretariat of the UNFCCC states: "Population growth and land scarcity has encouraged the migration of more than 10 million Bangladesh natives to neighbouring Indian states during the past two decades. This migration has been exacerbated by a series of floods and droughts affecting the livelihoods of landless and poor farmers in this region. Land loss in coastal areas resulting from inundation from sea-level rise as a result of climate change is likely to lead to increased displacement of resident populations."[25]

Cross-border movement between West Bengal and Bangladesh must be understood within the context of historical and cultural connections between the people, the porous nature of borders, economic opportunities, and many other factors. It is also noteworthy that migration from Bangladesh to the northeastern states of India (e.g., Assam) is a factor in the violence in that region. Thus while we need not securitize the issue wholesale, we cannot turn a blind eye to this aspect. The adaptive capacity of the Indian state where Bangladeshi migrants are most likely to move is not high.[26] In response to the ongoing and potential migration, the Indian government has started building an 8-foot-high barbed wire fence along the length of its 2,500-mile frontier with Bangladesh. John Ashton, the U.K. foreign secretary's special representative for climate change, feels that the fencing "is partly to prevent migration from Bangladesh as rising sea levels brought about by 'catastrophic climate change.'" This move has not only irked Bangladeshis but is also criticized as "a 'close the castle gates' phenomenon that can easily slip into a 'barbarian at the gates' mentality, which would be both counterproductive and lead to mounting radicalisation."[27]

Evidently, the possible (and ongoing) migration issue due to environmental stress will complicate the relationship between Bangladesh and its neighbors, especially India. The country is almost entirely surrounded by India, and it is unlikely that India will commit aggression against Bangladesh. Having said that, one cannot discount the possibility of increased tension between Bangladesh and India in the future. As indicated above, the tension may mount from a number of issues, including undocumented migration and water sharing. The situation could be further exacerbated by contested claims to offshore resources, particularly gas and oil fields in the Bay of Bengal. This is equally true with regard to the Bangladesh–Myanmar relationship. Because Bangladesh is facing resource constraints and contemplating the exploration of these fields, the tension seems to be growing. The demarcation of the maritime boundary and the configuration of the Exclusive Economic Zone have yet to be finalized. Bangladeshis allege that India has begun exploration in some of the contested oilfields. Similarly, Indian and Myanmar sources claim that Bangladeshis are in violation of maritime boundaries.[28]

Although the impact of climate change will vary across social groups, there is no ethnic dimension to this impact, thanks to the cultural homogeneity and absence of regional parties in Bangladesh. Political parties in Bangladesh are organized on a national basis, which has both positive and negative implications. The positive aspect is that these issues are not going to have an impact on governance at the local level—the threat of secession, for instance, is nonexistent. On the other hand, if a political issue emerges out of these problems, it will be a national issue and will be played out in the national political arena, which means that it will disrupt the entire nation at once.

Climate Change and Governance

The disruptive possibilities of climate change, both internally and externally, may weaken the capacity of the Bangladeshi state in many ways. They may also strengthen its authoritarian tendencies. The possible weakening of the state's capacity lies with its inability to address the grievances of various segments of society and to deliver services to deprived sections. The possibility of strengthening the authoritarian tendency, however, is dependent on the role of the political parties and the functioning of some relatively fragile democratic institutions. If sociopolitical instability threatens the political system and the political parties fail to contain it within manageable levels, a resort to extraconstitutional measures by the military is a very likely scenario. The political events leading to the declaration of a state of emergency on January 11, 2007, and the installation of a military-backed interim government are a clear indication. In this instance the military initially enjoyed popular support and launched an anticorruption reform process, but the initiatives began to falter after a year.[29] When the caretaker regime handed over power to the elected regime, it was evident that these initiatives would soon be scrapped.

The environmental scenarios discussed in the previous sections indicate the possibility of further instability in Bangladesh in coming decades unless some structural reforms take place and the adaptive mechanism is enhanced significantly. But the changes will take place in a gradual manner, allowing the state (and other political forces) time to adapt to them. The adaptive capacity of the state may not be great, but it should not be underestimated. Furthermore, there are forces that would work to forestall the collapse of the state. These forces include the growing middle class, the entrepreneurs who have benefited from the state's interaction with the global economy, and the NGOs, to name but a few. The strong military will also serve as a deterrent to state failure. The complete failure of the Bangladeshi state, therefore, is very unlikely; but the combination of limited resources, simmering discontent among the common people, and the possibility of radicalization owing to the climate changes may accentuate the crisis of governance.

Meeting the Challenges

In general, Bangladeshis demonstrate remarkable resilience in the face of natural calamities. Due to the recurrence of floods and localized storms, many communities have developed coping mechanisms. Although these mechanisms are far from institutionalized and have very little to do with the government, their presence and efficacy should not be ignored. The local people's resilience comes from their unique cultural ethos and generational experience. The most important elements of these mechanisms are a sense of community and the presence of social capital (i.e., social networks and trust). These qualities have served as means to adapt to postdisaster situations. They are yet to be utilized as a remedy for anticipated long-term changes, and no plan currently exists to incorporate these resources into the planning process. Among available institutional mechanisms, grassroots organizations such as NGOs have the potential to play a pivotal role. They can institutionalize indigenous knowledge and serve as a link to wider plans. In recent years the state, through various agencies, particularly the disaster management ministry, has been engaged in capacity building. Furthermore, there is growing awareness of the impact of climate change among the wider population.

To meet the challenges of climate change, the building of institutional capacity at the local level should receive the utmost priority. Essential steps to enhance existing local capacities include economic diversification to reduce dependence and vulnerability to shocks; the development of stable, organic, and effective democratic institutions, and the growth of crosscutting civil societies. The National Adaptation Program of Action, formulated in 2005, is a step toward this end.

The climate changes hypothesized by existing mainstream earth science are not seen by any social or interest groups to be in their best interests. There is no expectation that these changes will provide any benefits to any groups. Although some parts of the country may be hit harder than others, the impact is likely to be too great to remain local or to be exploited for political gain. Some political groups, however, are oblivious to these developments. This is due in large measure to a lack of knowledge, not because they consider these potential changes beneficial.

Impacts of Climate Change on Foreign Policy

Until recently, environmental concerns, particularly about the potentially devastating impact of climate change on Bangladesh, remained contained within academic and activist circles. Government officials at best paid lip service to the problem. Now this situation has begun to change. But unfortunately, these issues have yet to be envisaged within a broader context, and policymakers have yet to comprehend the intrinsic relationship between climate change and the political situation, let alone the role of climate change vis-à-vis security.

This is not to suggest that the government of Bangladesh is not aware of the potential devastating scenarios discussed in this chapter. Instead, it now recognizes that the country cannot meet these challenges alone. The heightened role of Bangladesh in various international forums on environmental issues, including the Copenhagen climate summit in December 2009, is a testimony to this understanding. This is a result of sustained efforts of various civil society organizations and environmental activists. The need for engagement with international bodies is now emphasized in the official discourse. However, this has not yet been followed up with concerted, multilateral strategic planning.

The country's foreign policy, since its dramatic shift in 1975 from a pro-Soviet to a pro-Western stance, has maintained consistency. The military regimes until 1990 and the elected civilian governments since 1991 have worked closely with Western governments, including the United States, at the bilateral level and in international forums. Engagement with Western governments is not a matter of dispute. Yet the government and the Bangladeshi people are extremely sensitive to anything that smacks of external intervention, however well-meaning the action may be. The populist nature of governments locally, and latent anti-Western sentiment among the masses, always prevent the government (and political parties) from being too forthcoming on engagement with the West. Bangladesh, in recent decades, has become less dependent on Western sources for its economic progress, which makes it difficult for international actors to influence domestic politics. Overt external pressure from the international community, especially from the United States and other Western nations, is certain to cause resentment and may have other unintended (and counterproductive) consequences.

On environmental issues, particularly on climate change, activists and a large section of the academic community in Bangladesh are of the opinion that the country should take a tougher stance in putting pressure on Western nations, particularly the United States, to comply with international standards. It is often argued that the rich countries must pay for their failure to address the root cause of the problem. The Climate Change Adaptation Fund is seen as a good step in this regard, but it is also argued that "the adaptation is not a substitute to the mitigation. The primary commitment of the industrialized country is to mitigate (i.e., reduce greenhouse gas)." The tone and tenor of the policymakers in Bangladesh can be gauged from the speech of the then–head of the caretaker government delivered at the United Nations on September 27, 2007. Fakhruddin Ahmed made a passionate plea to the international community for help, but he also demanded "justice": "I speak for Bangladesh and many other countries on the threshold of a climatic Armageddon." In March 2008, while visiting the United Kingdom, Ahmed commented: "There is every reason to feel angry and upset. . . . The least developed are suffering the most. It is unfair. We are suffering the most from climate change, but we did not contribute [to it] at all. We are prepared to do our part, but we require, and demand, access to a large amount of investment, resources and technologies that will be needed to adapt."[30]

Similar sentiments have been expressed by the Bangladeshi prime minister, Sheikh Hasina, in her speech at the Copenhagen climate summit. She said that representatives from Bangladesh and other countries came to the beautiful city of Copenhagen with hopes of justice, of equity, and for a fresh start to ensure the common safety of humankind.[31]

The urgency and the latent impatience of the comments reflect the mood of a large segment of society, particularly after Hurricane Sidr. Editorial comments from two influential newspapers in the wake of the Bali conference in December 2007 are indicative of the position of the community at large. The daily *New Age*, in its special issue on climate change, said: "As we turn the corner to the Bali summit, this *New Age* issue on climate change is a charge sheet of crimes for the world leaders to consider."[32] Similarly, the *Daily Star* commented, "Bangladesh must make its points in a forceful manner. It goes without saying that Bangladesh—where disasters are seen to be caused by changes in climatic behaviour—has been at the receiving end without in any way having contributed to such changes."[33] During the Copenhagen summit, the newspaper once again highlighted the issue of "climate justice" in an editorial published on December 18, 2009.[34]

The link between ongoing climate change and impending catastrophe is now commonly understood. Atiq Rahman, one of the lead authors of the IPCC report, echoes the sentiment: "Climate change is an issue of justice since individuals who are born in the countries hit hardest by it are inherently worse off while those born in the developed countries enjoy a far better lifestyle at the expense of greater carbon emissions."[35] This has been the argument of Bangladeshi political leaders in past years. The demands that developed countries should divert at least 1.5 percent of their GDP to affected nations, and that Bangladesh and other most vulnerable countries should be provided with compensatory grants to meet the full cost of adaptation to climate change, stem from this position.

Conclusion

Climate changes anticipated in both the medium and the long terms place Bangladesh in a precarious situation and pose challenges to the nation's security at various levels. These challenges, both domestic and regional, have implications for the international community. Sea-level rise, severe storms, repeated floods, increased water salinity, and worsening water scarcity will directly affect the availability of food. In a country with a large poor population, the decline in food security is a recipe for political and social instability. The decline in food production will make the country dependent on the international food market, the volatility of which needs no elaboration after the events of 2008. Combined with fragile political institutions, a contentious political culture, and a continued preponderance of violence, the possibility of radicalization is very likely. Militant organizations will make use of these to their benefit. The huge pool of urban poor, particularly new migrants from rural areas, may serve as a reservoir of a disgruntled army.

The domestic problems of Bangladesh will spill over the borders to neighboring countries, particularly to India, through migration. But migration will not be the only factor to increase the tension between these two countries. Water scarcity has already proved to be an influential factor. Exploration of resources in the Bay of Bengal is the emerging issue of contention. The potential of these tensions to become sources of security threats looms large because they will also affect the domestic politics of Bangladesh and India. Anti-Indian feelings among the Bangladeshis will increase, which usually serves as a legitimating factor for the Islamists (and radical Hindutva supporters in India).

Despite the willingness of the government to work with the international community, the reluctance of Western nations to act expeditiously is helping to fuel anti-Western sentiment among Bangladeshis. There is a growing impatience among the members of civil society groups and people in general. As political parties have returned to power, populist rhetoric has begun to increase, and anti-American and anti-Western sentiment are finding a hospitable environment.

These challenges are not dramatic enough to enter the twenty-four-hour news cycle of the international media, and thus attract immediate attention; but they will steadily affect the Bangladesh's ability to survive and thrive. The failure or inability of a country on the front line of global climate change to protect itself will affect the outlook of other countries around the world, just as surely as carbon emissions thousands of miles away are affecting the lives of poor Bangladeshis today.

Notes

1. The outflow from Bangladesh is the third highest in the world after the Amazon and Congo systems. Annually, the three major rivers of Bangladesh drain 175 million hectares of land. Christian Aid, "Bangladesh: Erosion and Flood," in *The Climate of Poverty: Facts, Fears and Hope*, May 2006, 32–37, www.reliefweb .int/rw/lib.nsf/db900sid/RURI-6PUL34/$file/cha-gen-may%2005.pdf?openelement.

2. Bangladesh Bureau of the Census, *Bangladesh Census at a Glance*, www.bbs.gov.bd-dataindex-census-bang_atg.pdf.

3. "Bangladesh," *CIA World Fact Book*, www.cia.gov/library/publications/the-world-factbook/geos/bg .html#People.

4. H. E. Rashid, *Geography of Bangladesh*, 2nd ed. (Dhaka: Bangladesh University Press, 1991), quoted by Shardul Agrawala, Tomoko Ota, Ahsan Uddin Ahmed, Joel Smith, and Maarten van Aalst, "Development and Climate Change in Bangladesh: Focus on Coastal Flooding and the Sundarbans," Environment Directorate, Development Cooperation Directorate Working Party on Global and Structural Policies Working Party on Development Cooperation and Environment, Organization for Economic Cooperation and Development, 2003, 9, www.oecd.org/dataoecd/46/55/21055658.pdf.

5. S. Huq and M. Asaduzzaman, "Overview," in *Vulnerability and Adaptation to Climate Change for Bangladesh,* edited by S. Huq, Z. Karim, M. Asaduzzaman, and F. Mahtab (Dordrecht: Kluwer, 1999), 1–11.

6. Sven Harmeling, *Global Climate Risk Index 2010: Who Is the Most Vulnerable? Weather-Related Loss Events since 1990 and How Copenhagen Needs to Respond,* Germanwatch Briefing Paper, December 2009, www.preventionweb.net/files/11973_GlobalClimateRiskIndex2010.pdf.

7. United Nations Development Program and United Kingdom Department for International Development, *Bangladesh: Reducing Development Risks in a Changing Climate,* March 2008, 4.

8. "World Bank Lists Most Vulnerable Countries to Climate Change," *One World South Asia,* July 9, 2009, http://southasia.oneworld.net/globalheadlines/world-bank-lists-most-vulnerable-countries-to-climate-change/.

9. Qazi Kholiquzzaman Ahmad, "Global Environmental Challenges: Local Dimensions: Focus on the Water Sector in the Eastern Himalayan Region," presentation at the IVE-TERI-IDDRI Conference on Energy, Environment, and Development, Bangalore, December 14–16, 2006.

10. Atiq Rahman, "Bangladesh Must Learn to Live Thru' Climate Change," *Daily Star* (Dhaka), March 29, 2008, www.thedailystar.net/pf_story.php?nid=29737. Rahman was the lead Bangladesh author on the IPCC.

11. For details of the Bangladeshi political landscape, see Ali Riaz, *Islamist Militancy in Bangladesh: A Complex Web* (London: Routledge, 2008), 7–28.

12. A category four tropical cyclone named Sidr slammed Bangladesh on November 15, 2007, killing at least 3,500 people, making millions homeless, and costing billions of dollars. According to a World Bank report, "the accompanying storm surge reached maximum heights of about 10 meters in certain areas, breaching coastal and river embankments, flooding low-lying lands and causing extensive physical destruction. The cyclone's winds of up to 220 kilometers per hour caused further destruction to buildings and uprooting of trees that in turn destroyed housing and other infrastructure inland." World Bank, "Cyclone Sidr in Bangladesh Damage, Loss and Needs Assessment for Disaster Recovery and Reconstruction," February 2008, mimeo. This was the third natural disaster the country faced in 2007. Previously, Bangladesh was devastated by two floods. Although the death toll was smaller than that caused by cyclone Marian in 1991, which cost 139,000 lives, the effect of cyclone Sidr was immense. Sunderban, the largest continuous mangrove forest of the world, bore the brunt. At least 31 percent of the forest, 19,000 square kilometers, was affected. For a preliminary assessment, see "Effect of Cyclone Sidr on the Sundarbans: A Preliminary Assessment," Center for Environmental and Geographic Information Services, Dhaka, November 2007, mimeo. Despite government and international efforts, many victims did not receive relief even after a year. The relatively low casualties caused by Sidr were due to massive evacuation efforts of local government agencies and NGOs. It is estimated that 650,000 people were evacuated from the path of destruction.

13. "Statement of Dr. Dipu Moni, the Hon'ble Foreign Minister of Bangladesh, at the Climate Vulnerable Forum, Maldives, 09 November 2009," www.climatevulnerableforum.gov.mv/?page_id=45.

14. Nadim Jahangir, "The Tipaimukh Dam Controversy," *Forum* 3, no. 7 (July 2009), www.thedailystar.net/forum/2009/july/tipaimukh.htm.

15. For detail of the treaty and the environmental impacts of the barrage, see *The Ganges Water Diversion: Environmental Effects and Implications,* edited by M. Monirul Qader Mirza (Dordrecht: Kluwer, 2004).

16. Sobia Nisar, "Indo-Bangladesh Water Issues," *Daily Mail* (Islamabad), January 5, 2006, http://dailymailnews.com/200601/05/dmcolumnpage.html.

17. "UK Should Open Borders to Climate Refugees, Says Bangladeshi Minister," *The Guardian* (London), December 4, 2009, www.guardian.co.uk/environment/2009/nov/30/rich-west-climate-change.

18. Lisa Friedman, "Bangladesh: Where the Climax Exodus Begins," *Environment and Energy,* www.eenews.net/special_reports/bangladesh/#previousjump.

19. IPCC, *Climate Change 2007: Impacts,* 48. Some analysts claim that in the medium term (i.e., 2030), climate change may not adversely affect overall rice production, but in the long term, food production will be significantly affected by water shortage. They argue that "the overall impact of climate change on the production of food grains in Bangladesh would probably be small in 2030. This is due to the strong positive impact of CO_2 fertilization that would compensate for the negative impacts of higher temperature and sea-level rise. In 2050, the negative impacts of climate change might become noticeable: production of rice and wheat might drop by 8% and 32%, respectively." See I. M. Faisal and Saila Parveen, "Food Security in the Face of Climate Change, Population Growth, and Resource Constraints: Implications for Bangladesh," *Environmental Management* 34, no. 4 (2004): 487–98.

20. Agrawala, Ota, Uddin Ahmed, Smith, and van Aalst, "Development and Climate Change," 21.

21. IPCC, *Climate Change 2007: Impacts*, 64.

22. Mahmuder Rahman has stated that "the Government of Bangladesh estimates that 30 million people are drinking water that contains more than 50 micrograms per liter of arsenic. However, up to 70 million people are drinking water that contains more than 10 micrograms per liter of arsenic, which is the provisional WHO guideline value. After a quick field survey in 2001, the government estimated that 40 to 50 percent of the estimated 10 million tube wells were contaminated with arsenic. In some villages that figure was as high as 80 to 100 percent. Now there is the problem that some tube wells that were not originally poisoned are becoming so." Mahmuder Rahman, interview, *Bulletin of World Health Organization* 86, no.1 (January 2008).

23. Masud Karim, "Bangladesh Faces Climate Change Refugee Nightmare," Reuters, April 14, 2008, www .reuters.com/article/environmentNews/idUSDHA23447920080414.

24. Ibid.

25. UNFCC, "Impacts, Vulnerability and Adaptation to Climate Change in Asia," Asian Workshop Background Paper, Beijing, April 11–13, 2007, 19, http://unfccc.int/files/adaptation/methodologies_for/vulner ability_and_adaptation/application/pdf/unfccc_asian_workshop_background_paper.pdf.

26. IPCC, *Climate Change 2007: Impacts*, 71.

27. Ben Vgel, "Climate Change Creates Security Challenge 'More Complex Than Cold War,'" Janes.com, January 30, 2007, www.janes.com/security/international_security/news/misc/janes070130_1_n.shtml.

28. Ian A. Blakeley, "Bangladesh: A Natural Gas Perspective," *PESGB Monthly Newsletter*, October 2006. For an Indian perspective on the politics of energy, see Srinjoy Bose, *Energy Politics: India–Bangladesh–Myanmar Relations*, New Delhi Institute of Peace and Conflict Studies Special Report 45, July 2007, www .energiasportal.com/download/117/. In November 2008, the dispute escalated and a naval showdown ensued. Naval ships from Bangladesh and Myanmar faced off in the Bay of Bengal, and the standoff continued for couple of days. In 2009 Bangladesh referred the issue to the United Nations for arbitration under the UN Convention on the Law of the Sea. In early 2010, there were some positive developments, as both countries agree to resolve the dispute through negotiations. On maritime boundary and security issues, see Abu Syed Muhammad Belal, *Maritime Boundary of Bangladesh: Is Our Sea Lost?* Bangladesh Institute of Peace and Security Studies Policy Brief (Dhaka: PBIPSS, n.d.), http://bipss.org.bd/download/mb_bd.pdf.

29. Leading up to the declaration of emergency, the country descended into chaos, with violence, general strikes, and transport blockades called by the major political parties demanding reform of the election commission, corrections in the voters' list, and the depoliticization of the administration. The violence that gripped the nation for months, disrupted public life, caused damage to public property and enormous economic losses and above all, the deaths of innocent people in clashes between law enforcement forces and political activists. On January 12, 2007, the military-backed caretaker government took over in Bangladesh promising sweeping reforms to the political system and the building of institutions necessary for sustainable democracy.

30. As quoted by John Vidal, "Remote Control," *Guardian*, Manchester, March 26, 2008, www.guardian .co.uk/environment/2008/mar/26/bangladesh.

31. "Poor Must Get Full Cost of Adaptation," *Daily Star* (Dhaka), December 17, 2009, www.thedailystar .net/newDesign/news-details.php?nid=118033.

32. "Living Climate Change," *New Age*, Special Issue on Climate Change, (December 1, 2007, www.newage bd.com/2007/dec/01/climatechange07/climatechange07.html.

33. "Climate Change Conference in Bali," *Daily Star* (Dhaka), December 2, 2007.

34. "Bangladesh Leader's Call in Copenhagen," *Daily Star* (Dhaka), December 18, 2009, www.thedailystar .net/story.php?nid=118110.

35. "Bangladesh Must Learn to Live Thru' Climate Change," *Daily Star* (Dhaka), March 29, 2008.

9
Russia

Celeste A. Wallander

The Russian Federation occupies a vast northern territory that is unlikely to suffer the most serious negative effects of anticipated climate change. Geology, geography, and climate may even combine to make Russia not merely a survivor but even a beneficiary of the environmental changes that elsewhere will be experienced as deterioration. Rising global temperatures are more likely to reduce than to worsen the stresses and constraints of daily and commercial life in high northern latitudes. An increase in average temperatures may also make more of Russia's territory suitable for development and human habitation, and it should reduce the costs required to heat inhabited regions. Though Russia is normally thought of as remote and landlocked, if it were to face warmer Arctic conditions, it could find itself with one of the world's longest commercially viable coastlines and with access to polar shipping routes that would facilitate its trade with Europe, Asia, and North America. Its European, Arctic, and Pacific coastlines are neither especially low lying nor densely inhabited, so a rising sea level is unlikely to flood significant developed areas. The predicted effects of changes in temperature and rainfall on Russian agricultural production are not negative, and may even be beneficial on balance.[1]

Even more dazzling, the thawing of the globe's northern regions could unlock Russia's vast known reserves of oil, natural gas, and other natural resources. The country is already the world's largest producer of natural gas, and it is currently matching Saudi Arabia in oil

Russia

production. Its potential has been limited, however, by the inaccessibility of much of its untapped reserves in cold and remote regions. Rising temperature could unlock this wealth, and navigation through the Arctic Ocean could make its petroleum reserves exportable at low cost. If one also takes into account the carbon resources under the Arctic Ocean, which may constitute as much as one-quarter of the world's total petroleum reserves, and to which Russia may succeed in laying a substantial claim, the economic growth and rising global influence that Russia has enjoyed during the past decade may prove to be merely the prelude to a new era of Russian power and prosperity.[2]

Yet Russia faces serious challenges that will determine whether it prospers or suffers as a result of climate change. Its response to those challenges will in turn have implications for its own security and that of the wider world. There is, first of all, the question whether climate change globally, and especially across the rest of Eurasia, will create stresses that affect Russia indirectly. Climate change may position Russia to produce more oil and natural gas, but climate-induced economic dislocation that entails a turning away from carbon-based energy might nevertheless undermine a fundamentally vulnerable Russian economy. Similarly, part of the country's strength lies in its geopolitical reach, a feature of its size and location; but its looming continental presence also makes it vulnerable to climate-change-induced social or political instability around its extensive and porous borders.

Russia's political system has a mixed record at best in coping effectively with complicated problems of the kind that will dominate environmental politics going forward. Russia may be poised to profit from global warming, but the state will need to plan strategically for investments in infrastructure, education, health, and resource management if the profits are to be realized. The initial effect of the melting of Russian permafrost, for instance, will be a marked deterioration of all the existing infrastructure that was built in the expectation of its permanence. Only after that difficult transition is weathered, and a new and more appropriate infrastructure is created, can the potential benefits of a warmer north be realized by the Russian economy as a whole.[3] State and society together will also need to cope with climate-driven migration and economic dislocation around its periphery, which will pose security challenges in themselves. Russia's track record in coping with such problems is not good, and though there could be improvements during the next two decades, the system that exists today affords few grounds for optimism.

Russia is unlikely to be a flashpoint for climate-induced instability or conflict, but even so, in its case the potential for high-impact effects warrants consideration even of unlikely outcomes. Broadly speaking, the country is dissatisfied with its place in world affairs—and it is arguably the largest, potentially the richest, and certainly the best-armed of all the states about which this might be said. To the extent that climate change creates dislocations or uncertainty in the management of international security and of global markets, Russia can be expected to attempt to seize whatever opportunities it perceives as a result. It will be the unforeseen consequences of such efforts (and perhaps especially of failed efforts and squandered opportunities) that pose the most substantial security risks to Russia, its neighbors, and its trading partners and political rivals around the world.

The Sources of Russian Power

Russian power, apart from the nuclear arsenal inherited from the former Soviet Union, rests chiefly on its commodities-export-driven economy and its geopolitical position. Its economy has been one of the most successful in the world in recent years, though like other developing nations it has suffered severely from the global economic slump that began in late 2007. Its annual growth rates ranged from 4 to 8 percent between 1999 and 2007, and its GDP per capita quadrupled.[4] Its government revenues and defense spending have expanded accordingly. In 2000 the Russian federal government's revenues were $24.9 billion (14.9 percent of GDP), budget expenditures were $26.7 billion (16 percent), and defense expenditures were $4.6 billion (about 3 percent). Russia's currency reserves were $13.3 billion, its external debt was $166 billion, and debt servicing cost it $9.2 billion (nearly one-third of its expenditures). By 2008 federal revenues had risen to $383.5 billion (21.8 percent of GDP), budget expenditures were $273.5 billion (15.6 percent), and defense expenditures were $36.8 billion (3.9 percent). By 2008 Russia held $600 billion in currency reserves, and it was no longer subject to International Monetary Fund constraints. The Russian government had sufficient resources to leverage external investment by private European business interests, and to offer official aid to its neighbors, such as the $2.1 billion package of grants and loans that led the government of Kyrgyzstan to close the American airbase at Manas (later reopened as a transit center).[5]

This growth was dependent on Russia's resource wealth. Both private consumption and public programs of social welfare and infrastructure maintenance are critically dependent on oil price levels, and thus they tend to suffer when those levels fall below the range of $60 per barrel.[6] Price levels are especially important because Russia's growth during the past decade has not necessarily been accompanied by an increase in energy production. Russian annual natural gas production remained basically flat, at about 650 billion cubic meters, in the post-Soviet era. Oil production fared better; due to improved business practices and the relative security of property rights in the late 1990s, as well as the foreign investment and technology that allowed privatized companies to go back to wells previously deemed exhausted by inefficient Soviet extraction practices, Russia increased its oil production from the post-Soviet low of 6 million barrels per day in 1999 to nearly 10 million in 2006 (equivalent to Saudi Arabia's), of which 5 million were exported.

The available efficiencies in current oil fields are now fully realized, however, and Russia's oil production growth rates stalled in 2007–8. The country's oil and natural gas industry has been investing only about half the amount of capital required to sustain current growth rates. And while the Russian state was seeking to limit and control foreign investment in its energy sector, the country's energy companies went on a buying spree, taking on substantial foreign debt and buying assets in other sectors or in downstream energy facilities abroad. What they did not do was invest in the country's capacity to produce new sources of wealth—in the energy sector or otherwise—in the coming years.

The resources created by its economy funded defense spending, attracted foreign direct investment, increased nonenergy trade with Europe, strengthened the Russian state internally, built societal support for the Putin regime, created investment funds that Russian companies and the state could use to buy energy and nonenergy assets in Europe and Eurasia, and funded official media outlets and public diplomacy ventures to spin and shape information with respect to a host of target issues (NATO, missile defense in Europe, the war with Georgia, and the American military presence in Eurasia). Russia's energy wealth removed vulnerabilities that might have constrained its leaders' foreign policy options; and though one might expect a reversal of economic fortune to have the opposite effect, that is not necessarily the case. There is no rule that says economic hardship leads to a more cautious policy.

As was mentioned above, the effects of the global economic recession have been palpable in Russia, as in all other countries whose economies are critically dependent on the prices of primary commodities. Russia has been hit by capital outflows and constrained credit, and though the massive currency reserves that it built up during the run-up in energy prices through mid-2008 have helped it to mitigate the effects of the global financial crisis to some extent, the effects of the subsequent decline have nonetheless been severe. The World Bank estimates that the Russian economy contracted at the rate of 4.5 percent of GDP in 2009, a number that may prove too conservative, and it predicts further growth in capital flight as investors seek more secure markets.[7] Unemployment is rising because major growth industries—particularly construction—have seen a virtual halt because of liquidity problems and declining consumption. The Russian government was projected to run a budget deficit of about 8 percent in 2009. About $135 billion in Russian banking and corporate debt fell due in 2009, of which perhaps 10 percent was expected to default.[8]

Russia is far from alone in facing serious consequences from the global financial crisis and recession. However, because of its undiversified economy, its dependence on foreign credit, and its long track record of inadequate infrastructure investment, its is more vulnerable than other major economies to difficult economic times. Its difficulties have been exacerbated by the official reaction that recent troubles have inspired. The World Bank reports that foreign investment in Russia's energy sector has already been declining as the Russian state has sought to increase its control of that sector, particularly following the enactment of its "strategic sector law" limiting foreign investment.[9] With Russian oil production growth now shrinking and the country unable to meet its contractual commitments for natural gas deliveries without its own purchases and redeliveries of Central Asian natural gas, Russia's preeminence as Europe's major supplier of oil and natural gas is reaching a turning point. Either Russian investors themselves will have to make serious strategic investments to develop new fields or foreign investors will have to be allowed to make the massive and complicated investments necessary to develop new oil and gas fields.[10]

The greater confidence and initiative that Russia has demonstrated in its conduct of foreign affairs in recent years is in many respects a derivative of high energy prices. Those prices in turn increased the state's capacity to generate massive strategic investments to develop its energy reserves. Both these developments can be seen as symptomatic of a healthy, or at any rate a rapidly growing, global economy. If climate change exacerbates global economic dislocation, either by depressing oil and gas prices directly or by decreasing consumption because environmental externalities get priced into energy costs, Russia's winning position as a holder of those energy reserves may not be fully realized.[11]

Economic hard times may cause Russia to delay a costly conventional military reform program aimed at shrinking and modernizing the officer corps, and shifting its force structure away from large formations designed for massive ground operations in Europe and toward the more mobile combined arms units needed for security contingencies in Eurasia. Such a delay would further enfeeble a Russian military struggling to perform modern military missions. If the defense budget shrinks substantially as a consequence of Russia's current financial troubles, its modest long-term strategic nuclear replacement program may have to be slowed further. This may increase Russia's incentives for nuclear arms control in order to gain greater stability of expectations about U.S. capabilities relative to Russia's (an argument Russian officials make consistently in private). On the other hand, a Russia with an unreformed and aging conventional military force may feel that it will need to rely even more heavily on its nuclear deterrent to underwrite its foreign policy. Fiscal constraints can help to make the case for relying on arms control to manage the balance of power, but even so, economic conditions that exacerbate military weakness are likely to feed insecurity.

Russia's energy wealth has made it more powerful because of the country's connection to global energy markets. But this very interdependence creates a vulnerability to the economic effects of global climate change. And insofar as this change reduces demand for (and thus the price of) energy, or creates incentives to shift energy consumption away from fossil fuels, it will constrain Russia's power and security in dealing with its complex and changing Eurasian neighborhood. Its dominant size and position have afforded it increasing leverage throughout Eurasia in the past decade and given its leaders substantial influence in Europe, the Caucasus, Central Asia, and the Far East. The necessary conditions for sustaining this position are stable borders and settled populations. If Russia's Eurasian neighbors become the victims of climate-induced agricultural or demographic crises, their populations may seek refuge under Russian protection. Assuming that Russia's future leadership retains its present character, its reaction to such a challenge is necessarily uncertain.

Russia's Political Institutions and Leadership

It is difficult to know what the Russian political system will look like in twenty years. The system lacks a firm institutional (much less constitutional) base, so that, since the collapse of the Soviet Union (and indeed for some time before), much of its form and function have depended on the preferences and personalities of leading individuals. Broadly speaking, Russian politics is a form of authoritarianism rooted in patron–client relationships that enable the political leadership to exercise power without accountability to the country's citizens. Vladimir Putin's leadership has successfully eliminated pluralism in the country, a condition that *has* been institutionalized to some extent, and thus seems certain to continue for some time. The media is state-owned, in the hands of a Kremlin-friendly businessman, or without access to national broadcast outlets. The lack of a professional independent media capable of reporting on government deeds or misdeeds makes it essentially impossible for Russian citizens to hold their government accountable for policy and performance. The atmosphere of intimidation is such that even independent-minded journalists tend to practice self-censorship.

Yet while Russia has shed the nascent pluralism of the 1990s, it has not reverted to the totalitarianism of the high Soviet era nor even to the repressive authoritarianism of the 1970s and 1980s. The country today functions through a complex system of patronage and selective threat, in which ideology plays no part, and which does not require the kind of comprehensive social control that has been exercised in the past. Limits on contestation extend beyond political competitors or the media. Civil society exists in organized form only to the extent that its activities and objectives are nonpolitical. Business interests are active in the country's politics, but not as independent actors. Instead, business interests and prominent individuals embody and enforce the demands of the political leadership, lending the entire system a corporatist character.[12]

Succession is the most serious challenge for authoritarian regimes in which power and wealth depend on patronage ties, which is why the future of such systems is especially hard to foresee. When the head of the system changes, the new head is likely to bring his clients in to run the show and access the wealth, threatening to dispossess the current winners. Putin finessed this problem temporarily by ensuring that the March 2008 presidential election was won by Dmitry Medvedev, who owed his position entirely to his predecessor and mentor; but the artificiality of the resulting arrangement merely illustrates the nature of the problem in the long run.

The danger is that economic troubles may lead Russia's citizens to question the deal that was forged between state and society during the decade of rapid economic growth that may now be coming to an end. If declining global energy consumption leads to difficult times at home, the government may become less popular and have fewer resources to make the case that it is delivering its part of the social bargain. Many governments face unpopularity, social demands, and economic contraction—and climate change will stress and challenge all of them. Russia is not unique, but it is distinctive in the degree to which its leadership has justified its power and its demand for obedience to a highly educated, modern, and urban citizenry based on what it can deliver in strictly material terms, rather than on the historical or constitutional legitimacy of its rule. Despite its image of strength and central control, Russia is a fundamentally brittle political system, not a strong state. The complex, long-run, slowly unfolding challenges likely to be posed by climate change are precisely the kind for which the primitive politics of patronage are unlikely to offer effective answers.

Vulnerabilities and Coping Mechanisms

Russia has three primary sociopolitical vulnerabilities: an economy excessively dependent on exporting raw materials, with an eroding industrial base and little new development in technological innovation or globally competitive sectors; a state that is good at executing one or two high-profile tasks but incapable of everyday governance and problem solving; and a backlog of unmet social needs with crisis points looming, including a declining and aging population, and shrinking human capital.

The Economy

Insofar as the legitimacy and popularity of the current regime are based upon its economic success, so that Russian citizens now compare their current lifestyles and future expectations favorably with conditions in the past, the country's stability rests first of all on oil prices that remain above about $60 per barrel. Because energy prices have lately risen, in large measure because of growing energy demand from Asian economies—themselves dependent on global demand for finished goods—a global economic downturn that brought oil prices below the $60 range for an extended period would create serious problems for Russia's political leadership.

Russia has enormous untapped reserves of natural gas and oil, and the efficient extraction of those resources is widely regarded as the key to the country's economic future. There is considerable recurring speculation in the Russian press about the economic benefits to Russia from global warming, because rising temperatures will melt Siberia's permafrost and the Arctic ice and make the exploitation of currently inaccessible northern oil and gas fields possible. All this may well be true, but it is essential to realize that this sort of scenario only works if the environmental changes that melt the permafrost and the ice do not also lead to a dramatic decline in general global economic activity.

If one of the effects of global climate change is economic dislocation in a form that suppresses energy demand and thus energy prices, climate change could fundamentally destabilize Russia's economy and political system, even if its direct and immediate effects in Russia itself prove to be relatively benign. Thus, for instance, the long-run warming of Siberia would facilitate Russia's ability to increase oil and gas production. But the latter process will probably not unfold rapidly enough to matter in the period that immediately concerns us—the next twenty years or so—when global environmental politics may well reach a kind of tipping point with respect to the consumption of fossil fuels, setting in motion new technological and pricing trends in the energy sector that once again leave Russia out in the cold.

Even in the short run, tapping new energy wealth beneath a melting Siberian permafrost cannot begin without an adequate provision of roads, infrastructure, and sustainable human habitation. Challenges in this area will remain formidable; rising temperatures will not turn the frozen tundra of Siberia into a garden, after all, but into something like a vast, semifrozen bog. The current Russian government has for the most part failed to invest in transportation infrastructure outside high-profile locations such as Moscow and Saint Petersburg. Though climate change in resource-rich areas may increase the incentives to do so on the margins, it will remain a difficult and expensive problem, requiring a high level of effort and investment of resources. More broadly, it would imply (and perhaps require) a change in relations between the center and the periphery of Russian government and society. Regional and local leaders in Russia's Siberian and Far Eastern regions are notoriously corrupt and ineffective even by Russian standards, and they have remained so despite Putin's stated desire to encourage development and settlement in Russia's eastern regions. It might also require a change in relations between Russia and the larger world, which will have to provide a major share of the financing and the advanced technology required to extract the mineral wealth from a swampy Siberia. The Russian government is currently very selective in welcoming foreign investors, most especially if such investment would lead to a substantial claim on physical infrastructure or assets inside Russia. Recent experience suggests that it is at least as likely that warming in Siberia will create an unrealized opportunity rather than a source of new economic growth.

Governance

As always in Russia, questions about what the future holds inevitably come back to questions about the state and its leadership. The Russian government has failed to cope structurally with fundamental social problems—including the provision of social services, education, and health—preferring instead to rely on easy fixes and ad hoc solutions calculated to increase support for the leadership and dampen social demands, at the cost of postponing a genuine reckoning with whatever the problem may be (during which time it generally gets worse). Such habits are not merely symptomatic of shortsightedness or incompetence. They are also a sign of how worried the government is about the subterranean social discontent arising from deteriorating living standards. Such concerns are, if not exactly new, then certainly greater than in the past. Russia is, historically, a country in which the state has perennially made enormous demands upon the people's patience and fortitude. The fact that its current leadership is less confident than its predecessors of its ability to demand sacrifice and forbearance from the nation is a sign of that political brittleness that was referred to above, and that may easily be mistaken for hardness or strength.

In creating a political system without effective competing political parties, and in which all decision making comes from the center, Russia's leadership has ensured that all bad news eventually ends up at the feet of the president and his associates. In eliminating or intimidating the independent media, it has also eliminated important sources of potential information, a critical requirement in educating public opinion about the emerging effects of climate change. Climate politics, in Russia and elsewhere, is necessarily going to be about the development of a durable public consensus that can support policies tailored for the long run. But such policies have had no place in the Russian regime up to now.

Corruption is, in its way, a natural expression of a government's anxiety about its social base, and in Russia corruption has reached astonishing proportions. The overall capacity of the Russian state to perform the basic functions of government has not improved since the collapse of the Soviet Union, and has probably eroded. Certainly it must be regarded as highly vulnerable to Katrina-like episodes, in which the consequences of a natural disaster are both amplified and politicized because of the government's incompetent reactions. The instincts of both central and local governments, if confronted by progressively rising coastal water levels or by flooding associated with more numerous severe weather events, would almost certainly be to deny the early signs of crisis until it is more than too late. Societal watchdogs like independent media and environmentally focused nongovernmental organizations (NGOs) have virtually no ability to monitor conditions and report on trends or problems in a way that would contribute to a constructive government response. Any such warnings would be most likely to be met with accusations of disloyalty or efforts to weaken the state. Local leaders have no incentive to call Moscow to alert it to problems, because the response will be punishment. Nor are local leaders accountable to their citizens, because they are either appointed directly by the Kremlin or attain office by means of elections engineered from Moscow.

The primary source of legitimacy for the current Russian government is that it provides decent economic performance and basic security. Russians are acutely aware of the incompetence and corruption of their government, but the alternative to supporting the state would be, in their minds, chaos and a loss of international standing and prestige, which Russians profess to value more highly than democratic governance and civil liberty.[13] By a kind of cultural inversion that is by no means unique to Russia, loyalty to the Motherland and its defense against outsiders may actually be strengthened by a sense of the regime's incapacity, which signals that in the end the its real strength lies within society after all.

The Erosion of Social Resilience

Given the likely ineffectiveness of government agencies in the face of climate-induced crises (or even the slow accumulation of climate-induced pressure for new policies), the question naturally arises as to the resourcefulness and resilience of Russian society. Russia's population numbers more than 140 million, a number currently declining at the rate of about 700,000 per year. If this trend were to continue, by 2030 Russia's population could fall as low as 125 million. This would reduce the demand for critical environmental resources such as water, and it would also reduce the human capital available for coping with more complex challenges. Russia's demographic decline, combined with the poor health of its working-age population, means that the country is not well positioned to respond to the social policy challenges arising from climate change in the coming decades. With fewer healthy people in the age cohort of eighteen to thirty-four years, Russia will have fewer capable men and women to serve in its military, security services, and police. Its human capital—which as of today remains impressive for a country of its developmental and income level—is a wasting asset that was created in the Soviet period and is not being renewed. There are some successes in the area of advanced training in finance and business, but Russia's primary and secondary education systems have collapsed.

The core challenge of 2030 may simply be that Russia's people—confronted with what by then will likely be incontrovertible evidence of a need for systematic adaptive responses to climate change, whether to mitigate its effects or capitalize on them—may be embittered, frightened, and in severe need reassurance and leadership from their government. For the reasons outlined above, they are unlikely to get it. Citizens often forgive their leaders for mistakes and inadequacies, but they need mechanisms for safely, legitimately, and effectively registering their discontent and demands. These mechanism do not exist in Russia at present, having been systematically suppressed, and there do not seem to be any forces at work within Russian society that are capable of creating them outside the framework of government.

For the most part, as has already been suggested, one would have to look outside the Russian state for hopeful indicators of reserves and resilience. The country's business leaders are increasingly well trained, not only in American and European universities but also in Russian advanced business and economic schools. The problem is not the leadership capacity of such individuals but the severe political constraints under which they must operate. If such constraints were loosened, the business community could very well serve as a resource for coping with climate change challenges. But if what remains of the private sector continues to fall progressively under the domination of the state, one can expect those skills to be lost as business leaders learn to function under state control.

Similarly, in the short to medium terms, Russia's scholars and intellectuals could still serve as a resource for knowledge, strategic analysis, and problem solving. As with the business community, however, the longer the intellectual elite has to trim its sails to survive in world of acceptable ideas and civic participation as defined by the Kremlin, the more this reserve of human capital will erode. Already social scientists and historians are shaping their scholarship to fit the Kremlin's approved script, and natural scientists are limiting their inquiries and contacts for fear of running afoul or the state's security services. By 2030, the country's intellectual and scientific leadership will have passed to those who advanced because of their conformity to Kremlin-accepted ideas, not genuine scholarship.

If current trends continue, Russia's society and NGO community will be of no use for coping with climate change challenges. As the Kremlin has cut off foreign funding and created funds for approved NGOs instead, NGOs with a true watchdog mission and with genuine independent expertise based upon knowledge and science are becoming impossible to sustain. This is a primary avenue by which the Kremlin is weakening and not strengthening the nation's capacity to cope with the entire range of challenges it faces in the coming decades.

Climate-Induced Migration

Russia is unlikely to experience significant out-migration as a consequence of climate change, though it may well have to cope with internally displaced populations leaving the low-lying areas of the Baltic Sea region or southwestern Siberia. It might well become a destination for people displaced by climate-induced economic dislocations along its periphery, though the welcome that awaits them will not be encouraging. The collapse of social services and the contraction of living standards that the Russian population faced in the 1990s exacerbated their latent xenophobic prejudices and antiforeigner suspicions, which have not abated as the economy has improved. On the contrary, suspicion of outsiders has been encouraged by the Kremlin as part of its campaign to control terrorism. The Kremlin-created youth movement Nashi (Ours) is explicitly antiforeigner, and its members have at times embarked on missions in major Russian cities to root out foreigners from the Caucasus and Central Asia. These habits of mind and officially sanctioned social practices are likely to be especially troubling in the event of climate-induced migrations or other forms of economic turmoil throughout Russia's "near abroad," a region more likely than Russia itself to feel the direct negative effects of rising temperatures, declining availability of water, and so on. Russian xenophobia is easily heightened in periods of fear, and it would most certainly be activated in the event of substantial migration, especially from the southern regions of the former Soviet Union.

If one were looking for potential sources of violent conflict connected with climate change, in fact, it would be most likely to arise as a consequence of an influx of climate-spurred refugees from Central Asia or the Caucasus into European Russia. Such people have already been systematically stigmatized as a consequence of the Russian version of the "war on terror," and the social practices that have been encouraged in this context could easily be extended to others, and probably would be. Immigration to European Russia by Caucasians or Central Asians would arouse and lend credibility to Russia's cryptofascist parties and movements, including Nashi, the Liberal Democratic Party, and the National Bolshevik Party. Although the first two of these (at least) are subject to Kremlin control and manipulation, it is by no means certain that such groups do not have genuine bases of support that will draw their own conclusions in the event of widespread social fear of an influx of southern immigrants. There have been numerous scattered instances of violence against such individuals. Significant refugee movements could quite easily spark widespread social violence in major Russian cities, to which the government would then be compelled to react in some fashion.

Perhaps more dangerous in the long run would be the movement of Central Asian or Caucasian populations into the North Caucasus. If there is one issue the Russian public really does expect the government to handle without question, it is preventing a resurgence of Caucasian-based violence. If climate-induced immigration were concentrated in Russian urban centers, anti-immigrant attacks would be high profile in media terms but of limited effect internationally. If climate immigrants from the Caucasus and Central Asia were concentrated in Russia's North Caucasus (or Caspian) region, the effects of anti-immigrant efforts and violence would be potentially far more destabilizing. This region is already Russia's most violent, overlaying the Russian Federation's most diverse and underdeveloped regions. The North Caucasus comprises hundreds of ethnic, linguistic, and religious groupings, and it has been the locus of a surge in low-level yet significant violence since the second Chechen war in the 2000s. Although the Putin-dominated state seems to have stabilized violence in the region through a combination of better counterinsurgency tactics, reliance upon local leaders (sophisticated in some cases, simply brutal in others), and an enormous infusion of resources, the relative stability thus obtained remains distinctly fragile. An influx of southern immigrants as a result of climate stresses is likely to reignite the violence that Russia faced in this region in the last decade. If there is a plausible path to real power for fascists / National Bolsheviks in Russia, it is via anti-immigrant social tensions exacerbated by deliberately or inadvertently inflammatory state actions. The pressure of climate change on Russia's neighbors may thus offer a plausible, if somewhat circuitous, path to the otherwise unlikely radicalization of Russian politics.

The worst-case scenario of climate-induced violence involving Russia would certainly arise from conflict with China. Experts on Russia's Far East are pessimistic about the region's future and about the coherence and capacity of the Russian state in the region. There is, indeed, a kind a cottage industry in Russia involving speculation over whether Russia's far eastern provinces will break off from the Russian Federation, what China might do to facilitate this development, and how both states would react if it did happen. Such a possibility is given a low level of probability by regional experts, who find it difficult to come up with a plausible scenario by which China would make either sudden or incremental claims on Russia's Far East, or by which Russia's Far Eastern political leaderships and citizens might not resist such an encroachment. At present Chinese traders and laborers, numbering roughly 1 million a year, move back and forth throughout the region as seasonal labor. So far they have shown no inclination to remain in Russia. Were they to do so, however, the reactions of both governments are necessarily uncertain. Here again, the pressures of climate change on populations in Central and Eastern Asia emerge as the most plausible triggering forces for otherwise unlikely but inherently very dangerous events.

Climate Politics and Engagement with the West

On the whole Russian attitudes toward climate politics are dominated by a concern that nothing be done to hamper continued economic growth. In this respect the outlook of

Russia society, at least, has resembled that of the developing world, which regards climate change as a problem created by rich people, and accordingly theirs to solve. In reality, of course, Russia is the third-largest producer of greenhouse gases on Earth, a position that is bound to expose it to international pressure to adopt mitigation standards akin to those employed in Europe and the United States. And it has the economic incentives to do so: Russia is a major producer of greenhouses gases largely because of the extraordinary inefficiency of its industries, which produce more than twice the global average of carbon emissions per unit of productivity, and more than three times that of the average for the European Union.[14] Given that the fundamental driver of Russian economic performance is the exportation of energy, there is every reason to wish to conserve as much of it as possible for that purpose.

Crafting policies to accomplish this is another matter. Russia is a participant in the UNFCCC, and a party to the Kyoto Protocol of 2004. Its acceptance of Kyoto was regarded as a politically important step at the time, particularly in the wake of America's rejection of the protocol, and was instrumental in bringing it into force. In reality, Russia's adoption of Kyoto's standards was more symptomatic of cynicism than environmental awareness; Kyoto, which expires in 2012, is aimed at reducing national emissions by that year to the levels that existed in 1990, a fortunate benchmark for Russia, because it corresponded to the last full year of production under the Soviet Union. In 2004 Russian emissions were far below Soviet-era levels, and the expectation was that when the time came, Russia would be able to sell the difference (as is allowed under the protocol) to parties that were unable to meet their targets. Ironically, Russia's recent economic success has reduced (and may completely eliminate) its marketable emissions surplus, and put it in the awkward position of either taking action toward meeting its obligations or explaining its failure to do so.

Current Russian climate policy, as articulated in a major statement promulgated in April 2009, was initially greeted with widespread satisfaction, if only because it contrasted markedly with the studied official silence of the previous five years and expressed renewed official recognition of the reality and significance of climate change. To the extent that its emphasis falls upon successful adaptation rather than active mitigation, however, this initial enthusiasm has cooled. Europeans, by virtue of their own dependency on Russia energy resources, would particularly like to see a greener attitude emerge in Moscow, and they have made a number of gestures inviting scientific and managerial cooperation toward that end.[15] In practice, however, their market dependency on Russian natural gas especially seems to be limiting rather than enhancing their influence. Russian climate policy at present does not include future targets for reduced carbon emissions, for instance, the kind of step that most environmentally focused NGOs would recognize as an elementary measure of political seriousness.

On balance, environmental politics do not loom large in Russian public life, except with respect to the melting of the polar ice cap, an issue to which a certain glamour attaches because its central dynamic appears to be conflict with West. Russians have come to expect a great deal from the retreat of Arctic ice, and they have little sense of the time and effort required, in this as in so many other areas of environmental policy, to realize whatever potential advantages may be in the offing. Their government has put forward extravagant claims, and it has made a striking gesture by way of dramatizing the issue and also of casting it in the familiar guise of rivalry with the West, whose policies are presented as yet another attempt to interfere in a part of the world where Russian interests should prevail by natural right. Russia's outlook has found a moderate echo in Europe, where talk of future potential conflicts over oil, fishing rights, and so on has become fairly commonplace even among those bearing considerable political responsibility—though this in itself is not a reason to suppose that the confrontational and even violent scenarios being envisioned are realistic.[16]

It is realistic to assume, however, that Russian foreign policy with respect to climate issues will share the confrontational qualities that have lately come to color its relations with the West, for as long as that coloration persist. The extent to which the politics of climate change is uniquely dependent upon effective multilateral cooperation makes the issue less rather than more attractive in Moscow, which can be expected to see climate issues not as a potential avenue for improving cooperation and engagement with the West but simply as another arena of rivalry and friction. In this respect, as in many others, Russian policies fuse the outlook and interests of the developed and developing worlds. It is easy to envision its policy evolving into an incongruous combination of attitudes, in which the need for technological cooperation with the West in pursuit of effective adaptation and modernization combines improbably with a desire to deflect responsibility for global problems onto those with whom it will in fact be seeking partnership.

The best means of moderating this corrosive process is to deepen contacts with Russia's business elites, whose current corruption and subservience to Moscow may prove subject to revision as global market conditions change—and particularly if they change in ways that more fully expose the state's economic incompetence. Similarly, Russia's Academy of Sciences remains a genuine professional association in which serious climate scientific research can still be done. It is not entirely unrealistic to hope that these communities, together with environmentally focused NGOs and some elements of the Russian media, may go into a form of political hibernation that allows them to survive even an extended period of political degeneration, so that they remain available to mitigate the damage once the Kremlin's current experiment in authoritarianism runs its inevitable course toward social and economic failure. If there is a best case scenario to be looked for in Russia's confrontation with environmental change, it may lie in the possibility that persistent pressure may be applied to a regime with which it is ill suited to cope, eventually breaking down its facade of iron-fisted competence so that something better can take its place. A catastrophe of any kind will almost certainly favor the dictators. It is in the slow unfolding of events, even bad and potentially dangerous ones, where the best hopes for modernity and democratization lie.

Notes

1. See appendix A. The issue of direct climate impacts on Russia is complicated by the degree to which climate change brings with it greater climate variability, and an increasing incidence of extreme weather events. Such variability is always bad in the agricultural sector of any economy, regardless of the whether the general trend may be judged favorably or not. See, e.g., Roderick Kefferpütz, "Climate Change as an Opportunity for Russia," *UNDP Newsletter on Development and Transition*, October 2008, www.developmentandtransition .net/index.cfm?module=ActiveWeb&page=WebPage&DocumentID=682.

2. U.S. Geological Survey, *Circum-Arctic Resource Appraisal: Estimates of Undiscovered Oil and Gas North of the Arctic Circle*, USGS Fact Sheet 2008-3049, 2008, http://pubs.usgs.gov/fs/2008/3049/.

3. See Sascha Müller-Kraenner, "Russia's Achilles' Heel: Climate Change," *IP Global* (German Council on Foreign Relations), Winter 2008, www.ip-global.org/archiv/volumes/2008/winter2008/russia---s-achil les----heel--climate-change.html.

4. See the series of World Bank Reports on Russia, prepared semiannually since October 2003, http:// web.worldbank.org/WBSITE/EXTERNAL/COUNTRIES/ECAEXT/RUSSIANFEDERATIONEXTN/0,,con tentMDK:20888536~menuPK:2445695~pagePK:1497618~piPK:217854~theSitePK:305600,00.html.

5. Ellen Barry, "Kyrgyzstan to Give U.S. 6 Months to Leave Base," *New York Times*, February 19, 2009, www.nytimes.com/2009/02/20/world/asia/20kstan.html?fta=y.

6. Luke Harding, "Russia Close to Economic Collapse as Oil Price Falls, Experts Predict," *Guardian* (Manchester), November 20, 2008, www.guardian.co.uk/world/2008/nov/20/oil-russia-economy-putin-medvedev.

7. In an interview on CNBC on June 2, 2009, Russian president Dmitri Medvedev proposed that Russia's GDP in 2009 would decline by "no less than 6 percent," and may reach 7.5 percent. The GDP decline for the twelve months ending in March 2009 was 9.8 percent. See "The Recession in Russia," *Stratfor Global Intelligence*, June 15, 2009, www.stratfor.com/analysis/20090612_russia_and_recession.

8. World Bank in Russia, *Russian Economic Report 18*, Moscow, March 2009, http://web.worldbank.org/WBSITE/EXTERNAL/COUNTRIES/ECAEXT/RUSSIANFEDERATIONEXTN/0,,contentMDK:20888536~menuPK:2445695~pagePK:1497618~piPK:217854~theSitePK:305600,00.html.

9. "New Russian Federal Law on Foreign Investment in Strategic Sectors," *Russia Update* (Hogan & Hartson LLP), May 23, 2008, www.hhlaw.com/files/Publication/99c462ba-19bf-466e-a0dd-af809f7fc218/Presentation/PublicationAttachment/152bef13-824f-4f7e-a390-37dc4f3cb6ae/Russia_Update_May2008.pdf.

10. Pierre Noel, *Beyond Dependence: How to Deal with Russian Gas*, European Council on Foreign Relations Policy Brief (London: European Council on Foreign Relations, 2008), 5–6, http://ecfr.eu/page/-/documents/Russia-gas-policy-brief.pdf.

11. See Daniel Moran and James A. Russell, "Introduction: The Militarization of Energy Security," in *Energy Security and Global Politics: The Militarization of Resource Management*, edited by Daniel Moran and James A. Russell (New York: Routledge, 2009), especially the discussion of "virtual peak oil," 3–4.

12. See Leon Aron, "21st Century Sultanate," *The American*, November 14, 2008, www.american.com/archive/2008/november-december-magazine/21st-century-sultanate; and Yuri Felshtinsky and Vladimir Pribylovskiy, *The Corporation: Russia and the KGB in the Age of President Putin* (New York: Encounter Books, 2009).

13. Amy Myers Jaffe and Ronald Soligo, "Energy Security: The Russian Connection," in *Energy Security and Global Politics*, ed. Moran and Russell, 123.

14. Michael Kozeltsev, "Working with Russia on Climate Change: Barriers and Opportunities for Enhancing EU-Rissia Dialogue," IES Autumn Lecture Series: The EU and the Fight against Global Climate Change, December 10, 2008, www.ies.be/files/repo/Michael_Kozeltsev_101208.pdf.

15. "Russia's Climate Policy Fails to Raise Hopes," EurActive.com, May 19, 2009, www.euractiv.com/en/climate-change/russia-climate-policy-fails-raise-hopes/article-182458.

16. See, e.g., Ian Traynor, "Climate Change May Spark Conflict with Russia, EU Told," *Guardian* (Manchester), March 10, 2008, www.guardian.co.uk/world/2008/mar/10/eu.climatechange?gusrc=rss&feed=networkfront.

10

Central Asia

Kazakhstan, Kyrgyzstan, Tajikistan, Turkmenistan, and Uzbekistan

Edward Schatz

Social scientists are unlikely to develop precise point estimates of the social and political effects of climate change. Although those who study the natural world can aspire to create a predictive science, scholars who consider the social world examine not rocks, genomes, and ecosystems (changeable only over the long term) but reflective and reflexive human beings who constantly revise and revisit their behaviors in often unpredictable ways. Much about the social effects of climate change depends on the ability of human beings to develop strategies to deal with climate-changing behaviors and adjust accordingly; these human propensities do not lend themselves to anything more than the roughest of predictions.[1] Nonetheless, the potential for climate change to create devastating social consequences recommends that we anticipate the basic contours of future scenarios, if only to prepare ourselves to react with rapidity when one of those futures is upon us.

This chapter considers the likely effects that climate change will have on regional security in the five countries of the former Soviet Central Asia—Kazakhstan, Kyrgyzstan, Tajikistan, Turkmenistan, and Uzbekistan—which are already facing notable risks of destabilization, some linked to the distribution of natural resources. It begins by providing a risk baseline, a broad estimate of traditional sources of insecurity that are, for present purposes, conceptually prior to the risks induced by climate change. It then turns to four clusters of climate-change-induced factors—the distribution of water resources, population displacements and refugee flows, the impact of variable harvests on food security, and the authoritarian exploitation of environmental issues—highlighting their added risk margin. Finally, it addresses possible Western responses to these factors. Although Central Asia does not face an inevitable environmental catastrophe, climate change contains the potential to combine with traditional risk factors to render the region deeply unstable.

Traditional Sources of Destabilization in the Former Soviet Central Asia

When the five Central Asian republics emerged from the Soviet collapse, they were considered by many to be at risk for serious instability. With the partial exception of Tajikistan, which experienced a brutal civil war from 1992 to 1997, the region has thankfully been much more stable than analysts expected. This happy fact does not, however, mean that risks are absent. On the contrary, possible destabilization could emerge from, inter alia,

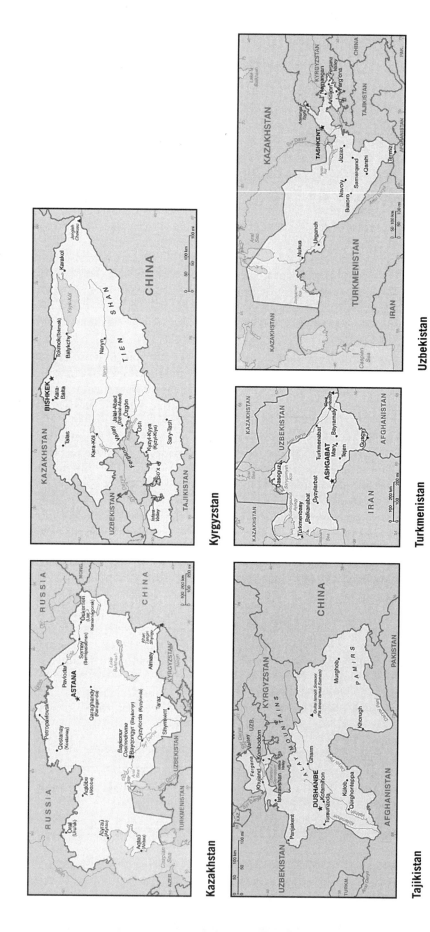

Kazakhstan

Kyrgyzstan

Tajikistan

Uzbekistan

Turkmenistan

tense interethnic relations,[2] intraethnic clan tensions,[3] the potential for Islamist militancy against the state,[4] a general decline in governance capacity,[5] and the aggressive foreign policies of particular states.

The presence of risk does not, of course, mean that Central Asia is in immediate "danger" or that it requires "rescuing."[6] The potential for conflict is mitigated, to varying degrees, by other factors, such as an active civil society (which is the case in Kyrgyzstan, and to a degree in Kazakhstan, but less so elsewhere), economic independence (the case to a degree in Kyrgyzstan and Kazakhstan), strong international engagement (e.g., a strong diplomatic corps, especially in Kazakhstan), and high levels of education and literacy (in a relative sense, this remains the case across the region).[7] These mitigating factors, which are present to varying degrees in the region, should enter into any evaluation of the prospects for its long-term stability. For the most part, however, these factors are unfortunately weaker than would be necessary to predict peaceful future trajectories for the region with confidence.

The lowest risk for serious instability based on traditional factors is in Kazakhstan.[8] By 2008 Kazakhstan enjoyed a relatively more liberal space for economic activity and for civil society groups, as well as an enormous patronage system based on oil receipts, which helped the regime to secure political loyalty and provided a relatively strong social welfare system (including education, old-age pensions, and rudimentary health care). Moreover, the economic fundamentals of Kazakhstan were relatively sound. The credit crunch that burst real estate and construction bubbles in Kazakhstan in 2008 did not risk creating a major or sustained macroeconomic downturn. Because politics inherently contains uncertainty, unfair elections, presidential succession, interethnic tensions in the labor market, and small-scale but increasing levels of Islamist militancy in the southern provinces include some risk of becoming more significant political problems. Yet Kazakhstan remains the Central Asian state where one would least expect major destabilization based on traditional factors.

Turkmenistan, the hardest authoritarian regime of Central Asia, in 2008 had a marginally higher risk for instability based on traditional factors. After the 2006 death of Saparmurat Niyazov, his successor, Gurbanguly Berdymukhammedov, engineered remarkable political continuity. Liberalization was minimal, although Niyazov's cult of personality was quietly dismantled and the new regime expressed increasing interest in opening hitherto closed Turkmenistan to economic, cultural, and educational ties abroad. Popular mobilization was practically nonexistent (except among Turkmen exiles in Moscow and Europe). Patronage via gas receipts and penetrative coercive structures helped maintain political quiescence. Although the potential for conflict was relatively low, the mitigating factors identified above were low as well. As a consequence, any hypothetical conflict (arising, for instance, from rivalry over presidential succession or tensions with Uzbekistan over water use) could prove difficult to contain.

Kyrgyzstan and Tajikistan shared a similar, and slightly higher, level of risk. After the ouster of the soft authoritarian president Askar Akaev in Kyrgyzstan in 2005, the "democrats" who replaced him returned to, and in some cases perfected, many of their predecessor's soft authoritarian practices.[9] Popular discontent and impatience were widespread. Mafia-related groups increasingly controlled economic affairs. Mitigating factors included a relatively vibrant civil society, especially in the north of the country; a strong memory of interethnic conflict in 1990 in the south and a popular desire to avoid repeating that experience; and strong international engagement by nongovernmental organizations (NGOs) and other governments. Serious potential flashpoints remained, however, including unfair elections, interethnic tension in the south, increasing Islamist militancy in the south, a weak state capacity in general, and a burgeoning tradition of potentially destabilizing street protests.

Tajikistan had a roughly similar profile as of 2008. Its civil war from 1992 to 1997 had fundamentally weakened state capacity. Although both elites and masses in 2008 were determined to avoid returning to such conflicts, clan-related and ethnic divisions lay beneath the surface, and the aging of the civil war generation risked renewing tensions. Islamists were active, and though most were nonmilitant, radical Islam increasingly served as a common language for decrying the authoritarianism of President Emomali Rahmon.[10] The factors that would mitigate emergent conflict were unfortunately weak. Education and literacy levels had plummeted as a result of the civil war and its aftermath. Economic activity beyond subsistence was limited and largely dependent on the state (and patronage structures within the state) or international donor agencies. International engagement continued at a medium level after the civil war, but international actors had come to view Rahmon as a guarantor of stability and therefore granted him virtual carte blanche to consolidate his authoritarianism and his structure of patron–client ties. A crisis could emerge from several sources, including presidential succession, a major deterioration of conditions in neighboring Afghanistan or Kyrgyzstan, tensions with Uzbekistan over water use, the increasing impatience of Islamists with authoritarian secularism, weak and variable harvests, and a weak state capacity across the board.

Uzbekistan, on the face of it a stable and conservative society, faced the highest risk of destabilization based on traditional factors. Like Turkmenistan, Uzbekistan was a hard authoritarian regime, but its population had a stronger potential to mobilize against the state. Transnational Islamist militants (especially the Islamic Movement of Uzbekistan and its offshoots) particularly targeted the Islam Karimov regime and identified Uzbekistan as a potential shari'a-based society, despite the fact that such ideas were fundamentally at odds with popular understandings of Islam in the region.[11] Tensions among ethnic Uzbeks, Kyrgyz, and Tajiks occurred, especially in the densely populated towns of the Ferghana Valley. Unfortunately, mitigating factors were weak. Civil society was not strong. Economic activity of any significance increasingly depended on patronage structures and links to the ruling elite. Uzbekistan kept all international actors (especially, but not exclusively, Western ones) at arm's length and, in the mid-2000s, evicted a whole range of NGOs that might have acted as watchdogs against regime-propagated abuses. Presidential succession, intra-elite factionalism, concern about the treatment of ethnic Uzbeks in neighboring Tajikistan or Kyrgyzstan, tensions with upstream Tajikistan and (in parts upstream) Turkmenistan over water use, and poor administration of the agricultural sector or low harvests in the generally fertile Ferghana Valley all brought additional risk of destabilization to Uzbekistan.

The Change Climate Change Brings

Some of climate change's anticipated effects will have a negligible direct impact on Central Asia. For example, the region is landlocked, so the inundation of coastal areas (widely associated with climate change in the Western popular imagination) is a nonissue. Other climate change effects, however, will likely have an impact on the region. In this section, I consider four probable areas of concern. In order of likely magnitude and concern, they are (1) locally significant shortages of water, which may dovetail with and therefore exacerbate preexisting tensions over water use; (2) immigration/refugee flows from neighboring states that could intensify preexisting resource scarcity; (3) local or regional food shortages resulting from expected fluctuations in harvests and food prices; and (4) an increased appetite for authoritarianism as a way to address problems of resource scarcity and goods' distribution.

Conflict over Water Distribution

Water is poorly distributed in Central Asia and is subject to constant manipulation and contestation. The downstream states (Kazakhstan, Uzbekistan, and Turkmenistan) require abundant water for agriculture, particularly for the water-intensive cotton production that dominates Uzbekistan and, to a lesser extent, Turkmenistan. Unless they change their agricultural practices, they have no choice but to rely on the upstream states (Tajikistan and Kyrgyzstan) to provide it. In the meantime, the upstream states are seeking to use water to generate electricity via hydroelectric stations. Without hydroelectricity, the upstream states are overwhelmingly energy poor and therefore reliant on energy-rich downstream states. Major interstate tensions escalated in the 1990s and 2000s over this issue,[12] and they will only be exacerbated further if water shortages prove significant or provide a pretext for aggressive state behavior (see appendixes A and B). Even if militaries show a disciplined commitment to avoiding conflict and remain uninvolved, conflicts that begin at the local level (e.g., disputes over irrigation) always contain the potential to escalate, given the intricate ways in which ethnic and clan communities are intertwined across state borders.[13]

Although the entire region is affected by this maldistribution of water resources, and though the vagaries of locally significant shortages cannot be predicted with accuracy, Uzbekistan is more likely than the other states to unleash preventive military action to improve its access to water, for several reasons. First, by regional standards, it has a significant military, including both ground and air forces. Second, it harbors grievances against and has border disputes with each of its neighbors.[14] Third, it is the most reliant of the states on downstream flows to irrigate its cotton monoculture, which generates much of its export earnings. In turn, the cotton sector is structured to maximize social control. For example, all farmers must sell all their cotton to the state at state-mandated prices, and they remain subject to exorbitantly high levels of indirect taxation.[15] Fourth, Uzbekistan's elite is prone to factionalization.[16] As a result of these conditions, if Uzbekistan experiences greater shortages of water for agriculture (or if its elite merely *fears* that it could face heightened shortages in the future), it could opt for preventive measures, such as those implied by the military maneuvers that simulated a capture of Kyrgyzstan's Toktogol Dam.[17]

Tajikistan and Kyrgyzstan have pursued massive hydroelectric projects to alleviate their reliance on oil and gas deliveries from neighboring states. In Tajikistan the main sites of such projects are the Panj and Vakhsh rivers. The latter has the Nurek Dam, which produces the overwhelming majority of the country's electricity. The Nurek faces a problem of induced seismic activity, the magnitude of which is debated. Engineers disagree about whether or not the dam is adequately reinforced to ensure its integrity and avoid the inundation of downstream settlements, including the capital city of Dushanbe. The Panj has the Rogun Dam, an unfinished project from the Soviet period for which additional Russian financing has stalled. Myriad other projects, including the Sangtuda-1 and Sangtuda-2 power stations downstream on the Vakhsh, are under construction, the former by Russia's United Energy Systems and the latter by Iran.

These projects in Tajikistan are designed to make the country a net energy supplier. Plans are afoot to provide Pakistan and Afghanistan with electricity from Tajikistan and Kyrgyzstan without relying on Uzbekistan's electrical grid, with financing in part from the

World Bank.[18] The concern is that these projects run the risk of souring relations with downstream Uzbekistan. Uzbekistan, in the meantime, is unlikely to sit idly as Tajikistan develops its hydroelectric sector. Rumors persist that Uzbekistan's Karimov promised closer ties with Russia in exchange for Russia's aluminum company, RusAl, putting Uzbekistan's concerns about the height of the Rogun Dam in play as part of its negotiations with Tajikistan.[19]

Kyrgyzstan's situation is roughly similar. Major dams line the Naryn River, which feeds the Syr-Darya downstream in Uzbekistan's portion of the Ferghana Valley. The Toktogol Dam is the largest, generates the most electricity, and is the closest to the Uzbek border. In early 2008 Kyrgyzstan was studying the best way to resurrect two hydroelectric station projects at Kambarata, which would supply energy for the northern part of the country. It is widely known that hydroelectric stations have become enormous opportunities for rent-seeking officials.

Lessening Kyrgyzstan's and Tajikistan's reliance on expensive energy inputs is, of course, necessary; of ongoing concern, however, is the likely reaction from downstream states, particularly Uzbekistan. The construction of these projects must be accompanied by confidence-building measures on both sides, as well as strong steps to ensure compliance in emerging interstate water-use regimes, which in the past have lacked serious enforcement capacity. As I suggest below, third parties involved in the region should do everything possible to ensure that long-term stability is taken into consideration at the *design* stage of these hydroelectric projects, regardless of whether they are pursued by private- or public-sector actors.

Although water shortages are likely to affect interstate and intercommunal relations negatively in the region, the opposite possibility—locally significant flooding—should not be overlooked. Given Kyrgyzstan's and Tajikistan's steep mountain topographies and abundance of glaciers at the high elevations, warming temperatures could generate glacial lake outburst flooding, potentially inundating downstream population settlements and perhaps generating destabilizing refugee flows. The design of hydroelectric projects should also take into account this possibility. One study expects an overabundance of water with climate change as the region's glaciers melt, followed by shortages and greater seasonal and annual variability.[20]

The possibility of glacial lake outburst flooding suggests a broad need for the region's states to develop their capacities for rapid emergency response, both individually and in concert through multilateral organizations such as the Shanghai Cooperation Organization. And Central Asian states are not prepared—as was illustrated, for instance, during the exceptionally cold winter of 2007–8, when electricity supplies and distribution infrastructure in Tajikistan proved inadequate, leading to a health and humanitarian crisis. Tajikistan received emergency assistance from various United Nations agencies to address the disaster. It is reasonable to assume that the inability of the Tajikistani state to deal with this particular crisis is similar across the board of potential crises, and that the region's other states, particularly Kyrgyzstan, are also unprepared.

The Absorption of Refugees and Migrants

Increasing migration and refugee flows are possible from Afghanistan and China. Already in the 1990s and 2000s Afghanistan sent out significant waves of refugees during its civil conflict. Chinese migrants have tended to be motivated by a search for economic opportunity, settling especially in Kazakhstan and Kyrgyzstan. Improved road links between Central Asia and China have begun to accelerate both legal and illegal migration. If Chinese coastal areas are threatened by rising sea levels, this would create population pressures that would presumably be resolved via some combination of internal and external migration (see appendix B and chapter 2 in this volume). The significance of these refugee and migrant flows from Afghanistan and China will depend upon their intensity and scope, along with the preparedness of Central Asian states to absorb newcomers. Areas where unemployment is high but the local population is hesitant to emigrate to seek employment are particularly at risk. This is the case for the Ferghana Valley, which, if its towns cannot easily absorb newcomer Chinese or Afghanis into the local labor markets, could see rising tensions. Everyday resentment against ordinary Chinese was already palpable throughout the region by the early 2000s. The cities of Osh (Kyrgyzstan) and Khujand (Tajikistan), regional centers of commerce increasingly linked to China via transport corridors for the trade in inexpensive manufactured goods, are particularly vulnerable to these pressures. Absent serious measures, they will tax the capacity of the already-weak states of Tajikistan and Kyrgyzstan. State collapse cannot be caused by these migrations, but, in combination with other factors that weaken state capacity, such pressures could be noteworthy.

In the 2000s each of the region's states enjoyed factors that mitigated migration's effects. For example, in Tajikistan, such stresses were partially alleviated by the willingness of Tajiks to emigrate to other countries for work; an estimated 800,000 to 900,000 Tajiks (out of a population of about 7 million) worked in Russia in 2008.[21] Likewise, an estimated 500,000 Kyrgyz lived abroad in the late 2000s. In Kyrgyzstan (and, to a lesser extent, Kazakhstan), a relatively vibrant civil society that generally inculcated norms of tolerance and peaceful coexistence tended to allow for some absorption of newcomers.[22] The willingness of the Kurmanbek Bakiyev regime to follow the trend throughout much of the former Soviet space by consolidating its control over NGOs and other parts of civil society was, in this and other respects, troubling. In Uzbekistan and Turkmenistan, population stresses were partially alleviated by the closed nature of each state's labor market.

It is important to emphasize that these factors were all linked to circumstances that cannot be expected to continue indefinitely. Russia, the main destination for Tajiks and Kyrgyz seeking itinerant work, has an ambivalent relationship with its foreign-born workforce, so scenarios in which Russia becomes an inhospitable destination for Central Asians are not hard to imagine. Likewise, one can expect a deterioration of civil society in Kyrgyzstan and Kazakhstan as state restrictions on NGO activities increase. Potentially, this could serve to change societal norms, intensify resource competition, and generate xenophobia. Finally, with an eventual change of regimes in Tashkent and Ashgabat, Uzbekistan and Turkmenistan may open their labor markets, thus creating more significant magnets for migrants and refugees.

Food Shortages

Although aggregate agricultural productivity is expected to increase in the region with changing temperatures, this is not the whole story (see appendix A). Russia and Kazakhstan developed the mechanized and semimechanized processes and effective, largely market-based distribution systems to supply food throughout the region. Uzbekistan had the potential to do the same, but its engagement in market mechanisms remained partial. As with water resources, the concern was less with the overall volume of food and more with its maldistribution. Particularly intense climate-related vagaries can increase year-to-year variation in harvests. The significance of this risk will depend on the ability of the Central Asian states to avoid becoming overwhelmingly reliant on such imports.

For example, Tajikistan relied strongly on subsistence agriculture in the 2000s, but if it comes to rely on imported foodstuffs from Russian and Kazakhstani producers, it will become vulnerable to supply-chain disruptions related to yearly fluctuations in harvests. The same is true for Kyrgyzstan. If such disruptions have locally concentrated effects, they will generate tensions in complexly interrelated ethnic and clan communities. We should not expect wholesale state collapses because of food shortages, but a significant erosion of state capacity remains an important theoretical possibility.

Authoritarian "Solutions"

Even in the advanced industrial democracies, calls have increased for greater state regulation to generate desirable environmental outcomes. In these contexts, such regulations will likely have little impact on the essential features of democracy and market economy. By contrast, in unconsolidated states, whether authoritarian or nominally democratic, anxieties surrounding increasing environmental threats likely will generate calls for strong-arm solutions. One paradoxical but nevertheless serious risk arising from climate change in Central Asia is that its consequences, though real enough, may be exaggerated for political effect.

The exaggeration of the scope and disruptive potential of "threats" is not new to Central Asian elites. The Karimov regime in Uzbekistan, for example, expertly exploited the issue of Islamic militancy to create a climate of fear and justify regime-sponsored repression in the name of combating terrorism.[23] Karimov was simply the most vocal of the "defenders" of Central Asia against the so-called Islamic threat; all the region's leaders used this specter to political effect.

If claims of an environmental threat can be similarly exaggerated, the manufactured crisis can shore up the kinds of repressive apparatuses that have characterized large parts of Central Asia. In turn, two additional risks would emerge from the resort to authoritarian "solutions." First, the regimes could adopt strongly anti-Western rhetoric, if the causes of environmental despoliation can be plausibly linked to Western sources (e.g., a disproportionate Western contribution to greenhouse gas emissions or an ecological emergency resulting from Western companies' oil and gas exploration). This possibility is strengthened by the synergy between environmentalism and anti-imperialist nationalism that emerged in the late Soviet period.[24] Second, a turn to greater authoritarianism brings repression that can inflame and embolden Islamic militants and other regime critics. Whether a further authoritarian turn would become destabilizing in the short term is unclear, but its longer-term effects on the social and economic stability of the region can only be negative.

Possible Western Responses

In and of themselves, none of these anticipated effects is likely to be catastrophic, but in a region with myriad intertwined traditional risk factors, the add-on potential for destabilization deserves focused Western attention. Western actors should consider four targeted strategies to minimize the effects of climate change on Central Asian stability.

First, the development of hydroelectric capacity in upstream Kyrgyzstan and Tajikistan must be pursued under conditions that will enhance long-term stability for the entire region. Crucially, hydroelectric projects must be a part of meaningful, enforceable bilateral and multilateral agreements on water use. To date, a variety of water-use treaties have been concluded, but their terms have been routinely violated by officials and ordinary people. A significant regional capacity to monitor compliance in any emerging water-use regime needs to be developed. Moreover, those engaged in upstream hydroelectric development must think of dam design as more than about maximizing power generation; certain designs can help to build the confidence even of downstream states that water will be available for their pressing needs. For example, "run-of-the-river" hydroelectric dams, as opposed to normal "reservoir" dams, are excellent designs for minimizing the ability of upstream parties to withhold water flows. Although Central Asia's mountain topography may make the run-of-the-river design impractical in some places, it may nonetheless be considered a possibility for new projects, given its potentially salubrious political and social effects. Put most generally: Given that there are technical choices involved in dam design that may help or hinder stability in the region, it is in the interests of the West to encourage and facilitate the choices that are most conducive to regional stability.[25]

Second, though any water-use regime should consider the water-use needs of the downstream states, the fact remains that the downstream states' overreliance on water-intensive agriculture is the heart of the problem. Uzbekistan should be encouraged to diversify its economy away from the current overwhelming emphasis on cotton cultivation. If the downstream states find it easier to meet their water-use needs, the likelihood that they would attempt to secure access to water by force would be radically reduced. Shifting Uzbekistan away from cotton is a long-term project. Not only is it a valuable cash crop on international markets (generating roughly half the country's export earnings), but in Uzbekistan its cultivation is a crucial part of a system of social and political control of rural areas.[26] The diversification of Uzbekistan's economic profile is thus linked to larger questions of political liberalization.

Could Uzbekistan realistically shift away from cotton cultivation? It would be folly to expect that this would happen smoothly or rapidly.[27] Nonetheless, the fact that the Uzbek government was able to achieve wheat self-sufficiency after the Soviet collapse suggests a significant margin for maneuver, given the political will.[28] Moreover, the Uzbek government has underexplored the potential of alternatives, such as traditional tourism and ecotourism, and the global price for cotton has generally been declining, with rising consumer preferences for synthetic materials and increased yields among the most advanced producers, such as those in the United States.[29]

At a bare minimum, Uzbekistan needs to take seriously the problem of inefficient irrigation systems in its river basins. As Spoor details: "Whereas in other major cotton-producing countries like the USA or Egypt water-saving techniques have been important in improving the efficiency of irrigation systems, they have not been a priority in Central Asia. The seepage from major canals and field channels seepage is [sic] enormous, and substantial losses are incurred when transporting water in open canals in the desert temperatures of Central Asia."[30] The Uzbek elite should be encouraged to anticipate that water-saving measures will become ever more crucial as hydroelectric projects are pursued upstream and the region potentially experiences a trend toward water scarcity.

Third, outside and regional actors should assist Tajikistan to develop its agricultural base beyond a subsistence level. The fact that bountiful seasonal produce rots on the fields suggests that Tajikistan could produce top-quality juices, canned goods, and other products. Foreign investment has been meager, and indigenous market know-how is limited. Although multinational companies may find Tajikistan's market to be small, Kazakhstan is in a good position to offer direct investment to develop Tajikistan's agricultural sector. Western actors could encourage Kazakhstan to develop its investment strategy for the region, in part centering on reducing Tajikistan's vulnerability to food insecurity.

Fourth, all the region's states should work to improve border policing. The European Union and the United States, among others, have already been involved in bolstering the interdiction capacity along the region's interstate borders, especially since 2001, with a view to improving their ability to absorb migrants who otherwise could present population and resources pressures that markets and states could not bear. Accommodating new migrants is a security issue, especially for the Ferghana Valley regions of Tajikistan, Uzbekistan, and Kyrgyzstan.

The (in)stability of Central Asia does *not* rest upon the vagaries of climate change, but climate change could contribute to destabilization by other factors. Luckily, the kinds of steps that would be needed to combat climate change are low cost and would address traditional risk factors as well. Even if the predicted natural effects of climate change do not come to pass, a stable water-use regime, a shift away from cotton production, a bolstered ability to contend with new population pressures, and greater food security are all essential public goods for the entire region—a region that will remain of crucial importance for global security as well.

Conclusions

In this chapter I have argued that the risks of destabilization in Central Asia will likely be elevated, given the climate changes that are anticipated for the region. Because of these elevated risks and their potential political and social consequences, outside actors (particularly, though not exclusively, Western ones) must find ways to address the issues of water distribution, the ability to absorb migrants and refugees, the variability in food availability and prices, and the possibility that climate politics may contribute to an authoritarian drift.

Any number of variables could thwart the accuracy of these predictions. Although natural scientists can aspire to develop a consensus about the scope, magnitude, and particular manifestations of climate change, social scientists can only hazard educated guesses based on a close knowledge of individual contexts. Whatever intellectual hazards are inherent in making such predictions, they are far outweighed by the real-world hazards of turning a blind eye to the issue of climate-related insecurity.

Notes

The author thanks Nikola Milicic for his excellent research assistance.

1. See Bent Flyvbjerg, *Making Social Science Matter: Why Social Inquiry Fails and How It Can Succeed Again* (Cambridge: Cambridge University Press, 2001).

2. David D. Laitin, *Identity in Formation: The Russian-Speaking Populations in the Near Abroad* (Ithaca, NY: Cornell University Press, 1998); Michele E. Commercio, *Russian Minority Politics in Post-Soviet Latvia and Kyrgyzstan: The Transformative Power of Informal Networks* (Philadelphia: University of Pennsylvania Press, 2010).

3. Kathleen Collins, *Clan Politics and Regime Transition in Central Asia* (Cambridge: Cambridge University Press, 2006); Edward Schatz, *Modern Clan Politics: The Power of "Blood" in Kazakhstan and Beyond* (Seattle: University of Washington Press, 2004).

4. Igor' Rotar', *Pod zelenym znamenem islama: Islamskie radikaly v Rossii i SNG* (Moscow: Assosiatsiia issledovatelei rossiiskogo obshchestva XX veka, Nezavisimaia gazeta, AIPO-XX, 2001).

5. Mark R. Beissinger and Crawford Young, eds., *Beyond State Crisis: Post-Colonial Africa and Post-Soviet Eurasia in Comparative Perspective* (Washington, DC: Woodrow Wilson Center Press, 2002).

6. See the insightful special issue of *Central Asian Survey* 24, no. 1 (March 2005), which addresses the "discourses of danger" surrounding Central Asia.

7. Following McMann, I have in mind the ability of individuals to thrive economically without relying on the state or other dominant structures (e.g., a dominant ethnic group). See Kelly M. McMann, *Economic Autonomy and Democracy: Hybrid Regimes in Russia and Kyrgyzstan* (Cambridge: Cambridge University Press, 2006).

8. The estimates are relative, not absolute. Further research is required to know how these estimates compare to those outside Central Asia.

9. Edward Schatz, "The Soft Authoritarian 'Tool Kit': Agenda-Setting Power in Kazakhstan and Kyrgyzstan," *Comparative Politics* 41, no. 2 (January 2009): 203–22.

10. See Kirill Nourzhanov, "Saviours of the Nation or Robber Barons? Warlord Politics in Tajikistan," *Central Asian Survey* 24, no. 2 (2005): 109–30; and Collins, *Clan Politics*.

11. Adeeb Khalid, *Islam after Communism: Religion and Politics in Central Asia* (Berkeley: University of California Press, 2007).

12. International Crisis Group, *Central Asia: Water and Conflict*, International Crisis Group Report 34, May 30, 2002, www.crisisgroup.org/home/index.cfm?id=1440.

13. Jeremy Allouche, "The Governance of Central Asian Waters: National interests versus Regional Cooperation," *Disarmament Forum* 4 (2007): 45–55. On the complexities of overlapping communities and identities in the Ferghana Valley, see John S. Schoeberlein-Engel, "Identity in Central Asia: Construction and Contention in the Conceptions of 'Ozbek,' 'Tajik,' 'Muslim,' 'Samarqandi' and Other Groups" (PhD diss., Harvard University, 1994). For one account of the potential to escalate, see Sam Nunn, Nancy Lubin, and Barnett R. Rubin, *Calming the Ferghana Valley: Development and Dialogue in the Heart of Central Asia* (New York: Council on Foreign Relations and Century Foundation, 1999).

14. International Crisis Group, *Central Asia: Border Disputes and Conflict Potential*, International Crisis Group Report 33, April 4, 2002, www.crisisgroup.org/home/index.cfm?id=1439.

15. Sanjar Djalalov, "Indirect Taxation of the Uzbek Cotton Sector: Estimation and Policy Consequences," in *The Cotton Sector in Central Asia: Economic Policy and Development Challenges*, proceedings of a conference held at the School of Oriental and African Studies University of London, London, November 3–4, 2007, edited by Deniz Kandiyoti (London: University of London, 2007), 90–101.

16. Alisher Ilkhamov, "The Limits of Centralization: Regional Challenges in Uzbekistan," in *The Transformation of Central Asia*, edited by Pauline Jones Luong (Ithaca, NY: Cornell University Press, 2004), 159–80.

17. International Crisis Group, *Central Asia: Water and Conflict*.

18. See "S 2010 goda Kirgiziia i Tadzhikistan budut postavliat' elektroenergiiu v Afganistan i Pakistan," December 20, 2007, www.ferghana.ru/news.php?id=7994.

19. Alexander Sadikov, "Tajikistan's Ambitious Energy Projects Cause Tensions with Uzbekistan," Eurasianet.org, October 4, 2006, www.eurasianet.org/departments/insight/articles/eav100406.shtml.

20. World Bank, *Drought: Management and Mitigation Assessment for Central Asia and the Caucasus*, Report 31998-ECA, March 11, 2005, 60, www-wds.worldbank.org/external/default/main?pagePK=641930 27&piPK=64187937&theSitePK=523679&menuPK=64187510&searchMenuPK=64187511&cid=3001&entit yID=000310607_20061213143452.

21. "Natsional'nyi institut migratsii pomojet tadjikskim gastarbaiteram v Rossii," REGNUM Informatsionnoe agenstvo, April 22, 2008, www.regnum.ru/news/990623.html.

22. On the context in which nascent civil society developed in the 1990s in Kyrgyzstan, see John Anderson, "Creating a Framework for Civil Society in Kyrgyzstan," *Europe-Asia Studies* 52, no. 1 (January 2000): 77–93.

23. Since September 2001, the United States has afforded states claiming to be fighting Islamic militants wide latitude to operate against anyone suspected of being a militant, whether or not they pose the threat ascribed to them. See Khalid, *Islam after Communism*.

24. Jane I. Dawson, *Eco-Nationalism: Anti-Nuclear Activism and National Identity in Russia, Lithuania, and Ukraine* (Durham, NC: Duke University Press, 1996).

25. Thanks to Matthew Keefer (personal communication) for his insights on the range of technical choices available.

26. See *Cotton Sector in Central Asia*, ed. Kandiyoti, 1–11 and passim.

27. Max Spoor, "Cotton in Central Asia: 'Curse' or 'Foundation for Development?'" in *Cotton Sector in Central Asia*, ed. Kandiyoti, 54–74.

28. Uzbekistan engineered a dramatic increase in wheat production, from 1 million tons in 1991 to 5.2 million tons in 2004, enormously increasing the land dedicated to wheat and partially opening up the sphere to market principles. See Djalalov, "Indirect Taxation."

29. Richard Pomfret, *The Central Asian Economies since Independence* (Princeton, NJ: Princeton University Press, 2006), 145.

30. Spoor, "Cotton in Central Asia," 71.

11

The European Union

Chad M. Briggs and Stacy D. VanDeveer

In recent years U.S. and European Union officials as well as nongovernmental analysts have devoted increasing attention to assessing and debating the ramifications of climate change for national and international security. At the same time, the EU and the United States have spent the better part of the past decade disagreeing about the Kyoto Protocol and a host of associated issues and mechanisms for addressing global greenhouse gas emissions. The EU has developed relatively ambitious policy goals on this score, and it has limited the growth of carbon dioxide emissions more effectively than has the United States. The EU and many of its member states have initiated numerous national and supranational programs aimed at meeting Kyoto commitments, together with a range of other EU policies motivated by a concern for global warming; several EU member states are thus struggling to bring down emissions and meet relevant targets. Through 2010, the United States, for its part, had few serious national greenhouse gas mitigation or adaptation policies to its credit, despite several robust state and local initiatives that might provide examples of how to proceed.

Differences between Washington and Brussels on climate change policy since the 1990s have been stark, sometimes described as symbolic of a deep "climate divide" across the Atlantic.[1] The EU's ratification of the Kyoto Protocol was greeted with celebrations in Brussels and many other European cities.[2] In contrast, the announcement of the United States' withdrawal from Kyoto was greeted with dismay and anger in Europe. President George

The European Union

W. Bush's first visit to Europe in June 2001, to attend an EU summit, was met by public protests, as fifteen thousand people demonstrated in the streets of the host city, Gothenburg, Sweden. Although those protestors certainly had agendas that extended beyond climate change, it is clear that the Bush administration's approach to climate change contributed to a significant cooling in transatlantic relations. The different climate change and energy policy priorities of the Barack Obama administration, conversely, suggest that transatlantic climate relations can be improved with greater U.S. engagement in international climate-change-related negotiations. Yet a long list of policy differences remains, and a transatlantic climate change accord did not emerge in the first year of the new U.S. administration or at the global climate summit in Copenhagen in December 2009.

On each side of the Atlantic, much greater attention has been paid by high-level analysts to climate change and security connections in recent years. In the United States, the post-2005 period witnessed increased policymaker and analytical attention to the connections between climate change and security, including a number of high-profile reports from Washington-based think tanks, a National Intelligence Assessment, the scenarios work done by the U.S. National Intelligence Council, and strategic scanning and foresight projects developed by the Energy and Environmental Security Directorate at the U.S. Department of Energy.[3] European governments have also started addressing the security dimensions of climate change in recent years, with notable contributions of the Swedish and German governments.[4] The European Commission, under the direction of High Representative Javier Solana, released a paper on climate and security in March 2008 with follow-up recommendations in December of that year.[5] Both these U.S. and EU efforts build on the analysis of environment and security connections done in the 1990s, when these linkages last received sustained policymaker attention on both sides of the Atlantic.[6]

European Climate Concerns

Climate change and its impacts hold serious ecological, economic, political, and social consequences for European states and societies during the coming decades.[7] Yet, though climate change impacts are likely to have a number of negative ramifications, they seem unlikely by themselves to significantly destabilize the social order in any EU member countries. Climate security concerns at the senior levels of European leadership tend to focus on impacts external to the European continent, whose resulting second- and third-order effects may be anticipated to affect Europe. Within Europe, the most severe expected impacts until 2030 include extreme weather events (intense droughts, flooding, and acute summer temperature increases), increased seasonal climate variability, an increased risk of forest fires, coastal and river flooding, a reduction of winter precipitation and snowpack, and human health effects due to changing pests, pathogens, and extreme weather. These are likely to harm agriculture and winter tourism, and to entail significant and growing costs for doing repairs and providing an emergency response during extreme weather events, along with adaptation generally.

In some states the immediate social and ecological challenges may nevertheless prove quite serious. Across the Mediterranean region, for example, water scarcity may worsen substantially. In Poland, where a larger number of citizens rely on the agricultural sector, and which has several million people in the coastal zone, climate impacts may be significant for large portions of the citizenry. Large variability in seasonal conditions and the damage this might do to the agricultural sector, in conjunction with a higher frequency of extreme drought and flooding events, holds the potential to increase rural poverty and further accelerate urbanization in an already highly urbanized society.

Recent Central European examples that may be illustrative include a number of comparatively severe, even catastrophic, flooding events that occurred in the region between 1997 and 2002. Though numerous environmentally oriented actors interpreted these in terms of climate change, this is not how the flooding is generally perceived in the region.[8] Although the Polish and Czech governments were the subject of considerable criticism in the wake of these flooding events, the resulting strain posed no threat to social stability. In contrast, those living in and caring for the coastal regions in the Netherlands and the United Kingdom are more acutely aware of the relationship between systemic climate triggers and coastal storms or flooding, and of the long-term risks of sea-level rise. Such concerns help motivate current efforts at climate mitigation.[9]

In comparison with many other nations and regions discussed in this volume, widespread violence or instability due to climate change remains highly unlikely in European countries. Increased migration from rural and/or flood-prone areas is more likely, and it may arise as part of a general movement of population toward the greater economic opportunities of Western Europe. Such intra-European migration, whether motivated by ecological or economic motives, can largely be accommodated under existing political structures. Only Poland, with a population of about 40 million, would be likely to produce a sufficient number of such migrants to cause negative policy reactions. Because the number of Poles working in Western Europe is already controversial in states such as the United Kingdom and France, any substantial increase may have political ramifications there and elsewhere in Europe. Still, these appear quite unlikely to rise to the level of serious social cleavages, much less organized violence. A much more likely response among less wealthy EU member states and citizens would be increased demands on EU and Western European state budgets for financial assistance in the wake of extreme events. It is reasonable to assume that, as the costs of adaptation to climate change increase—in the form of infrastructure investments, flood protection, agricultural subsidies, resettlement, and so on—so too will demands that those costs be borne disproportionately by the EU's richer states. In this respect European adaptation to climate change may be more immediately linked to the future of the global economy, European defense and social policy spending, and the financial markets than to the stability of European society.

In comparison with many other regions, state failure as a result of climate change appears highly unlikely in Europe during the next twenty-five years. Nevertheless, some predicted climate impacts and their economic ramifications present challenges sufficiently serious to increase the chances of political paralysis in some EU member states. This may be particularly true in some recently democratizing states, where political parties remain very weak, small, and short-lived, and where urban/rural and class divisions frequently result in divided governmental authority. As such states confront growing global economic uncertainty, acute concerns over local conditions may severely hamper the ability of relatively fragile governments to implement effective responses to long-term climate-related risks. The fact that risks and impacts are most likely to be felt by different social groups than those that will be required to pay for their mitigation will likely exacerbate this problem within European states, as it may do among EU members.

Nevertheless, European states are generally well equipped, financially and administratively, to respond both to climate-related emergencies to longer-term trends. Even in the transition states, such capacities have grown over time.[10] This growing capacity results from numerous factors, including the political and economic transitions away from state socialism, capacity building assistance from Western countries, and the integration of Central and Eastern Europe transition states into the EU, NATO, and other Western European and/or transatlantic institutions. Civil society is developing as well, albeit more slowly than many analysts would like. The European Union, and the growing integration with other European countries that it represents, is a source of tremendous public-sector capacity across the continent. For example, the EU institutions helped organize timely assistance to the Polish and Czech states during and after the floods of the last decade. They also worked to build emergency response capacities across the continent and to expand knowledge and preparedness programs related to climate change.

European Climate Concerns: Common Foreign Policy and Security Approaches

European climate-related security concerns, like those in the United States, often focus on the impact of climate change in the developing world, and on the need to mitigate whatever troubles may arise from that quarter. Although climate change poses significant risks to vulnerable infrastructure and health within Europe, the security concerns as defined by the EU and its member states have tended to look outward. Broadly speaking, the primary areas of concern are environmental migrants from neighboring regions; the impacts of water, food, and energy insecurity on the stability of extra-European states; sea-level rise; and changes to the geopolitics of the Arctic.

Threats to the availability of water and food among Europe's less-developed trading partners—as a consequence of the increasing severity and frequency of drought, the spread of desertification, or rising sea level—are a particular concern. The increasing pressures that such environmental changes would place upon populations in North Africa, Central Asia, and the Middle East would act as "threat multipliers" of the kind hypothesized by the Center for Naval Analysis report on similar indirect threats to the interests of the United States.[11] If such pressures were to become sufficiently severe to create acute instability or state collapse, the cascading effects of such events would likely spill over European borders.

The potential impacts of environmentally induced migration are arguably a more sensitive topic in Europe than similar concerns in the United States.[12] Illegal immigration into the "Schengen Zone" (or into the United Kingdom and Ireland) has a relatively low threshold for political reaction, and small numbers of migrants from North Africa have been known to make European-wide headlines.[13] Existing tensions between immigrant populations and European nationalist sentiments may be worsened if climate-related migration from the South or East increases. Within neighboring regions, migration is well recognized as a potential trigger for violence or instability, with larger regional security implications for Europe. The EU has recently funded projects to determine the causes and likelihood of environmental migration, with driving forces linked to the underlying vulnerabilities in affected societies.[14]

For EU members such as Sweden, Denmark, Germany, Norway, the United Kingdom, and the Netherlands, integration of climate threats with overseas development aid is crucial to increasing the adaptive capacity of affected regions.[15] This includes assistance to shift economies to low-carbon models of development, as well as adaptation measures for what may be inevitable climate-related environmental changes. These EU member states are also among those that remain the most involved in global climate change negotiations, including those associated with the meeting of the Conference of the Parties to the UNFCCC in Copenhagen in December 2009 and the UN-sponsored talks that continue in that summit's wake.

The pan-European climate security strategy proposed by the European Commission involves closer dialogue with vulnerable regions in the European neighborhood, including the Middle East and North Africa, Sub-Saharan and Central Africa, and Central Asia, and the development of early warning systems to detect processes that may lead to larger security concerns.[16] Such an early warning network would attempt to identify key environmental vulnerabilities in advance of conflict or acute instability so that appropriate action can be taken to reduce such risks or mitigate the environmental changes that are producing them. This approach reflects European attitudes toward human security. From a European perspective, military options are viewed as irrelevant to climate-related instability. Nor are they available either substantively or politically to most European states. The long-term commitment of NATO forces to Afghanistan merely heightens the attraction of preventative or palliative action in lieu of direct intervention.

Changes to the Arctic have a particular relevance in Europe, given that Sweden, Finland, Denmark, and Norway have territorial jurisdiction over substantial Arctic regions. Temperature changes in the higher latitudes are occurring at a much faster pace than further south, and climatological and ecological systems in the far north exist closer to crucial systemic tipping points.[17] Average changes of a few degrees Celsius can lead to the melting of large areas of permafrost or glaciers, with impacts that are already visible in terms of Greenland ice sheet melting or the loss of summer sea ice in the Arctic Ocean. The observed and anticipated changes have two potential types of impacts for security. The first type of impact is that the loss of summer sea ice, which may completely disappear within the next decade if present trends continue, creates new geopolitical space where previous access had been severely limited. The opening of the Northwest Passage can create new territorial and resource access disputes between Europe (including Greenland), Canada, the United States, and Russia, along with new concerns about pollution from shipping through ecologically fragile regions.[18] Although unlikely to result in open conflict, such shifts are significant security concerns for the Nordic states of Europe, and they may place an additional strain on long-established transatlantic security relationships and increasingly stretched defense budgets.

The second type of impact of Arctic climate change is more physical but has far more serious security implications. Many of the most powerful climate feedback mechanisms that have been identified so far lay in the Arctic. The accelerated melting of the Greenland ice sheet may have effects on the thermohaline circulation of ocean currents, which can result in very acute and sudden changes in global climate.[19] Little is known about the speed with which such effects may develop. An increased melting of permafrost or warming of Arctic waters can result in large releases of methane, a powerful greenhouse gas whose increased presence in the atmosphere could act as a positive feedback mechanism for accelerated warming.[20] The melting of marine methane clathrate deposits could release at least 100 times more carbon-equivalent emissions than Europe produces every year, with far-reaching consequences for acute climate change and impacts.[21] The geographical location of Europe gives many states a more sensitive perspective on such issues than those of more southerly neighbors.

Transatlantic Climate Politics

The United States and the European Union are responsible for the vast majority of all historic anthropogenic greenhouse gas emissions. They also constitute a substantial fraction of the global economy. As a result, if climate change and its effects are to be addressed, they both face the simultaneous tasks of reducing greenhouse gas emissions, adapting to climate change at home, and responding to the growing need for adaptation and mitigation assistance around the globe. All three challenges, and their associated costs, hold the potential to have an impact on security-related planning, budgets, and risks.

The United States and the EU remain two of the world's three largest annual emitters, having lately been overtaken by China, whose historical contributions to global greenhouse gas emissions has until recently been negligible (see table 2.1 in chapter 2 of this volume). The United States, with a population of more than 300 million (about 5 percent of the world's population) emitted 21 percent of global greenhouse gases in 2000. The fifteen members of the EU (the EU-15), which were originally bound by Kyoto Protocol reduction commitments, emitted about 12 percent. By 2008, the EU-15's greenhouse emissions were about 6.7 percent below 1990, while the emissions for the whole, now-enlarged EU-27 were down by 10.7 percent.[22] Although this remains short of the EU's Kyoto commitment, it stands in stark contrast to the 14.7 percent growth in U.S. emissions between 1990 and 2006.[23] These very different post-1990 trajectories result, in large measure, from the different energy and climate policy responses adopted on each side of the Atlantic. European officials expanded energy efficiency and renewable energy investment incentives and emissions reduction policies, while the U.S. federal government failed to do so. U.S. per capita emissions are well above twice those of the EU-15, and almost four times those of an average Chinese citizen.

The EU's climate change policies have been shaped by a combination of domestic policy measures, bargaining among EU member states, and the efforts of the European Commission and the European Parliament to stimulate policy change.[24] A shifting group of EU states took the lead in developing progressive national environmental policies and promoting a more aggressive EU climate change policy.[25] In turn, EU climate change policies have accelerated the development of national policies, even in the pioneer states.[26] The Commission also played a central role in working out an EU negotiating strategy for Kyoto and in developing a burden-sharing agreement for implementing emissions reductions, which set individual targets for each EU member state based on their different economic and geographic profiles. However, member states differ greatly in their progress toward their targets.

Since 2004 twelve more states have become EU members: Poland, Hungary, the Czech Republic, Slovakia, Estonia, Latvia, Lithuania, Slovenia, Cyprus, Malta, Bulgaria, and Romania. New EU entrants were added to the burden-sharing arrangement, even as their accession increased incentives for EU members to invest in greenhouse gas reduction measures in these neighboring countries, and to link the joint implementation mechanism to the EU's greenhouse gas trading scheme. Enlargement has thus increased the ability of the EU to assert global leadership on climate change.[27]

Continued EU action on climate change is guided by a regional political consensus that global average temperature increases should not exceed 2 degrees Celsius. In 2007 the European Commission proposed a cut in EU greenhouse gas emissions by 20 percent below 1990 levels by 2020, in an effort to contain global average temperature increase to no more than 2 degrees Celsius.[28] The leaders of all twenty-seven EU member states accepted this plan. The EU has since gone beyond this, adding a goal of achieving a 20 percent share of renewable energies in EU energy consumption by 2020. The combined goals are now referred to as the "20 20 by 2020" policy.[29] Since 2007 EU officials have pushed other developed countries to accept a similar policy. The EU Commission further stated that the EU would accept a 30 percent reduction below 1990 levels by 2020, if this goal was accepted by other industrial countries, including the United States.

The EU continues to support a legalistic approach to environmental policy based on multilateral legal agreements. The Bush administration, in contrast, championed the idea of voluntary partnerships and programs. Transatlantic tensions around climate change issues remained quite high throughout the Bush administration, as manifest in a host of forums and across a wide range of issues.[30] The 2008 Group of Eight Summit achieved a rare transatlantic agreement around the idea that greenhouse gas emissions must decline by half by 2050—the first time President Bush accepted a long-term numerical reduction goal. However, no baseline year was established, and no shorter-term goals or actions were included in the declaration.

European policies have nevertheless been influential in the United States, as a host of municipal and state efforts have been shaped by European experiences and programs.[31] By 2007 California and EU climate policies were on several similar paths. Both EU and California state policies include short- and long-term caps on greenhouse gas emissions, efforts to establish low- or zero-carbon fuel standards, renewable portfolio standards, existing or developing cap-and-trade schemes for utilities and some industrial emissions, large-scale energy-efficiency programs and goals, and frequent calls for more stringent U.S. federal policies. All these efforts have attracted the attention of policymakers, private-sector actors, and nongovernmental organizations across North America, Europe, and parts of Asia.[32] The arrival of the Obama administration breathed new life into transatlantic climate change cooperation, but a long list of substantial differences persist between European and U.S. domestic policymakers, and in their negotiating positions at the 2009 Copenhagen negotiations. These differences, and the lack of U.S. support for any kind of binding agreement to follow the Kyoto Protocol, contributed substantially to the failure of the Copenhagen summit to produce agreement about global climate change institutions. U.S. and EU policies differ on how much and how quickly greenhouse gas emissions should reduced, and about how much international financial and other types of assistance should be made available to developing countries, along with a host of other issues. A complaint that one hears frequently from European climate change policymakers and scientific and technical experts is that U.S. policymakers simply do not take the catastrophic risks associated with climate change as seriously as do many in Europe.[33]

Because of the political and economic importance of Europe and North America, and the fact that they are large emitters of greenhouse gases, transatlantic relations are of significant importance for global climate change cooperation. In fact, no effective global climate change regime can be built without transatlantic cooperation. Certainly other large emitters in the developing world, including China, would be hard-pressed to make any meaningful mitigation commitments absent transatlantic agreement and, likely, incentives funded, at least in part, by such an agreement. Moreover, it is unclear that the EU is on track to reach its various policy goals for 2020,[34] or that the United States will seriously attempt to reduce its emissions during the same period. States and societies on both sides of the Atlantic will need to undertake difficult actions to address climate change causes and threats. Enhanced transatlantic cooperation could greatly facilitate their efforts, but such an agreement is by no means a foregone conclusion.

Climate Geopolitics and Transatlantic Connections

Climate change impacts, and their potential significance for transatlantic and global politics, raise a host of issues of broad concern to U.S. policymakers. Although the list of such possibilities is nearly endless, four merit special attention. First, because the United States has contributed significantly to the climate change problem and has, at least through 2010, been unwilling to incur mitigation costs or to help finance adaptation costs abroad, the risk that U.S. policies, social practices, and economic interests may be blamed for negative climate change impacts by political actors elsewhere, including Europe, remains significant. Perceptions of injustice remain crucial factors to environmental security assessments.[35] A greater willingness to engage in climate change mitigation and adaptation on the part of the United States may address this, but the general point holds in the coming years. As impacts become more apparent, calls for the United States to undertake more mitigation and to finance more mitigation and adaptation abroad are likely to become more insistent. This raises the prospect that other governments and international organizations may seek to link issues of ostensibly greater importance to the United States, such as terrorism and international money laundering, to its cooperation with respect to climate policy.

The perception that the United States is largely indifferent to climate change beyond its borders is well established in the minds of many citizens and political actors around the world already. It will be difficult to dislodge, especially if blame for the 2008–9 global financial crisis is also attributed to the United States, or if the American response to that crisis is thought to display the sort of indifference to global opinion that has also characterized its climate policy. Furthermore, if the EU or the Chinese can sustain or increase the assistance they are able to offer the developing world with respect to climate issues, this may give them a new and potent form of "soft power" leverage vis-à-vis the United States. European policymakers are already aware that their larger foreign assistance budgets offer such opportunities. Chinese assistance to a number of African states, particularly as it relates to Chinese access to oil and other resources, also illustrates the potential importance of this dynamic.

Second, the combined adaptation and mitigation costs in Europe and for other high-income U.S. allies like Japan, Australia, and New Zealand are liable to rise substantially over the next twenty-plus years. Given this, and combined with the costs of aging populations, prospects for greater European or Japanese defense spending—a U.S. foreign and security policy priority—dim further. These cost and spending issues raise a larger point about the ability of high-income societies to sustain even relatively low levels of foreign assistance in the face of growing domestic climate change costs. One of the paradoxes that surround climate politics is that an issue that would seem to place an absolute premium on international cooperation may well inspire increasingly self-regarding policies instead. Similarly, the security implications of high-consequence, low- (or unknown-) probability events—such as changes in marine currents and accelerated glacial melting—may need to be assessed beyond their direct material effects. There is good reason to suppose that, in Europe and elsewhere, the lion's share of blame for such developments, if they occur, will be laid at the feet of the United States.

Third, as climate change impacts become more apparent and costly across Europe—from its polar regions to it water-scarce Mediterranean climes—EU members might be expected to enhance their support for strong EU global climate change leadership. Up to now Eastern and Southern European member states have often supported EU climate policies somewhat reluctantly. As climate impacts become more severe, it seems likely that public and official opinion in these states will become significantly more stringent and demanding, if only because their reliance on outside assistance for mitigation and adaptation will be high. The general pattern in the EU is that its domestic and foreign policy agenda is driven by those members with the greatest stake in particular issues. If impacts accelerate, EU states as a group may be expected to push harder for non-EU states to do more to mitigate climate change.

As a related point, if the United States remains a policy laggard with respect to climate change, the most adversely affected EU members might be expected to become less amenable to cooperation with the United States on a variety of other issues. At some point the climate change issue may become sufficiently important in Europe to inspire more "issue linkage" strategies toward the United States and other major greenhouse gas emitters that are not perceived to be doing their share to address the problem. If the gap between U.S. and European environmental policies remains wide, calls to reduce cooperation on issues that American policymakers care most about—security and trade, for example— could become increasingly difficult for European governments to resist.

Fourth, if climate-related economic and social problems do emerge in states to the east and south of the EU's borders, European political actors at various levels of authority are likely to call for greater securitization/militarization of those borders. Such calls can be expected to focus on Russia and North Africa in particular, and to further complicate Russian, and possibly Turkish, relations with Western Europe, NATO, and the United States.

Conclusions: Transatlantic and Global Climate Geopolitics

With respect to climate policy, the United States is generally unwilling to act globally before it has acted domestically, or to employ international standards as leverage for shifting domestic policy toward more aggressive action.[36] Nevertheless, innovative policies developed at the local and regional levels, and in the private sector, may develop sufficient political momentum to push U.S. national policy closer to that of Europe.[37] If this occurs, the EU and the United States may find it easier to cooperate in addressing environmental threats.

Nevertheless, transatlantic climate politics appear likely to remain contentious. The EU appears poised to sustain global leadership on the issue, possibly adding to the global sense that the United States is to blame for accelerating climate change and energy-related inequalities. Even as American opinion begins to catch up, European opinion is liable to surge ahead, owing to its more direct exposure to a variety of environmentally engendered risks. The long-term influence of economic conditions, and of agreements over mitigation/ adaptation financing, remain to be seen. Although the implications of climate change in Europe do not at present appear to exceed the adaptive capacity of European society and institutions, they are serious nonetheless, and they are likely to increase in cost over time. It seems certain that the politics of climate change will retain a prominent place in European public life. The implications of climate change for a host of states close to Europe are adding to a growing sense of concern there, which in turn will react upon European relations with the United States.

Notes

1. Miranda Schreurs, *Environmental Politics in Japan, Germany, and the United States* (Cambridge: Cambridge University Press, 2002); Miranda Schreurs, "The Climate Change Divide: The European Union, the United States, and the Future of the Kyoto Protocol," in *Green Giants? Environmental Policy of the United States and the European Union*, edited by N. J. Vig and M. Faure (Cambridge, MA: MIT Press, 2004); Miranda Schreurs, "Environmental Policy-Making in the Advanced Industrialized Countries: Japan, the European Union and the United States of America Compared," in *Environmental Policy in Japan*, edited by H. Imura and M. A. Schreurs (Cheltenham, U.K.: Edward Elgar, 2005); Miranda Schreurs, "Global Environment Threats and a Divided Northern Community," *International Environmental Agreements: Politics, Law, and Economics* 5 (2005): 349–76; Miranda Schreurs, Henrik Selin, and Stacy D. VanDeveer, "Conflict and Cooperation in Transatlantic Climate Politics: Different Stories at Different Levels," in *Transatlantic Environment and Energy Politics: Comparative and International Perspectives*, edited by Miranda Schreurs, Henrik Selin, and Stacy D. VanDeveer (Farnham, U.K.: Ashgate, 2009); Joshua Busby and Alexander Ochs, "From Mars and Venus Down to Earth: Understanding the Transatlantic Climate Divide," in *Climate Policy*

for the 21st Century: Meeting the Long-Term Challenge of Global Warming, edited by D. Michel, (Washington, DC: Center for Transatlantic Relations, Johns Hopkins University School of Advanced International Studies, 2004); Joseph F. Jozwiak Jr. and Patrick M. Crowley, "Comparing the EU and NAFTA Environmental Policies: Comparative Institutional Analysis and Case Studies," in *Crossing the Atlantic: Comparing the European Union and Canada*, edited by G. Bouchard, P. Bowles, and W. Chandler (Farnham, U.K.: Ashgate, 2004); and Loren R. Cass, *The Failures of American and European Climate Policy: International Norms, Domestic Politics, and Unachievable Commitments* (Albany: State University of New York Press, 2006).

2. Sabine Castelfranco, "US Rejection of Kyoto Protocol Brings Protest in Italy," Voice of America, February 15, 2005; Henrik Selin and Stacy D. VanDeveer, "Global and Continental Governance Challenges and Opportunities," in *Greenhouse Governance: Addressing Climate Change in America*, edited by Barry Rabe (Washington, DC: Brookings Institution Press, 2010), 313–35; Stacy D. VanDeveer and Henrik Selin, "Multilevel Governance and Transatlantic Climate Change Politics," in *Greenhouse Governance*, ed. Rabe, 336–52.

3. Kurt Campbell, ed., *Climate Cataclysm: The Foreign Policy and National Security Implications of Climate Change* (Washington, DC: Brookings Institution Press, 2008); Center for Naval Analysis Corporation, *National Security and the Threat of Climate Change*, 2007, http://securityandclimate.cna.org/report/; Thomas Fingar, "National Intelligence Assessment on the National Security Implications of Global Climate Change to 2030," Statement for the Record, House Permanent Select Committee on Intelligence, House Select Committee on Energy Independence and Global Warming, June 25, 2008, www.dni.gov/testimonies/20080625_testimony.pdf; National Intelligence Council, *Global Trends 2025: A Transformed World* (Washington, DC: National Intelligence Council, 2008), www.dni.gov/nic/PDF_2025/2025_Global_Trends_Final_Report.pdf; and Peter Schwartz and Doug Randall, *An Abrupt Climate Change Scenario and Its Implications for United States National Security*, Global Business Network Report, 2003, www.gbn.com/articles/pdfs/Abrupt%20Climate%20Change%20February%202004.pdf.

4. Swedish International Development Cooperation Agency, *A Climate of Conflict*, February 2008, www2.sida.se/shared/jsp/download.jsp?f=A+Climate+of+Conflict.pdf&a=36114; German Federal Ministry for Economic Cooperation and Development, *Climate Change and Security: Challenges for German Development Cooperation*, 2002, www.crid.or.cr/digitalizacion/pdf/eng/doc17309/doc17309-0.pdf; and Schubert, H., J. Schellnhuber, N. Buchmann, A. Epiney, R. Grießhammer, M. Kulessa, D. Messner, S. Rahmstorf, and J. Schmid, *World in Transition: Climate Change as a Security Risk*, Report from German Advisory Council on Climate Change (Earthscan: London, 2008), 144, www.wbgu.de/wbgu_jg2007_engl.html.

5. European Commission, *Climate Change and International Security*, paper from High Representative and European Commission to European Council, March 14, 2008, www.docstoc.com/docs/6483935/climate-change-and-international-security; and European Commission, *Climate Change and Security: Follow-up Recommendations by EUHR Solana*, December 18, 2008, www.europa-eu-un.org/articles/en/article_8382_en.htm.

6. See Geoffrey D. Dabelko, "Tactical Victories and Strategic Losses: The Evolution of Environmental Security" (PhD diss., University of Maryland, College Park, 2003).

7. Joseph Alcamo, Jose M. Moreno, and Bela Novaky, "Europe," in IPCC, *Climate Change 2007: Impacts*, 541–80; and Schubert et al., *World in Transition*.

8. More than fifty structured interviews conducted between 2003 and 2006 by Stacy D. VanDeveer, JoAnn Carmin, and a team of research revealed that very few public officials or civil society actors viewed the increased incidents in flooding since the 1990s as a product of climate change; project funded by National Science Foundation grant 0303720.

9. Netherlands Environmental Assessment Agency, *The Effects of Climate Change in the Netherlands*, 2005, www.mnp.nl/bibliotheek/rapporten/773001037.pdf.

10. JoAnn Carmin and Stacy D. VanDeveer, eds., *EU Enlargement and the Environment: Institutional Change and Environmental Policy in Central and Eastern Europe* (London: Routledge, 2005).

11. Center for Naval Analysis, *National Security and Climate Change*.

12. European Commission, *Climate Change and International Security* and *Follow-Up Recommendations by EUHR Solana*; Stacy D. VanDeveer, "Environmental Security: Conceptual Contestation and Empirical Relevance in the Mediterranean," in *Security and the Environment in the Mediterranean: Conceptualizing Security and Environmental Conflicts*, edited by H. G. Brauch et al. (Berlin: Springer-Verlag, 2003), 455–67.

13. "France Warns over African Migrants," BBC, July 10, 2006, http://news.bbc.co.uk/2/hi/africa/5164514.stm.

14. Fabrice Renaud, Janos J. Bogardi, Olivia Dun, and Koko Warner, *Control, Adapt or Flee: How to Face Environmental Migration?* (Tokyo: United Nations University Press, 2007), www.reliefweb.int/rw/lib.nsf/db900SID/PANA-7D5DMM?OpenDocument.

15. Margaret Beckett, "Foreign Policy and Climate Security," speech by the U.K. foreign secretary, Berlin, October 24, 2006, summarized at www.politics.co.uk/news/foreign-policy/international-development/debt-and-debt-relief-in-developing-world/climate-change-serious-threat-global-security-$455615.htm; German Federal Ministry for Economic Cooperation and Development, *Climate Change and Security*; and Norwegian Foreign Ministry, *A More Ambitious Norwegian Development Policy*, February 17, 2009, www.norway-un.org/News/Latest+news/ambitiousdevelopment_170209.htm.

16. European Commission, *Follow-Up Recommendations.*

17. R. B. Alley, J. Brigham-Grette, G. H. Miller, L. Polyak, and J. W. C. White, *Past Climate Variability and Change in the Arctic and at High Latitude*, U.S. Geological Survey Synthesis and Assessment Product 1.2, January 16, 2009, www.climatescience.gov/Library/sap/sap1-2/default.php.

18. Alan Dupont, "The Strategic Implications of Climate Change," *Survival* 50 (2008): 29–54; Cleo Paskal, *How Climate Change Is Pushing the Boundaries of Security and Foreign Policy*, Briefing Paper (London: Chatham House, 2007), www.chathamhouse.org.uk/research/eedp/papers/view/-/id/499/.

19. R. B. Alley et al., "Abrupt Climate Change," *Science* 299 (2003): 2005–10.

20. K. M. Walter et al., "Methane Bubbling from Siberian Thaw Lakes as a Positive Feedback to Climate Warming," *Nature* 443 (2006): 71–75.

21. Natalia M. Shakhova et al., "Why Should the East Siberian Shelf Be Considered a New Focal Point for Methane Studies in Terms of Global Climate Change?" *American Geophysical Union Fall Meeting Abstracts*, http://adsabs.harvard.edu/abs/2008AGUFM.U23D0081S.

22. European Environment Agency, *Greenhouse Gas Emissions Trends and Projections in Europe 2009*, EEA Report 9/2009, www.eea.europa.eu/publications/eea_report_2009_9.

23. U.S. Environmental Protection Agency, "Inventory of U.S. Greenhouse Gas Emissions and Sinks: 1990–2007," http://epa.gov/climatechange/emissions/usinventoryreport.html

24. Paul G. Harris, *Europe and Global Climate Change: Politics, Foreign Policy and Regional Cooperation* (Cheltenham, U.K.: Edward Elgar, 2007).

25. Miranda Shreurs and Yves Tiberghien, "Multi-Level Reinforcement: Explaining EU Leadership in Climate Change Mitigation," *Global Environmental Politics* 7, no. 4 (November 2007): 19–46.

26. Mikael Skou Andersen and Duncan Liefferink, eds., *European Environmental Policy: The Pioneers* (Manchester: Manchester University Press, 2000).

27. Carmin and VanDeveer, *EU Enlargement and the Environment.*

28. Commission of the European Communities, *Limiting Global Climate Change to 2 Degrees Celsius: The Way Ahead for 2020 and Beyond*, January 10, 2007, http://eur-lex.europa.eu/LexUriServ/site/en/com/2007/com2007_0002en01.pdf.

29. European Commission, *Follow-Up Recommendations.*

30. Schreurs, Selin, and VanDeveer, "Conflict and Cooperation in Transatlantic Climate Politics."

31. Selin and VanDeveer, "Political Science and Prediction"; and Henrik Selin and Stacy D. VanDeveer, *Changing Climates in North American Politics* (Cambridge, MA: MIT Press, 2009).

32. Selin and VanDeveer, "Political Science and Prediction"; VanDeveer and Selin "Global and Continental Governance Challenges and Opportunities."

33. This perceived difference of outlook among policymakers on either side of the Atlantic was a topic of discussion at, for example, workshops and conferences including the following: "A Low Carbon, Alternative Fuels Future: Perspectives from Europe and the Americas," Florida International University, Miami, March 13, 2009; "Governing the Climate: Lessons from the National Conference on Climate Governance," Woodrow Wilson International Center for Scholars, Washington, January 12, 2009; "Global Jean Monnet Conference 2007: The European Union and World Sustainable Development," Brussels, November 5–6, 2007; and "Climate Change: Science, Politics and the Management of Uncertainty," Merton College, Oxford, September 21, 2007.

34. European Environment Agency, *Greenhouse Gas Emissions Trends.*

35. Chad Briggs, "Post-Conflict Environmental Health Risk: The Role of Risk Analysis in Foreign Policy," in *Energy and Environmental Challenges to Security*, edited by S. Stec and M. Baraj (London: Springer, 2009).

36. Elizabeth R. DeSombre, "Understanding United States Unilateralism: Domestic Sources of U.S. International Environmental Policy," in *The Global Environment: Institutions, Law, and Policy*, edited by R. Axelrod, D. Downie, and N. J. Vig (Washington, DC: CQ Press, 2005); and Selin and VanDeveer, "Political Science and Prediction."

37. Selin and VanDeveer, "Political Science and Prediction."

12

Turkey

Ibrahim Al-Marashi

Greenpeace, the international environmental activist organization, is leading an effort to reconstruct Noah's Ark in the mountains of eastern Turkey. According to the Bible, the Ark, which Noah built to save humans and animals alike from a great flood, is believed to have finally rested on Mount Ararat in eastern Turkey.[1] Although global warming is unlikely to lead to a new flood of biblical proportions, the location of the Ark's final resting place is symbolic of Turkey's centrality to the interaction of climate change, international security, and regional stability. Turkey straddles the Balkans and the Caucasus, the Black, Aegean, and Mediterranean seas, and the Upper Mesopotamian and Tigris basins. Whatever impact climate change may have in Turkey by 2030 is certain to have repercussions for the region and the entire world.

This chapter assesses the effects on anticipated climate change in Turkey, based on three principal sets of sources. First, it examines one of the most widely read daily newspapers in Turkey, *Sabah* ("Morning"). The role of the news media in shaping perceptions of global warming is substantial. According to one source, "News reports on such matters may influence public and elite opinion and, in democracies, the policy positions of national governments. The extent to which press coverage of global warming issues varies from country to country may affect international agreement or disagreement on collective efforts to solve these problems."[2]

Turkey

This chapter argues that *Sabah* and other major media outlets will help to set the policy agenda in this area, by bringing global warming to the public's attention and educating the public about the issue at the same time, a trend that will inevitably impose itself on official opinion and the policymaking process.

The second set of sources are concerned with the military. Throughout the development of the Turkish Republic, the military played a surprisingly important role in this area, as it did in many other aspects of Turkish domestic politics. The Turkish military is a much-esteemed public institution, and it is widely considered the ultimate guarantor of the state's secularism. A note on the military's website, for instance, was enough to communicate its institutional displeasure when the religious, conservative Adalet ve Kalkinma Partisi (AKP, or AK Party) in Turkey tried to nominate Abdullah Gul for the nation's presidency in the spring of 2007. The military carried out what was termed as an "e-coup" by trying to block this motion. Although the military's power has been challenged by the AKP government, understanding how the Turkish military conducted political communication is crucial to understanding how the perceptions of climate change in Turkey would affect the state's perspectives on internal and external security.

There is every reason to suppose that the military will play a decisive role in public and intragovernment debates that arise about climate change. Indeed, it is already doing so. The worst-case scenarios that this chapter surveys have already been communicated to the Turkish public by the military. Along with the private media, the Turkish military plays an independent, agenda-setting role by making public assessments of the effects of climate change and by using the mass media to circulate these assessments.

Finally, the third set of sources this chapter analyzes are publications issued by Turkish ministries in collaboration with international organizations such as the United Nations. The Turkish government has been proactive in assessing the risks of climate change, and it has issued reports that aim to educate the Turkish public and to alert the international donor community to the threat posed by global warming to a country whose political and geographic position ensures that its problems cannot remain of merely local concern.

Climate Impacts on Sociopolitical Change

Serious environmental stress in Turkey in the upcoming decades may lead to both civil conflict and external aggression, in addition to population movements. Internally, rising temperatures will have their most adverse affect on the southeastern part of the country, where temperatures already tend to be the highest. This strip includes the low-lying areas running from the base of Diyarbakir, Urfa, and Mardin to the Silopi regions bordering Iraq and Syria. These areas are inhabited by concentrations of Turkey's Kurdish population. The fighting between Turkish security forces and the Kurdistan Workers' Party (Partiya Kar-kerên Kurdistan, PKK) in the 1980s contributed to internal displacement and the migration of rural populations to cities such as Istanbul, Ankara, and Izmir. These internally displaced persons would often take their sympathies or loyalties to the PKK with them to these urban centers. Thus, continued rural–urban migration due to climate change has the potential of aggravating political tensions within Turkey's major cities.

Populations displaced by global warming have been referred to as "climate refugees." The sociologist Thomas Faist refers to climate refugees in the Turkish context: "The difference between good and bad political management can be seen in Turkey. In the west, where land reform has been under way for decades, the farming sector has flourished; people can support themselves and export their products. In eastern Turkey, where most farmland still belongs to a handful of huge landlords, productivity is low, poverty is high, and many people are leaving for the cities."[3] According to Faist, western Turkey is in a better position to mitigate the effects of climate change than is the east.

Climate change may lead to both local conflicts over land and rural–urban migration, due to the collapse of agricultural and nonindustrial economies in the periphery. The "fight" element is especially likely arise due to water scarcity, resulting in competition for water between agriculturalists in the interior, who need water mainly for irrigation, which competes with the fast-growing domestic and industrial demand for freshwater in the coastal zone.

Geographic Vulnerabilities

The low-lying coastal areas of Turkey running from the Aegean to the Mediterranean will be affected by climate change, in particular the tourist sectors, which will suffer from rising water levels. Here again, population can be expected to move from the scenic rural coastal areas to the urban centers, as coastal inhabitants seek better employment.[4]

The Turkish debate on geographic vulnerabilities to climate change is mixed. Some Turkish media exaggerate or sensationalize this issue to attract audiences, while others fail to address it at all due to their dependency on advertising revenues from Turkish petroleum and auto industries. The *Sabah* newspaper has performed a watchdog function in this regard, by using climate change as a means to criticize the AKP government for its alleged inaction in responding to failing crop yields.[5] Other articles have criticized the Turkish political parties' neglect of global warming in their electoral campaigns in the summer of 2007.[6]

At times *Sabah* exaggerates the issue. In one article in September 2007, the author quotes the Quran as foretelling a water conflict between different states in 2010.[7] For the most part, however, *Sabah* has represented a responsible voice of criticism and information. It has, for instance, examined scenarios under which specific geographic areas are likely to suffer due to climate change. One article considered the impact of the drop in the level of lake waters on water life, and it noted that one result would be a reduction in the supply of freshwater fish, which provide a livelihood for local fisherman. Coastal areas, which are home to more than 30 million people, would also suffer. On the seas, climate change is expected to affect the availability of seafood, disrupting the economies of twenty-seven urban centers located on the coast. By way of illustration the article notes how climate change has already disrupted the movement of anchovies to the Black Sea, which are decreasing due to changes in atmospheric rhythms, winds, and currents.[8]

Other studies examine how increasing erosion has led to a substantial loss of topsoil, reducing agricultural output and raising the price of food.[9] The Turkish corn, barley, rye, oats, and sunflower harvests have decreased dramatically on a yearly basis due to increasing temperatures, longer summers, and decreasing water supplies. The production of olives and grapes on the Aegean Coast has also suffered, as has water-intensive cotton production.[10] Recognizing these regional variations is especially important because they are likely to be concealed within national averages; precipitation decreases along Turkey's Aegean and Mediterranean coasts, for instance, are believed to be linked with increases in the Black Sea and Caucasian coastal regions.[11]

A UNDP study, conducted in collaboration with the Turkish government and released in 2007, made several dire assessments of the ramifications of climate change.[12] According to the study, a rise in sea level is expected to result in erosion, flooding, the inundation of coastal lowlands, and salt water intrusion. A sea-level rise averaging 4–8 millimeters a year will be sufficient to threaten fertile land in settled areas, along with engineering structures in coastal areas, including roads. Its preliminary assessment of the impact of coastal erosion along Turkish shorelines places about 6 percent of Turkey's gross national product at risk of total loss, and estimates that about 10 percent of gross national product would be needed for environmental adaptation and protection nationwide. Istanbul's major freshwater lakes—including Lake Terkos near the coastline of the Black Sea, and the Haliç estuary that separates Istanbul's historic old town from its business district—are all vulnerable to salinization. Cultural and historical sites along the Bosporus in Istanbul would also be affected by the projected rise in sea level.

The same UNDP-Turkish joint study estimates that 86.5 percent of Turkey's total land area is vulnerable to desertification, and 73 percent of its arable land is at risk of erosion, land degradation, and desertification. Southeastern Anatolia and the interior, which are already relatively arid, are more prone to desertification, followed by the Mediterranean and Aegean regions. Desertification is also confirmed in a report on the Middle East by the U.S. National Aeronautics and Space Administration (NASA), which predicted that current global warming trends will turn most parts of Turkey into a desert. The report forecast that rising global temperatures will result in increased forest fires and droughts. The NASA report uses 2030 as a future date to project other effects in Turkey. By 2030 NASA predicts that most parts of Turkey will experience temperatures averaging around 40 degrees Celsius.[13]

Because water will most likely become a scarce commodity, one cannot discount the emergence of an illegal trade in water supplies. A black market water economy could emerge with nonstate actors managing exports and cross-border smuggling of siphoned water. Such a market could lead to a violent struggle for the control of these black market networks. The Iran/Iraq/Turkey border remains porous, allowing unregulated traffic from Iran into Turkey and vice versa. On the basis of Turkey's attempts to control the smuggling of narcotics across its borders, one can assume that it will find it equally difficult to eradicate professional networks seeking to smuggle water out of the country.

Institutional Capacity and Climate Science in Turkey

The Turkish authorities are aware of the challenge posed by climate change, and they have thus taken preliminary steps to address it. In February 2007 Turkey's legislature, the Grand National Assembly, began articulating a state-sponsored initiative to research the effects of global warming on the nation. Efforts have been made to develop legal frameworks to address climate change and to conserve water through better pipeline systems. The Assembly also established a Research Commission, to consist of fourteen members of Parliament who will provide input as the country attempts to adapt to climate change.[14] In the same month, the minister of the environment and forestry, Osman Pepe, the minister of energy and natural resources, Hilmi Guler, and the minister of agriculture and village affairs, Mehdi Eker, prepared a plan of action to deal with global warming.[15]

The Ministry of the Environment and Forestry has also worked with the UNDP to assess the effects of climate change. It has sought to educate the public and raise awareness in Turkish society about climate change, and to encourage the adoption of environmentally friendly technologies. The Turkish ministries of energy, transportation, agriculture, and health have also coordinated with the Ministry of the Environment in these efforts. Recommendations have been made to establish a Scientific Steering Committee on Climatic

Issues and a mechanism for monitoring the National Action Plan. The Ministry of the Environment also plans to establish a Coastal Zone Department for Environmental Impact Assessment and an Authority for the Protection of Special Areas.

The Regional Environmental Center (REC) Turkey has also initiated a Climate Change Project.[16] The Ministry of the Environment and Forestry has supported the REC under Article 6 of the UNFCCC, to provide education, training, and public awareness on a national level. The REC publishes Çevre, a climate-change-related publication in Turkish.[17] It has also held workshops and technical training courses for government, scientific, and business institutions. The REC also provides information on climate change activities in Turkey on its website (www.iklimnet.org).

Other organs affiliated with the state have been tasked with studying climate change. The Supreme Council for Science and Technology, the highest scientific decision-making institution in Turkey, has held meetings to study global warming, scenarios of climate change, and scientific and technological research programs to deal with the issue.[18] TUBITAK, the government academic funding body, has provided funds to conduct several studies on climate change in Turkey.[19] The Technical University in Istanbul and Dokuz Eylül University in Izmir, both state owned, have been conducting research on best- and worst-case scenarios for a 2 to 3 degree Celsius increase in temperature. Other studies have analyzed the emergence of diarrhea as a summer illness in Turkey. Because water shortages are a symptom of climate change, inhabitants of the interior have been using unhealthy water that has proven to be especially dangerous for children.[20] The Turkish Union of Chambers and Commodities has also urged industrialists to cooperate with the UNDP's efforts to adopt more stringent environmental standards, an effort related to ensuring that these companies can meet EU environmental standards in the future.[21]

In judging the likely efficacy of all these efforts, however, some account must be taken of the tendency toward venality that afflicts Turkey's public institutions. Climate change may result in increased corruption in the ministries that deal with water and environmental resources (the ministries of energy, transportation, agriculture, health, and environment and forestry). Projects designed to rehabilitate the nation's hydropower infrastructure may be hampered due to bribery and kickbacks prevalent in the various ministries responsible for this effort. In the worst case, scarcity of water could result in a clientist shadow state within the Turkish government, based on the patronage of water resources. Local parties or government could also allocate water based on patronage as a way to cement the loyalty of local constituents. Both phenomena have ample historical precedent and could become prevalent in Turkey regardless of the strength of the state at the center. Private networks to distribute water resources based on tribal or business connections may also emerge to compete with those based upon public institutions.

International Cooperation

International cooperation in environmental matters is generally not seen as a challenge to sovereignty in domestic Turkish politics. Turkey is a member of the Organization for Economic Cooperation and Development and of NATO, and a prospective member of the European Union, and thus it is amenable to cooperative international efforts.

Turkey delivered its First National Communication pursuant to the UNFCCC in February 2007, detailing how the country is affected by climate change, especially in the form of water resource shortages and drought, and related effects on the agricultural and health sectors of the economy. The study evaluated alternative energy scenarios, preventive measures to reduce greenhouse gas emissions, and how to raise public awareness. More than twenty Turkish research institutes and a hundred researchers and experts contributed to the preparation of the report.[22] Turkey is also a signatory to the United Nations Convention to Combat Desertification.[23] It has cooperated with the IPCC and has allowed it to raise awareness of global warming among the Turkish public.

The Turkish REC has implemented climate awareness projects with the support of national and international partners and international donors. The REC in Turkey is part of a network of seventeen country offices of the REC for Central and Eastern Europe, headquartered near Budapest. The REC was established in 1990 by the United States and the European Commission as an international nongovernmental organization that coordinates its efforts with local nongovernmental organizations, governments, and businesses, as it seeks to affect and support environmental decision making.

In April 2008, Spain and the UNDP granted $7 million to develop Turkey's capacity to counter global warming. Hasan Sarikaya, undersecretary of the Environment and Forestry Ministry, said that the funding would go to projects such as integrating developments regarding climate change to all areas, generating adjustment projects in pilot regions and then spreading them throughout the country, and enhancing the capacity to adapt to climate change.[24]

Nevertheless, Turkey has not signed the Kyoto Protocol, and it did not adopt a greenhouse gas emissions limit or reduction target in the first commitment period of the protocol. Its Research Commission was merely "expected to encourage a new Turkish approach to the first and consecutive commitment periods of the Kyoto Protocol."[25] In 2006 the minister of the environment and forestry, Osman Pepe, said that Turkey had not signed the protocol because "we are not the responsible side of the global warming. When you examine the carbon dioxide emission in developed countries, it is seen that Turkey is very much innocent on that. So Turkey can't sign the Kyoto Protocol like some African countries which immediately signed it."[26] In 2008 his successor, Veysel Eroglu, said that Turkey was not against the fundamental principles of the protocol: "We say Turkey can be a party of Kyoto. Talks are under way." He further added, "We support them but we also want global justice." When referring to "global justice," he added, "if a measure will be taken, then the whole world should take it. It is wrong to exclude several countries from the list, and impose obligation to some other countries."[27]

Security Scenarios

Turkey's overriding security concern has always been to preserve its territorial integrity. A constant worst-case scenario is the potential emergence of a separatist Kurdish entity in the southeastern part of the country. The first serious challenges to the Turkish state in its formative years were rebellions among its Kurdish population, which have been echoed in the recent fifteen-year struggle with the PKK.

The challenges posed by climate change connect to Turkey's traditional security preoccupations primarily via its fight against the PKK. That struggle is contingent on an issue that is rarely analyzed in connection with this conflict—water. The emergence of the PKK in 1984 can be partially attributed to efforts by Saddam Hussein and Hafiz al-Asad to support it as a means of putting pressure on Ankara to increase the downstream flow of the Tigris and Euphrates rivers, on which Iraq and Syria are highly dependent.[28] Iraq's policy of supporting the PKK in the past was tied to Baghdad's need to maintain some kind of leverage over Turkey in negotiations over water rights.

Iraq's support for the PKK ended after the fall of the Ba'athist government in 2003, and Syria's ties with the PKK were severed in 1998, when the Turkish military mobilized along the border. Under the AKP government, relations between Ankara and both Baghdad and Damascus have improved substantially.

As both Turkey's and Iraq's populations grow, the ability to manage their shared water supplies becomes more important. The Tigris and Euphrates rivers originate in Turkey, and as a result Iraq is dependent on its northern neighbor for maintaining the flow of both rivers. The Turkish Southeastern Anatolian Dam Project (known as GAP, its Turkish acronym) has caused some concern in Iraq, because it enhances Turkey's ability to control both rivers.[29] By 2030 diplomatic tensions, and even a military confrontation over these water resources, remain a possibility in the future, regardless of who governs in Baghdad.

A United Nations report, issued in March 2007, refers to the water wars in the Middle East, and it warns that the water available per person in Turkey in 2025 will be below 1,000 cubic meters per year, making Turkey "a poor water country." The report also warns that Turkey's dams may be targeted by Iraq and Syria, which will suffer serious water shortages by 2040.[30]

One element that has to be taken into account in answering these questions is what might be called "hydronationalism." An example of hydronationalism in Turkey was afforded by the inauguration of the Ataturk Dam in 1992, at which former prime minister Süleyman Demirel declared, "The water resources are Turkey's. The oil resources are theirs [the Arabs]. We don't say we share their oil resources; and they cannot say they share our water resources."[31] In this instance, water as a resource underlined the ethnic- and identity-based differences between the Turks and Arabs. Relations in the twenty-first century have improved significantly between Ankara, Baghdad, and Damascus, and thus cooperation over shared water sources is a possibility. Yet at the same time, there is also a potential for a conflict over water resources to reemerge in the future.

In October 2005 Turkey's National Security Council, the supreme decision-making body regarding security affairs, included the water issue in the National Security Political Document, which stressed the importance of global warming. The Turkish military has conducted several studies on global warming, and it has come to the conclusion that "this will pose a potential threat to national security."[32] The Armed Forces General Staff headquarters and its Strategic Research and Study Center (known as SAREM) have been tasked with conducting studies on internal and external threats, which include global warming. The primary concern focuses on the demand for water and possible regional "water wars." According to these military studies, global warming "will have serious implications on security and 'water' may be a cause of war. Since then 'global warming' has been listed in the documents of the Turkish Armed Forces as a priority threat."[33]

The military also predicted by 2025 that Turkey, with the most water resources in the region, may be targeted by Syria and Iraq. This threat was highlighted in the July 2007 themed issue on climate change in Turkey's *Armed Forces Periodical* quarterly, published by the General Staff. One of the authors, the Middle East specialist Lieutenant Colonel Süleyman Ozmen, wrote an article titled "Effects of Global Warming on Security Policies of Turkey," dealing with a decline in living standards, increasing death rates and mass health problems, food shortages and an economic disaster, leading to general instability.[34]

To analyze the repercussions of environmental stress and conflict in Turkey, several possible scenarios can be offered. The first scenario is focused on the substate level and assumes that relations between Ankara and a Kurdish entity in northern Iraq have failed to stabilize by 2030. Almost any federal framework in Iraq would enhance the power of the

Iraqi Kurds, whose leadership likely will have some control over the Iraqi oil centers in Mosul and Kirkuk, and over the pipelines that run from there to Turkey, across the territory of the Kurdish Regional Government (KRG). Thus a dynamic emerges where Turkey is wary of an oil-rich Kurdish entity in the north of Iraq influencing separatist sentiment among their fellow Kurds across the border. As the supplies of water decrease in the north of Iraq, Turkey can control the flow of the Tigris River to pressure the KRG administration in Irbil. If the Kurdish parties seek full independence, which Ankara vehemently opposes, Turkey has the option of significantly reducing the flow of the Tigris. Although the KRG has access to the Zab River and its tributaries, its agriculture and water supplies are still primarily reliant on the Tigris, and hence it would have an incentive to stir up Turkey's Kurdish communities in retaliation.

Less alarming outcomes are also possible, however. It is not impossible that the common threat of water scarcity may bind the region together, as neighboring states seek to link themselves to Turkey's hydroelectric power resources in order to augment their own shrinking water supplies. Countries that may actively seek such links with Turkey include Iran, Syria, Jordan, Kuwait, Saudi Arabia, and the other Gulf countries. In the past the Turks have contemplated the construction of a water "peace pipeline" to Saudi Arabia, Kuwait, Bahrain, Qatar and the United Arab Emirates, as well as Jordan, Syria, Israel, and the Palestinian Authority, an idea whose revival might offer a path toward a more integrated regional order. Turkey's dependence on the Gulf for energy need not become the basis for some kind of resources standoff, as former prime minister Demirel's comments, quoted above, might suggest. It could instead contribute to the development of a mutually beneficial oil-for-water system in which all participants would have the kind of stake that would serve as a strong incentive for broader cooperation. Such optimistic outcomes are perhaps not the norm in the region, but they remain a possibility that should not be discounted out of hand.

Conclusion

The Turkish government has taken preliminary measures to deal with the actual and potential future effects of global warming, chiefly in cooperation with international organizations. These initiatives are still in the research phase, and it remains to be seen how effective the state will be in applying them in practice on a national basis. Turkish civil society has also been engaged in this effort. Public opinion and academia have both been mobilized to some extent, though not always in the direction of constructive solutions. The Turkish military, for its part, has adopted a relatively provocative stance in anticipating future threats arising from climate change, and it has generally placed its prestige on the side of worst-case scenarios and unilateral or autarkic solutions. Nevertheless, Turkey's relatively high level of international engagement, via its defense and economic ties to Europe and the United States, weigh on the other side of the questions, and ensuring that such engagement continues is probably the best way for the outside world to contribute to Turkey's search for solutions to its environmental problems. Only time will tell if Turkey will develop a regional, cooperative approach that will include its neighbors in a joint effort to mitigate global warming in the Middle East.

Notes

1. Daren Butler, "Noah's Ark Rebuilt to Show Climate Change Threat," Reuters, May 23, 2007, www .reuters.com/article/idUSL2242550220070523.

2. Brigitte L. Nacos, Robert Y. Shapiro, Natasha Hritzuk, and Bruce Chadwich, "News Issues and the Media: American and German News Coverage of the Global-Warming Debate," in *Decisionmaking in a Glass House: Mass Media, Public Opinion, and American and European Foreign Policy in the 21st Century*, edited by Brigitte L. Nacos, Robert Y. Shapiro, and Pierangelo Isernia (Lanham, MD: Rowman & Littlefield, 2000), 41.

3. Gerald Traufetter, "The Age of the Climate Refugees?" *Spiegel Online*, June 4, 2007, www.spiegel.de/ international/world/0,1518,druck-476062,00.html.

4. Istem Erdener, "Türkiye'de Küre Kaç Derece?" *Sabah* (Ankara), June 16, 2007, http://arsiv.sabah.com .tr/2007/06/16/ct/haber,FF25E82343F44700ABB2F8B479672C39.html.

5. Abdurrahman Yildirim, "Domates, biber apartmanlarını da görür müyüz?" *Sabah* (Ankara), 1 November 2007, www.merkezyayinholding.com.tr/2007/11/01/haber,CF571B39F78C47A5BF8364A64C62D063 .html.

6. Erdener, "Türkiye'de Küre Kaç Derece?"

7. "Küresel Isınma Mesajı," *Sabah* (Ankara), September 1, 2007, http://arsiv.sabah.com.tr/2007/09/01/ haber,6B5F7AFC0073498DB1E9905B5406DD99.html.

8. "Karadeniz Akdenizleşecek!" Ibid., May 18, 2007, http://arsiv.sabah.com.tr/2007/05/18/haber,90DDEF 811FA14689AAAC29A3C6B179C6.html.

9. Rahim Ak, "Küresel Isınma Artık Midemize Uzanıyor," ibid., September 17, 2007, http://arsiv.sabah .com.tr/2007/09/17/eko118.html.

10. Fatih Atalay, "Küresel Isınma Yüzünden Zeytin Ağacının Var Ve Yok Yılları Karıştı!" Ibid., June 24, 2007, http://arsiv.sabah.com.tr/2007/06/24/haber,43F2CFC9F6B2405D9475C4AE00A2AA48.html.

11. Ibid.

12. *New Horizons*, United Nations Development Program Turkey Monthly Newsletter, April 16 2007, www.undp.org.tr/Gozlem2.aspx?WebSayfaNo=866.

13. Evren Deger, "Army: New Threat Is Global Warming," *New Anatolian*, July 7, 2007, Open Source Center document GMP20070707744006.

14. "Turkey Moves to Address Climate Change," *Environmental News Service*, February 16, 2007, www .ens-newswire.com/ens/feb2007/2007-02-16-03.asp.

15. "Climate Changes, Drought and Water Management—Pepe: 'There Is No Drought Yet'"; and "Climate Changes, Drought and Water Management (2)—'Turkey Will Inevitably Sign the Kyoto Protocol in the EU Process,' Pepe," Anatolia News Agency, February 6, 2007.

16. Ibid.

17. Ibid.

18. "Supreme Council for Science & Technology to Discuss Global Warming," Anatolia News Agency, March 1, 2007.

19. Turkish Ministry of Environment and Forestry, *First National Communication to UNFCCC of Turkey on Climate Change*, January 2007, www.undp.org.tr/Gozlem2.aspx?WebSayfaNo=627.

20. "Küresel Isınma Akdeniz Havzasını Yakacak," *Sabah* (Ankara), October 17, 2007, http://arsiv.sabah .com.tr/2007/10/17/haber,6DCFEC76616B4A67A9F378F2475E99E9.html.

21. Caglar Guven, ed., *Climate Change & Turkey, Impacts, Sectoral Analyses, Socio-Economic Dimensions* (New York: United Nations Development Program, 2007); www.undp.org.tr.

22. Turkish Ministry of Environment and Forestry, *First National Communication*.

23. *New Horizons*, 16.

24. "Spain, UN Donate 7 Mln USD to Turkey to Fight Global Warming," Anatolia News Agency, April 3, 2008.

25. "Turkey Can't Sign Kyoto Protocol Immediately, Pepe," Anatolia News Agency, March 16, 2006.

26. Ibid.

27. "Environment Minister: Turkey Not against Kyoto Protocol," Anatolia News Agency, May 7, 2008.

28. Adel Darwish, "Troubled Waters in Rivers of Blood," *Mideast News*, December 3, 1992, www.mideast news.com/water004.html.

29. John Podesta and Peter Ogden, "The Security Implications of Climate Change," *Washington Quarterly* 31 (2007–8): 120–21.

30. Deger, "New Threat Is Global Warming."

31. Darwish, "Troubled Waters."

32. Deger, "New Threat Is Global Warming."

33. Ibid.

34. Evren Deger, "Military Issues Global Warming Warning," *New Anatolian*, September 10, 2007.

13

The Persian Gulf

Bahrain, Iran, Iraq, Kuwait, Qatar, Saudi Arabia, the United Arab Emirates, and Oman

James A. Russell

> Climate change is the defining human development issue of our generation. All development is ultimately about expanding human potential and enlarging human freedom. It is about people developing the capabilities that empower them to make choices and to lead lives that they value. Climate change threatens to erode human freedom and limit choice. It calls into question the Enlightenment principle that human progress will make the future look better than the past.
>
> *—United Nations Human Development Report 2007/2008:*
> *Fighting Climate Change—Human Solidarity in a Divided World, 7*

Nothing green grows on its own. As visitors to and inhabitants of the nations ringing the contemporary Persian Gulf can attest, the region's vast efforts to make its environment appear to be anything other than dusty and sand colored have required extraordinary interventions by its governments to create an artificial and human-made world that allows its inhabitants to escape from the effects of the inhospitable physical environment. The tree-lined highways and flower-dotted promenades of the cities on today's Persian Gulf are all testaments to long-standing efforts to mitigate the impact of an environment that is hotter and drier than anyplace on Earth. States have gone to extraordinary lengths in this quest, and the work continues unabated. For example, Saudi Arabia was recently reported to be carrying out only the latest in a series of experiments that will seed clouds with calcium chloride and silver iodide in an attempt to generate sorely needed rain over its arid landscape.[1]

As the world slowly but inexorably turns its attention to the daunting challenges of climate change and the specter of a future in which environmental issues may exert an increasing impact on regional security and stability, the Persian Gulf region offers lessons on the problems and prospects of adapting to and mitigating the effects of an already-hostile environment. Although the steel-and-glass towers of Dubai, Riyadh, and Doha represent the envy of developing countries around the world, their continuing viability to some extent depends on the planet's continuing environmental and economic folly. These cities live off of the ongoing expansion of world petroleum markets—markets that are dumping carbon emissions, largely from the developed world, into the atmosphere at ever-increasing rates. These emissions must be controlled if the world is to credibly address the inexorable march of climate change. The challenge of climate change is thus inextricably intertwined

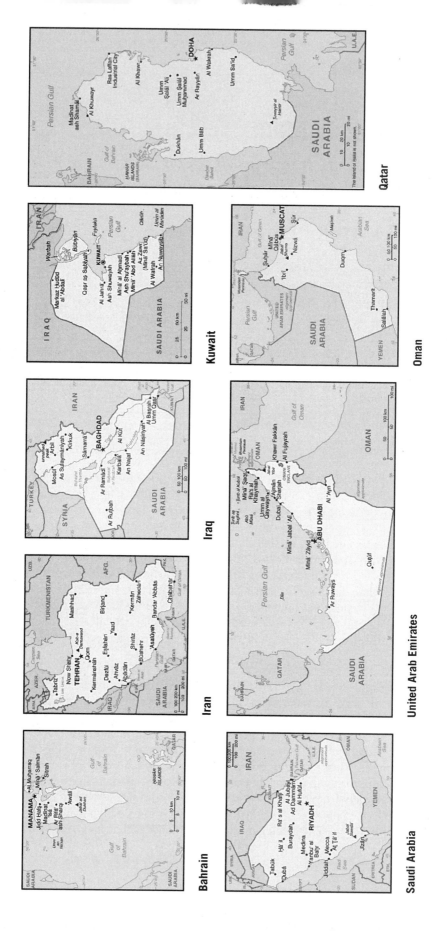

Qatar

Kuwait

Oman

Iran

Iraq

United Arab Emirates

Bahrain

Saudi Arabia

with the functioning of world petroleum markets, on which the states of the Persian Gulf depend for their environmental mitigation and adaptation efforts. If these efforts fail or are compromised, the societies in the Gulf region will be affected in a host of negative ways. Regional stability may surely be a casualty of this process.

This chapter assesses the challenge posed by climate change and environmental security to the Persian Gulf region, focusing on the states surrounding the Gulf: Iran, Iraq, Saudi Arabia, and the smaller states of Bahrain, Kuwait, Qatar, the United Arab Emirates (UAE), and Oman. Each of these states faces similar challenges. All largely depend upon oil and gas markets to fuel their economic growth. These markets have also provided these states with the means to delay political reforms while they maintain anachronistic forms of government. The cascades of cash dumped into the coffers of the region's elites by their customers have, in turn, allowed these elites to buy off their populations with expensive and inefficient subsidy and welfare programs—not to mention air conditioning. In the Gulf, the security of the governing elites is thus inextricably intertwined with the orderly functioning of international energy markets. These market functions must be addressed as a global consensus slowly coalesces around courses of action to address climate change. The chapter summarizes the challenges facing the region's ruling elites as they seek to continue their grip on political power while simultaneously dealing with the politics of climate change and the severe environmental stresses throughout the Gulf region.

Environmental Stress and Climate Change

That environmental stresses strike hard in the Persian Gulf states as they do throughout the wider Middle East is not in question. All statistical indicators suggest that the Persian Gulf is one of the world's hottest, most water-starved environments. With the exception of Iraq and Iran, which have somewhat better access to freshwater than their neighbors, most Gulf states suffer from acute freshwater scarcity, defined by the World Bank as access to less than 1,000 cubic meters a year per capita. These scarcities promise to become more acute as the world's average temperature increases, and the demand for freshwater grows along with the region's population. Domestic water demand is projected to double in the Gulf region by 2025, and the demand for water required for industrial uses will increase threefold over the period.[2] As indicated in table 13.1, the baseline of renewable freshwater available in today's Gulf states is already an environmental crisis.

The United Nations identifies freshwater scarcity as a critical risk factor in all societies, contributing to such systemic problems as poverty, unplanned urbanization, environmental degradation, and the fragmentation of government.[3] Stated differently, water security is now recognized as essential to societal growth, development, and stability.[4] Water scarcity will be perhaps the most serious of the direct environmental stresses in the Gulf region over the next twenty years. It is by any measure a systemic problem. The World Bank projects that by 2050, per capita annual water availability throughout the Middle East and North Africa will decline from today's average of about 1,000 cubic meters (i.e., the very margin of sufficiency) to 500 hundred cubic meters. Simultaneously, in 2050 the world's population is expected to reach 9 billion. Most people will in fact have sufficient water; per capita annual water availability will average 6,000 cubic meters per person, an order of magnitude greater than can be expected in the Gulf region.[5]

Total water demand is projected to increase in the states that belong to the Gulf Cooperation Council (GCC) by 36 percent during the next decade. In addition to systemic shortages, water scarcity in the Persian Gulf region promises to gather momentum as a result of

Table 13.1 Freshwater Availability in the Persian Gulf States (cubic meters per capita)

Country	Renewable Water
Bahrain	157
Iran	1,957
Iraq	2,652
Kuwait	7
Oman	550
Qatar	71
Saudi Arabia	96
United Arab Emirates	35

Source: UNESCO, "Total Actual Renewal Water Resources," 2006, in *Water in a Changing World,* United Nations World Water Development Report 3 (New York: United Nations, 2009), update at www.unesco .org/water/wwap/wwdr/indicators/pdf/Table_4.3_Updated.pdf. Information for Saudi Arabia is omitted from the 2009 report. That figure is taken from UNESCO, *Water: A Shared Responsibility,* Second UN World Water Development Report (New York: United Nations, 2006), 132–35, www.unesco.org/water/ wwap/wwdr/wwdr2/pdf/wwdr2_ch_4.pdf, and is for 2005.

persistent mismanagement by the region's states of their limited renewable water resources. It is a disheartening picture. In 2007 the GCC countries extracted 19.5 million cubic meters of freshwater from underground aquifers, while the recharge of these aquifers accumulated at the rate of only 4.8 million cubic meters. The Gulf states currently extract 91 percent of their total freshwater supply from these underground sources, with the remaining demand satisfied by desalinization and treated effluent.[6] This unsustainable practice has resulted in falling water tables, a deterioration in water quality, and saline water intrusion into the declining aquifers.[7]

Adding to this dismal situation, the GCC states have pursued a nonsensical and hugely inefficient policy of developing their own agriculture despite the inhospitable environment. A whopping 85 percent of the groundwater in the GCC states is used for agriculture, despite the fact that food can be imported more cheaply than it can be produced locally.[8] Moreover, the disproportionate investment of their limited freshwater assets in agriculture has had a negligible impact on their economies—contributing on average less than 1 percent of GDP throughout the region.[9] Saudi Arabia's particularly egregious agriculture program illustrates the point. During the 1980s, the Kingdom became the sixth-largest wheat exporter in the world (with production reaching nearly 5 million tons in the early 1990s) courtesy of nonrenewable groundwater provided through inefficient irrigation systems.[10] In belated recognition of this folly, Saudi Arabia announced plans in early 2008 to reduce grain production annually by 12.5 percent and halt all production by 2016.[11]

Another egregious example of Saudi Arabia's freshwater folly is represented by the Al Safi Dairy Farm, identified in the *Guinness Book of World Records* as the largest integrated dairy farm in the world. The farm, located about 60 miles outside Riyadh, covers 14 square miles and supplies approximately one-third of the country's dairy needs. The farm's 29,000 cows produce an estimated 122,000 gallons per day, with each cow needing up to 30 gallons of freshwater daily to drink and stay cool (a cow's normal body temperature is about the same as a human's) in temperatures that can reach as high as 45 degrees Celsius (115 degrees Fahrenheit) in the summer. Water for the entire operation is pumped from a depth of 6,000 feet underground. In addition to cooling down the cows, the water is used to grow all the food for the farm's four-legged inhabitants.[12] Clearly, projects like this must become a casualty of the region's dwindling freshwater supplies.

Table 13.2 Historic and Projected Population in the Gulf Region, 1950, 2000, and 2050 (millions)

Country	1950	2000	2050
Saudi Arabia	3.200	20.800	45.03
Iran	16.900	66.001	100.17
Iraq	5.300	25.020	61.90
Kuwait	.152	2.200	5.20
Oman	.456	2.400	4.60
Qatar	.025	.617	1.30
Bahrain	.116	.650	1.17
Totals	26.150	117.787	219.37

Source: United Nations Population Division, "World Population Prospects: The 2008 Revision Population Database," http://esa.un.org/unpp/.

As dire as the story of freshwater scarcity may be, however, all the Gulf states have taken dramatic steps to address chronic shortages by building desalinization plants. The region today boasts the most developed infrastructure for freshwater production in the world. However, desalinated water is extremely expensive to produce, costing on average between 50 and 60 cents per cubic meter.[13] The Gulf states today operate more than half the world's estimated 10,400 desalination plants, which produce more than 35 million cubic meters of water per day around the world. Saudi Arabia's Saline Water Conversion Company is the largest desalinated water company in the world, producing approximately 3 million cubic meters per day, along with 5,000 megawatts of power, representing 50 percent of the Kingdom's drinking water supply and 20 percent of its electricity. In March 2006 the company's governor, Fehied al Shareef, indicated that the Kingdom will need an additional 6 million cubic meters of water and 30,000 megawatts of power generation capacity to meet anticipated demand.[14]

The scale of the joint desalination–electrical power projects under consideration throughout the region is staggering. In August 2007 the French company Veolia Water Solutions and Technology announced that it had launched an $805 million project to build a desalination plant in Fujairah, UAE, that will produce 590,000 cubic meters of water per day when it is completed in 2010. The same company also received a $1.4 billion contract in June 2007 to build what will be the world's largest desalinization plant in Jubail, Saudi Arabia. It will produce 800,000 cubic meters of water per day.[15] In December 2006 Saudi Arabia began studying a potential $5.3 billion "Water Bank" project in Tihama that will add significant desalination capacity for the entire country.[16] Demand for desalinated water in the region is projected to grow at an annual rate of 6 percent and may require an investment of more than $100 billion in new capacity over the next decade to meet demand growth.[17]

The freshwater shortage is the Gulf states' critical environmental stress for the foreseeable future, and it becomes exacerbated by population growth and the economic growth that must sustain the region's burgeoning population. As indicated in table 13.2, the population of the Gulf region is expected to grow from 117 million in 2000 to 219 million by 2050, an increase of more than 85 percent.

Regional economic growth to accommodate this increased population must be fueled by world energy markets. Without revenues from these markets, the Gulf states face the prospect of declining per capita GDP, economic stagnation, and political uncertainty. The future for these markets, however, appears bright for the region's regimes as they seek to continue their rentier model of redistributing their energy market proceeds. The U.S. Energy Information Administration forecasts that the world will need 40 percent more oil than it is using today by 2030, when global demand is projected to increase from approximately 85 million barrels per day in 2008 to 118–120 million barrels per day. The United States is expected to need another 10 million barrels per day by 2030. Asia will be the Gulf's most important market over the period; its anticipated economic expansion will be fueled by increased Gulf production of gas and oil. Net oil imports in China and India are forecast to jump from 5.4 million barrels per day in 2006 to 19.1 million barrels per day in 2030. The production of oil and natural gas in the Gulf will need to double during the next twenty years if it is to keep pace with this anticipated increase in global demand. Gulf suppliers will face particular pressure after 2020, when those that do not belong to the Organization of the Petroleum Exporting Countries are projected to plateau in production output. The Energy Information Administration projects that the Persian Gulf share of global petroleum exports may reach 66 percent by 2025.[18]

The paradox of the Gulf states' situation is that their continued ability to adapt and mitigate the impact of environmental stresses for their growing populations depends on the functioning of markets that must somehow be artificially restrained if the world is to successfully regulate carbon emissions. This fundamental contradiction lies unaddressed by all major energy market participants. Both suppliers and consumers of fossil fuels continue to believe that the future will be like the past. That the Gulf states are proceeding down the road of planning their future based on the premise of continued unrestrained revenue growth is not in question. The recent past suggests their reasons for optimism. Revenues in the region delivered courtesy of the increase in oil prices have delivered a veritable waterfall of cash into the coffers of these states. According to a recent Kuwaiti economic report, regional oil revenues surged from $364 billion in 2007 to $636 billion in 2008.[19] However, the global financial and economic crisis saw Gulf oil revenues decline to an estimated $280 billion in 2009.

Despite the global economic slump of 2008–9, the future for economic growth and development looks bright for the Gulf states. As many regions staggered through the world's economic crisis during this period, Gulf economies continued to grow, albeit at slower rates. The Saudi American Bank estimated that the GCC economies grew by less than 1 percent in 2009, but it forecasts that growth will quickly rebound to 5 percent in 2010.[20] The future for economic growth and development looks bright in the Gulf states. The region today is among the fastest growing in the world. A reduction in oil revenues definitely has hurt the Gulf states, with GCC energy-related export income declining from $630 billion in 2008 to an estimated $209 billion in 2009. Contrary to popular perception, economic growth is not solely dependent on energy markets. Non-oil-sector growth is also an important factor in driving the growth of the Gulf economies.[21] The GCC has taken steps to open its markets to outside investors during the past decade and is becoming steadily more competitive relative to other global states, according to World Bank figures. In early 2010 the GCC states had $2.5 trillion in development and infrastructure modernization projects either planned or under way.[22]

Although the long term looks bright, however, a period of economic retrenchment is definitely under way. The bursting of the region's real estate bubble has littered Dubai's landscape with partially completed projects. Nearly half the UAE's estimated contracts, which are valued at more than $580 billion, have been either put on hold or canceled.[23] During 2010 the Al-Nahyans of Abu Dhabi provided cash injections of $25 billion to help refinance Dubai's estimated $80 billion debt. In early 2010 and in acknowledgment of this assistance, Dubai named its newest skyscraper (and the world's tallest, at 2,717 feet), the Burj Khalifa, in honor of its most important benefactor, UAE president Sheik Khalifa bin

Table 13.3 Gulf States' Ecological and Carbon Footprint per Capita, 2005

Country	Total Ecological Footprint (hectares per capita)	Carbon Footprint
Bahrain	N.A.	N.A.
Iran	2.7	1.66
Iraq	1.3	0.84
Kuwait	8.9	7.75
Oman	4.7	3.4
Qatar	N.A.	N.A.
Saudi Arabia	2.6	1.33
United Arab Emirates	9.5	7.82
United States	9.4	6.51
Global average	2.7	1.44

Note: N.A. = not available. "Carbon footprint" includes the direct carbon dioxide emissions from fossil fuel combustion, as well as indirect emissions for products manufactured abroad.
Source: World Wildlife Fund, *Living Planet Report 2008*, 32–40, http://assets.panda.org/downloads/living_planet_report_2008.pdf. See also ibid., chart, p. 14.

Zayed Al Nahyan. Despite the slowdown, the Emirate of Abu Dhabi continues to pursue a series of aggressive development projects. In early 2008 Abu Dhabi broke ground on Masdar City, a $22 billion project to build a 6-square-kilometer city with no carbon emissions at all.[24] Similar aggressive economic development is proceeding in Doha, fueled by export revenues from the North Dome natural gas field.[25] Other ambitious (and less "green") projects abound throughout the region. There are plans, for example, to position the region as a leading producer of aluminum in global markets. A series of huge, environmentally unfriendly aluminum smelter projects are under way in Kuwait, Qatar, Oman, and the UAE, which will boost their combined production to 1.8 million metric tons per year by 2010.[26]

Saudi Arabia has launched aggressive plans to build a series of new cities to provide housing and jobs for its growing population. The Saudi Arabian General Investment Authority has launched a massive development plan to build six new cities that it hopes will add $150 billion to the nation's economy by 2020, including housing for 4.3 million and 1.3 million jobs. The King Abdullah Economic City, Knowledge Economic City in Medinah, Prince Abdulaziz bin Mousaed Economic City in Hail (500 miles north of Riyadh), Jizan Economic City, and Petro Rabigh represent the cornerstone of the regime's attempts to build an infrastructure that can absorb the bow wave of population growth that will be breaking over the Kingdom in the next thirty years.[27]

Unsurprisingly, the rapid economic growth of the last five years has increased the environmental footprint of the region's populations. The World Wildlife Fund has developed an index to measure the demand a country places on the biosphere in the area of biologically productive land and sea required to provide the resources and absorb the waste of the world's population.[28] The index references the number of global hectares used per person for resource consumption in each country (table 13.3). The ecological footprint of the Gulf states, where it is known, is significantly higher than global averages, particularly in countries like Kuwait and the UAE. The figures indicate that the UAE boasts the world's largest ecological footprint on a per capita basis. Each citizen is using a whopping 9.5 hectares for resource consumption and waste absorption (edging out even the average American).

Environmental Vulnerability

The dire projections for the impact of declining access to freshwater are but one component in assessing the cumulative impact of climate change on the Gulf region states. Interestingly, the data summarized in appendix A of this volume do not indicate a "serious" societal vulnerability to projected increases in the world's temperature. Although the data show that Iran and Saudi Arabia will see continued significant shortages of potable water, these shortage are not deemed threatening to the social fabric of these countries. Another of the environmental phenomena commonly associated with climate change is the prospect of rising sea levels (appendix B). In contrast to Egypt, the Middle Eastern state most at risk from rising sea levels, none of the Gulf states in the CIESIN sample—Saudi Arabia, Iran, and Iraq—are assessed to have significant coastal populations that might be affected by dramatic rises in sea levels. That is not true elsewhere in the Gulf, however. Other studies have suggested that the island nation of Bahrain could lose up to 15 kilometers of coastline with significant increases in sea levels.[29] Moreover, the aggressive development of human-made islands off the coast of Dubai and the land "reclamation" projects in Qatar and Bahrain would certainly become more problematic in the face of rising sea levels.

Regional Stability and Climate Change

Rising average temperatures in and of themselves will not significantly affect the Gulf states: the region is already one of the hottest, driest places on the planet. The limitations of the physical environment have always proven to be a powerful systemic factor shaping Gulf state societies. As the Gulf societies have moved from agrarian to industrial economies and from a rural to urban populations, the region's ruling elites have devised sophisticated and expensive means to mitigate and adapt to the limitations imposed upon them by a hostile environment. The environmental adaptive capacities throughout the Gulf states today are the most advanced in the world, although Iraq and Iran have some catching up to do relative to the states of the Arabian Peninsula. Saudi Arabia, for example, has constructed an elaborate adaptive infrastructure at a cost of billions of dollars that have enabled the Kingdom to cope with environmental extremes. There is an admitted air of unreality to these measures, which have been taken without regard to cost or common sense. To survive, the region's regimes must continue to fund expensive and environmentally unfriendly programs to continue the process. Assuming that these societies can continue their unfettered investments in freshwater development, power generation, housing, and economic development, these efforts can continue as long as petroleum markets provide revenues that will allow them to do so. Continuing down this path may mitigate the prospect of internal instability. All the Gulf states thus find themselves in a series of difficult contradictions. They all rely on revenues from fossil fuels—whose consumption must be limited if the world is to successfully address the problem of greenhouse gas emissions.

The Saudis are frankly ahead of much of the rest of the world in their thinking about the politics of climate change in relation to their national interests; though their thinking is not necessarily to the liking of outsiders. The regime has spent much of the last fifty years investing in arguably the world's best-developed climate-related adaptation and mitigation infrastructure. An American football metaphor illustrates the royal Al-Saud family's approach. At a time when the rest of the world still has yet to arrange a huddle to call plays, the Saudis are already positioned in a preventive defense on the issue of climate change—a strategy to prevent the "Hail Mary" touchdown pass by accepting the need to give up the short pass and the run up the middle. They aim to avoid catastrophe, and they are not worried about losing ground as long as they do.

This approach to the politics of climate change has earned the Saudis the ire of environmental groups, who in 2006 rated Saudi Arabia the worst country in the world at addressing climate change.[30] At the December 2007 United Nations conference on climate change in Bali, environmental groups labeled Saudi Arabia the "fossil of the day" for its reluctance to constructively support the global climate change talks.[31] The Saudi approach to the issue seems encapsulated on the one hand by King Abdullah's November 2007 announcement that the Kingdom would spend $300 million to support climate change research, and on the other by the announcement, at the same time, that Prince Alaweel bin Abdulaziz al Saud was spending $300 million for an Airbus 380 flying palace.

Saudi Arabia, the United States, and China have united over the last eight years to water down findings of the Intergovernmental Panel on Climate Change.[32] The Saudi approach to the issue has been well articulated by Saudi oil minister Ali bin Abrahim al-Naimi, who, in objecting to attempts by the industrial world to restrain gasoline demand through higher taxes, told the United Nations General Assembly in September 2007 that "those industrialized nations are imposing more high taxes which are . . . providing direct and indirect aid for the industries of coal and nuclear energy which are the most polluting sources of climate and the global environment. . . . This affects growth rates in the world for demanding oil in the coming period and contributes to the negative impact on the march of development in our country." Naimi added that "the call for moving away from fossil fuel consumption as a way to address climate change is not a viable alternative. I can assure you that through the use of technology solutions that the world can continue to rely on oil."[33]

The Al-Saud family does not fear the impact of climate change on its own physical environs, which will change little if the world continues to heat up; but it does foresee disaster in the politics of climate change as the international community starts to grapple with the problem. The Saudi and Gulf states nightmare is global agreement on a system of market interventions that produces two outcomes: (1) a reduced demand for energy and (2) requirements that energy producers shoulder the costs for developing states that lack the resources to implement climate-related adaptation and mitigation measures of their own.

This is a strategy to hold off for as long as possible a system of global carbon taxes or other mechanisms to spread the costs of environmental adaptation and mitigation. The Saudis and their colleagues around the region look upon this outcome as inevitable, but the longer they can avoid dipping into their own pockets to contribute to global mitigation efforts, the better off they will be in developing their own systems and infrastructure. The Saudis are motivated by economic self-interest in this matter and, more broadly, by the recognition that the Kingdom's rentier state depends upon increasing amounts of cash to cope with the "traditional" sources of instability that confront the Kingdom—population growth, urbanization, unemployment, a lack of freshwater, and disruptive social movements that could spring from Saudi urban centers, to name but a few.

Saudi Arabia and the other Gulf states have paid close attention to a series of domestic stakeholders in cementing their hold on their respective countries. In Saudi Arabia, these stakeholders include the extended royal family; the religious establishment; the merchants of the Hijaz; the new caste of dissident religious clerics who wield influence in the Nejd; Shias in the eastern provinces, who are still second-class citizens in the Kingdom; and tribal and clan leaders throughout the peninsula, who have been indirectly integrated into the familial structure via marriage. Each of these stakeholders benefits in various ways from the system of economic and political patronage that permeates the governance of the Kingdom.

Internally, each ruling family in the Gulf states has constructed an elaborate system of political patronage and wealth redistribution in the form of free education, cheap gas and electricity, and government jobs for a mostly underemployed male population. Continued economic growth built on ever-expanding world demand for energy provides the means for them to continue this rentier system that keeps their friends happy and co-opts and buys off potential internal opponents.

The climate effects that may be realized by 2030 represent an appreciable additional risk factor for the Gulf states that could easily aggravate more "traditional" ones: population growth, unemployment, social movements that lead to political challenges to the regime, and urbanization. The Saudis and their Gulf state partners greatly fear the impact that climate change could have on the orderly functioning of global markets for petroleum, and they fear that the politics of the issue may result in market distorting forces that work against them. Both issues could lead to a drop in revenue and reduce the regime's ability to address traditional risk sources.

Climate Change and Systemic Risks to Regional Security

The analysis presented here suggests that regional security will not be seriously threatened by climate change per se through 2030, provided that the structure of global energy markets does not turn markedly against producer states, as a consequence of reactions to environmental stress elsewhere. Direct climate effects in the Gulf are not forecast to gain socially destructive momentum until the second half of the twenty-first century, far beyond the range of meaningful political or strategic analysis. The Saudis and the region's other regimes above all seek to ensure the security of the royal family and their continued political and economic ascendance. As was indicated above, we can expect all the Gulf states to act with alacrity in managing threats to the current structure of global energy markets. As long as the world demand for energy continues on its current path, the region's regimes are provided with the means to stave off stresses to the state stemming from environmental and climate-related forces.

Nevertheless, climate models do not account for disruptive, cascading events that can dramatically alter orderly political and economic interaction among global actors. In other words, the cumulative impact of climate change may produce unanticipated incremental changes that can materialize into much more serious problems. Surprises happen. Climate change *will* affect economic development around the world and will make it more difficult for various states, particularly in Asia, to sustain a predictable path of economic development. The continued economic expansion in Asia is vitally important to Saudi Arabia as a destination for its oil exports.

The analysis here is that the latent reserves of social and political resilience are proportionate to, and derivative from, the Gulf region's latent reserves of oil and natural gas. If the oil runs out or if markets fundamentally change due either to a sustained global economic slowdown or to successful energy demand mitigation efforts around the world, it is doubtful that today's residents of the Gulf will willingly and peacefully return to the Bedouin-type existence of their ancestors—a way of life that coped with the extreme environment before oil was discovered. The Saudis hope this will never happen, and their hope seems justified. The U.S. Geologic Survey estimates that the Kingdom may have as much as 1 trillion barrels in recoverable reserves of all kinds of oil. And no amount of demand mitigation measures will dry up the world's thirst for oil any time soon.

Saudi vulnerability stems from the functioning of international energy markets, which may be subject to change in light of the environmental or economic interests of other market participants. Such a scenario is not difficult to imagine if "peak oil" becomes a reality (or is simply perceived as such), or, alternatively, if the world's major oil-consuming states decide that the ever-increasing prices of oil are politically and economically unacceptable. In such an environment the Gulf states, and most particularly Saudi Arabia, would be subjected to intimidation and coercion by their major customers. To that extent, whatever potential for armed aggression may arise out of climate-related issues in the Gulf will not result from the socially disruptive effects of climate change, which all the Gulf states are at present well positioned to managed, but from a loss of confidence in international energy markets, and the reactions this may inspire among powerful consumer states.

The Outlook for Official Policies

As was noted above, the Gulf region's regimes will seek to mitigate developments in global politics that distort the functioning of international energy markets. To do this, the regimes must engage with a variety of international actors around the world, including both states and international organizations. This engagement is necessary to forestall the development of market-distorting forces and avoid being placed in the position of paying for climate-related adaptation and mitigation costs elsewhere. They are amenable to Western interests as a function of maintaining good customer relations, and because of their reliance on the military capabilities of the United States to maintain overall regional security. The Saudis have assiduously avoided offending the United States, while seeking to build close political relationships with the established European states as a counter to U.S. hegemony and as another source of protection against external threats. There is no reason to suppose that the Saudis will alter this approach in the foreseeable future, unless revolution from within topples the House of Al-Saud, replacing it with some kind of populist Islamic regime.

Conclusion

All the states in the Gulf region will face profound environmental stresses resulting from climate change in the coming decades. Nevertheless, the region's regimes are reasonably experienced at dealing with such stresses, and they have thus deployed elaborate and expensive climate-related mitigation and adaptation systems paid for by oil revenues. For example, the region is heavily investing in new freshwater capacity to address its shortfalls and prepare for the population growth that is expected during the next several decades. Unlike the oil wealth explosion of the early 1980s, whose windfall profits were mostly frittered away, this time the region's states are heavily investing in development projects at home, to build their infrastructures and promote some form of sustainable development.

Despite these prudent steps, however, the Persian Gulf region's regimes all remain vulnerable to fluctuations in global energy markets. A sudden drop in global demand for energy or a sustained drop in energy prices will negatively affect their ability to continue their mitigation and adaptation efforts, which are by and large of the have-your-cake-and-eat-it-too variety. The global politics of climate change threaten to change the dynamics of international energy markets in ways that would redound to the disadvantage of Gulf producers. The region's governments will thus continue to publicly embrace green development policies at home while joining together with other states to forestall a global system that will limit emissions and, hence, demand for energy. They will also seek to avoid schemes that distribute their wealth to the less-developed world to pay for the climate-related mitigation and adaptation efforts around which they have organized their own modernization efforts. Any global system that comprehensively addresses climate change will need to incorporate the needs and interests of the energy-producing states in the Gulf region. This will not be easy.

Notes

1. Samir Al Saadi, "Kingdom to Carry Out Another Cloud Seeding Experiment," *Arab News* 3, June 2008, www.arabnews.com/?page=1&section=0&article=110547&d=3&m=6&y=2008.

2. Nermina Biberovic, "Water and Agriculture Issues in the Gulf," *Gulf Monitor* 7 (Dubai: Gulf Research Center, January 2008, www.grc.ae/index.php?frm_module=contents&frm_action=detail_book&sec=Contents&override=Articles&PHPSESSID=dde75ec24ca921d7c05ac19d7e5dd977%20%3E%20Water%20and%20Agriculture%20Issues%20in%20the%20Gulf,%20in%20Gulf%20Monitor,%20Issue%20No.%207&book_id=54907&op_lang=en.

3. United Nations, *Water Hazard Risks*, United Nations Water Series 1 (January 2005), www.unwater.org/downloads/unwaterseries.pdf.

4. See World Water Council, *Synthesis of the 4th World Water Forum*, Mexico City, September 2006, www.worldwatercouncil.org/fileadmin/wwc/World_Water_Forum/WWF4/synthesis_sept06.pdf.

5. World Bank, *Making the Most of Scarcity: Accountability for Better Water Management Results in the Middle East and North Africa*, MENA Development Report (Washington, DC: World Bank, 2007), 5; http://web.worldbank.org/WBSITE/EXTERNAL/COUNTRIES/MENAEXT/0,,contentMDK:21244687~pagePK:146736~piPK:146830~theSitePK:256299,00.html.

6. This is as detailed by Mohamed A. Dawoud, *Water Scarcity in the GCC Countries* (Dubai: Gulf Research Center, 2007), 14.

7. Ibid.

8. Ibid.

9. Mohamed Bazza, "Policies for Water Management and Food Security under Water Scarcity Conditions: The Case of the GCC Countries," paper presented at Seventh Gulf Water Conference, organized by the Water Science and Technology Association, Kuwait City, November 19–23, 2005, www.fao.org/world/Regional/rne/morelinks/Publications/English/PoliciesforWaterandFoodSecurityintheGCCCountries.pdf.

10. World Bank, *Making the Most of Scarcity*, 12.

11. Andrew England, "Water Fears Lead Saudis to End Grain Output," *Financial Times* (London), February 27, 2008, www.ft.com/cms/s/f02c1e94-e4d6-11dc-a495-0000779fd2ac,Authorised=false.html?_i_location=http%3A%2F%2Fwww.ft.com%2Fcms%2Fs%2F0%2Ff02c1e94-e4d6-11dc-a495-0000779fd2ac.html%3Fnclick_check%3D1&_i_referer=&nclick_check=1. On the long-term trend in Saudi grain output, see Marianne Lavelle, "Beyond the Barrel," *US News & World Report*, May 21, 2008, www.usnews.com/blogs/beyond-the-barrel/2008/05/21/forget-saudi-peak-oil--worry-about-peak-grain.html.

12. The details given here are drawn from Craig Smith, "Al Kharj Journal: Milk Flows from Desert at Unique Saudi Farm," *New York Times*, December 31, 2002, http://query.nytimes.com/gst/fullpage.html?res=9A07E2DC153FF932A05751C1A9649C8B63.

13. Phil Dickie, *Making Water: Desalination-Option or Distraction for a Thirsty World?* World Wildlife Fund Global Freshwater Program, June 2007, www.waterwebster.com/documents/desalinationreportjune2007.pdf.

14. Javid Hassan, "Kingdom Leads in Desalination, but Needs More to Meet Demand," *Arab News*, March 22, 2006, www.arabnews.com/?page=1&section=0&article=79565&d=22&m=3&y=2006.

15. "Veolia Awarded Huge Desalinization Contract in Saudi Arabia," Cleantech Group LLC, June 28, 2007, http://media.cleantech.com/1392/veolia-awarded-huge-desalination-contr.

16. Mariam al Hakeem, "Saudis Consider $5.3 Billion Water Project," *Gulf News*, December 24, 2006, http://archive.gulfnews.com/articles/06/12/24/10091822.html.

17. Meena Janardhan, "Water Day-Gulf: Forced to Look beyond Desalinization Plants," Inter-Press Service News Agency, March 21, 2007, http://ipsnews.net/news.asp?idnews=37013.

18. The figures given here are drawn from *International Energy Outlook 2008*, Energy Information Administration, Department of Energy, www.eia.gov/oiaf/ieo/highlights.html.

19. John Isaac, "UAE Oil Income May Hit $110B," *Khaleej Times Online*, June 22, 2008, www.khaleejtimes.com/DisplayArticle.asp?xfile=data/business/2008/June/business_June706.xml&section=business&col=.

20. Saudi American Bank, *GCC 2010 Economic Outlook, February 2010*, www.samba.com/GblDocs/GCC_2010_Outlook_Eng.pdf. Also see World Bank, *Middle East and North Africa Region: 2009 Developments and Prospects* (Washington, DC: World Bank, 2009), http://siteresources.worldbank.org/INTMENA/Resources/MENA_EDP_2009_Overview_ENG.pdf.

21. Ibid.

22. Ibid., 9.

23. Paul Lewis, "Dubai's Six-Year Building Boom Grinds to a Halt as Financial Crisis Takes Hold," *Guardian* (Manchester), February 13, 2009, www.guardian.co.uk/world/2009/feb/13/dubai-boom-halt.

24. "Abu Dhabi Masdar Initiative Breaks Ground on Carbon-Neutral City," PR Newswire, February 9, 2008, www.anzses.org/files/Masdar.pdf.

25. Saudi-American Bank, *GCC Economic Outlook*, appendix 1, 18, provides supporting details.

26. "Gulf States Plan Higher Aluminum Output," *Engineering and Mining Journal*, September 2004, http://findarticles.com/p/articles/mi_qa5382/is_200409/ai_n21357315?tag=untagged.

27. Jad Mouawad, "The Construction Site Called Saudi Arabia," *New York Times*, January 20, 2008, www.nytimes.com/2008/01/20/business/worldbusiness/20saudi.html; Raid Qusti, "Saudi Arabia to Build Two More Economic Cities This Year," *Arab News*, April 27, 2009, www.arabnews.com/?page=6&section=0&article=95554&d=29&m=4&y=2007.

28. The World Wildlife Fund defines the environmental footprint as follows: "The footprint of a country includes all the cropland, grazing land, forest, and fishing grounds required to produce the food, fibre and timber it consumes, to absorb the wastes emitted in generating its energy uses, and to provide space for its infrastructure. People consume resources and ecological services from all over the world, so their footprint is the sum of these areas, wherever they may be on the planet." World Wildlife Fund, *Living Planet Report 2006*, 14, http://assets.panda.org/downloads/living_planet_report.pdf.

29. Mohamed A. Raouf, *Climate Change Threats, Opportunities, and the GCC Countries*, Middle East Institute Policy Brief 12, April 2008, www.mei.edu/Portals/0/Publications/CLIMATE-CHANGE-THREATS-OPPORTUNITIES-GCC-COUNTRIES.pdf.

30. "US, Saudi, China Rank among Worst on Climate Change: Group," Agence France-Presse, November 14, 2006, www.commondreams.org/headlines06/1114-01.htm. The report by the German environmental group Germanwatch rated Sweden as best, with the United States, China, and Saudi Arabia at the bottom of the heap.

31. "Saudi Arabia Tops the Roll of Dishonour," *OneWorld Net*, December 5, 2007, http://uk.oneworld.net/article/view/155885/1/.

32. This is as noted in "Billions Face Climate Change Risk," BBC, April 6, 2007, http://news.bbc.co.uk/2/hi/science/nature/6532323.stm.

33. These remarks are as reported by Andrew Leonard, "Don't Cry for Saudi Arabia," *Salon.com*, September 27, 2007, www.salon.com/tech/htww/2007/09/27/saudi_arabia_oil/.

14

Egypt

Ibrahim Al-Marashi

About a hundred and thirty thousand years ago the desert that currently envelops the Nile Basin was thought to be a savanna inhabited by life ranging from humans to rhinoceroses. Geoarchaeological studies of the Nile have reconstructed a history of climate change to explain how the inhabitants then responded to dramatic shifts in climate.[1] Such research demonstrates how human life in this area has proven resilient in adapting to climate change over the millennia. This chapter covers a less ambitious time span, and it seeks merely to estimate how climate change may affect Egyptian politics and society during the next twenty years or so. Climate change has already had an impact on Egypt, and also on socioeconomic dynamics in Africa, the Middle East, and the Mediterranean, underlying the need to conceptualize a framework of regional "climate security."

Egypt is home to approximately 80 million people, which makes it the most populous nation in the Middle East and North Africa. Its population has grown very rapidly in recent years—it has doubled since the country's current president, Hosni Mubarak, was first elected in 1981—a growth rate that official efforts to encourage smaller families has done little to check. Egypt's population is also densely concentrated in an elongated T, formed by the Lower Nile River and its delta, a glyph of green in an ocean of brown. These demographic and geographic facts are not easily amenable to amelioration by state intervention, and they constitute the basic realties by which the future impact of climate change will be shaped.

Egypt

Most centrally, rapid population growth will increase demand for water and energy resources, at the same time that rising temperatures throughout the Nile Basin may reduce available drinking water from the river. The concentration of Egypt's population and economic productivity in the Nile Delta means that a disproportionate share of Egyptians are liable to suffer as a result of even a moderate rise in global sea level. According to one Egyptian environmental scholar, Mohammed El-Raey, "Egypt is the third most vulnerable country in the world to climate change. It is only surpassed by Bangladesh and Vietnam."[2] The World Bank, for its part, has concluded that climate change would result in "catastrophic consequences" for Egypt.[3]

The Domestic Impact of Climate Change in 2030

Politically, Egypt is one of the world's more durable authoritarian states. Its present government originated in 1952, when the country's last monarch, Farouk, was overthrown by a clique of army officers under the leadership of Gamal Abdel Nasser. Nasser's presidency, which lasted from 1956 until his death in 1970, was marked by considerable turmoil and a good deal of warfare, chiefly versus Israel, against whose creation Egypt had taken up arms in 1948. It laid them down pursuant to the Camp David Accords concluded by Nasser's successor, Anwar Sadat, in 1978. Since then Egyptian politics have focused less on international adventurism than on the top-down promotion of economic and social development—a cause that, in Nasser's time, had often been driven into the background by his aspirations to lead the nonaligned world. Following Sadat's assassination by Islamist extremists in 1981, Egyptian politics acquired an almost dynastic stability, as personified by the Hosni Mubarak presidency. Whether that stability will survive the popular agitation that is roiling Egyptian politics in 2011 is, as of this writing, impossible to say.

Peace with Israel has allowed Egypt to draw ever closer to the United States, an association that has generally served its government well in terms of ensuring that it has the economic and military support required to sustain itself. Its ties with America have also heightened its visibility as a target of Islamist and other anti-Western opposition movements. Egypt is the birthplace of the Muslim Brotherhood (founded 1928), the largest Islamic political organization in the world, from whose roots many revolutionary Islamist movements have sprung. In Egypt the Muslim Brotherhood functions as a semilegitimate political party, the largest in the country, routinely harassed by the police but tolerated on the margins as an expression of popular political opposition and dissent.

Broadly speaking, the authoritarian and police instruments (and countervailing social and developmental policies) by which the present regime has maintained itself in power have been crafted to address the kinds of intrastate challenges the Muslim Brotherhood embodies. The central question with respect to climate change is whether these same tools and governing habits can be adapted to the quite different challenges that may arise from that quarter. Although a Western and technocratic education has become increasingly commonplace among the rising generation of leaders in the developing world, it is likely that future Egyptian governments will continue for some time to be dependent upon the same traditional instruments of power, above all the police and the army, that have supported the state since the reign of Farouk. There is nothing in the public record to suggest any impulse within Egypt's ruling elites to change this. One editorial on the relationship between President Mubarak and the public has proposed that he "has stayed in power by making an implicit contract with Egypt's masses: We provide food, you keep your noses out of politics. That deal is now fraying."[4] This comment undoubtedly simplifies the complex dynamics of politics in Egypt. But it does also illustrate how far the basic social contract that has kept the present regime in place has depended on its ability to wrest the basic necessities of life away from the forces of nature, in a landscape that promises to become increasingly unforgiving in the future.

About a third of Egyptians work the land. There is no question that their lives and economic circumstances are liable to become more difficult in the future. Moderate to severe climate change, in the form of rising temperatures, altered patterns of rainfall, and declining total freshwater discharge, will have far-reaching detrimental effects on agricultural production. In addition, rising sea levels will contribute to the salinization of groundwater along the low-lying northern coast and delta regions, and saltwater intrusion into coastal wells. Climate change has already resulted in a decrease in fishing, and it is expected to lead to more severe and frequent sandstorms, droughts, and flooding.[5]

Egypt is not, however, a primarily agricultural country. About half its population works in the service sector, of which the largest share, outside the government, is based on tourism. Tourism is a critical activity politically because its revenues, being drawn primarily from outsiders, are easy to tax and provide a disproportionate share of government revenues. Tourists bring the government about twice as much revenue as the Suez Canal, an income stream that is vulnerable in the long term to a range of environmental impacts.[6] Although Egypt is a hot country, it is not equally hot all year round, and it is cooler in the north, which is where tourists congregate. Rising temperature will reduce the duration of the relatively temperate tourist season and aid the spread of insect-borne diseases from Upper Egypt (i.e., the south), where they are currently confined, to areas of greater economic significance.

Agriculture (including fishing) and tourism are the two sectors of the Egyptian economy most directly vulnerable to the negative impacts of climate change. Whether these impacts will prove to be the seeds of politically disruptive social crisis is hard to say, though unlikely in the short run. Their effects, even if eventually severe, are certain to accumulate slowly, providing time for the political leadership to take what measures it can to provide relief for those most immediately effected. Crucial sources of government revenue, including foreign aid and rents from the exportation of oil (of which Egypt has a modest but nontrivial endowment) are not likely to be immediately reduced by climate change, so the government can be expected to retain its accustomed freedom of action to hold up its end of the social bargain described in the editorial quarter quoted above—at least for a while.

Egypt is not at present a major source of out-migration to other countries around the Mediterranean or in Africa. This is partly a reflection of the relative success of economic reform there since the 1990s, when a general campaign of economic modernization and privatization got under way. Nevertheless, several hundred Egyptians have perished in recent years attempting to transit the Mediterranean in makeshift craft, motivated by a desire to improve their economic prospects. Such incidents may well increase if the economics of Egypt's densely populated coastal zone deteriorate owing to a rising sea level, saltwater intrusion, and so on.

On balance, however, Egypt is more likely to be a destination for climate refugees from elsewhere in northern and equatorial Africa, where climate impacts will be at least as severe and government capacity is substantially lower. It is unlikely that climate change will make any of Egypt's immediate neighbors more attractive than Egypt itself as a destination for economic refugees; while Egypt's relatively greater capacity to provide social services, compared, for instance, with states like Sudan, may make it an attractive choice for outsiders. It goes without saying that the reception such environmental refugees receive will not be entirely welcoming. The potential troubles that would arise from a massive influx of refugees would exceed the well-proven capacity of the Egyptian state to maintain order.

Further reassurance in this regard arises from the fact that the kinds of issues likely to be spurred by climate change do not play particularly well into the hands of such organized political opposition as exists. Climate change is not an obviously fertile issue for groups whose main grievances have generally pointed toward a betrayal of traditional Islamic values, excessive deference to the West, and a general resentment of modernity. If anything, the arguments that dominate environmental politics, with their emphasis on science, technology, and economic adaptation, are likely to favor the kinds of academic and technical experts who already work for the government and speak through the government-controlled media.

Nevertheless, climate change could lead to disruptions in the lives of traditional communities. Feuds are likely to erupt between nomadic and pastoral groups, which have been a source of tension in Egypt since time immemorial. Areas will become uncultivable as crops are unable to bear the higher temperature, forcing farmers to relocate to marginal lands. Farming in these areas, where soils are already poor, would only contribute to their degradation.[7] Egypt's environmental agency has already predicted that climate change could have an impact on the population's general health. Such afflictions would include vector-borne diseases, physiological disorders, skin cancer, eye cataracts, respiratory ailments, heat strokes, and heat-related illnesses, which would cause a strain on the public health infrastructure.[8]

Sea-Level Rise and Coastal Urban Areas

Any scenario envisioning Egypt in 2030 must reckon with the substantial implications of a rise in sea level. More than 6 million Egyptians (9 percent of the population) are currently estimated to live in the 1-meter low-elevation coastal zone. Such people could be displaced and find their sources of livelihood disrupted by even the low end of currently forecasted changes in sea level. According to a report released by the Friends of the Earth Middle East at the Bali conference, climate change could displace 2 to 4 million Egyptians by 2050.[9] As the editor of this volume has noted, there is no aspect of climate science in which greater uncertainly prevails than sea-level rise, whose anticipated pace and extent have been subject to continuous revision in recent years and will probably continue to be so for the foreseeable future. The one thing of which we can be certain is that there are few states that are more vulnerable to this dimension of climate change, in relative terms, than Egypt.

Forty percent of Egypt's industrial activities occurs in and around Alexandria, the country's largest seaport and second-largest city. Multiple projections estimate that vast parts of this urban landscape could be inundated due to climate change. One Egyptian study cited computer simulations of climate change that estimated that a sea-level rise of only 50 centimeters would flood more than 30 percent of the city.[10] A professor of environmental studies at Alexandria University predicted that a 50-centimeter rise would affect tourist beaches and cause the flooding of coastal agricultural and industrial areas. He estimates that 194,000 jobs would be lost (151,000 in industry, 34,000 in tourism, and 9,000 in agriculture) and that 1.5 million persons would become environmental refugees.[11] The authorities in Alexandria are cognizant of the disruptive potential of a rise in sea level and are allocating $300 million to construct concrete sea walls to protect coastal areas.

Other affected areas would include smaller towns near the Mediterranean or Red Sea. In the town of Rashid (Rosetta) a rise of 50 centimeters in sea level would lead to the loss of its coastal areas, its historical Islamic monuments, and agricultural areas along with an estimated 30,000 jobs.[12] A system of walls has also been erected around this town to protect it against changes in water levels. Port Said, Port Fouad, Shark El-Tafria Port, and Sahl El-Tina, which are situated near the Mediterranean, Lake Manzala, and the Suez Canal, would also be negatively affected by a rise in the sea level.[13] However, according to one Egyptian geologist, Roshdi Said, a rise in the water level would not affect ship traffic on the Suez Canal, because the Egyptians placed mounds of sand on both sides of the canal during its construction that would allow them to absorb any increase in the water level.[14]

The River Nile

The impact of climate change on the River Nile is likely to be especially complex. Climatologists have predicted that rising temperatures will accelerate the Nile's evaporation process, thus reducing Egypt's already dwindling freshwater supplies. Not only would such a trend lead to a shortage of potable water; it would also affect the water available for irrigation and hydroelectric power. The Nile provides 95 percent of Egypt's water for these needs, in addition to other industrial activities.[15] Again, however, this is an area in which climate science appears to be especially uncertain. Some estimates anticipate increased rainfall patterns in the Ethiopian highlands and Lake Victoria, which serve as the headwaters of the Nile. Mohamed El-Raey has summarized a range of possible scenarios: on one side, a group of studies foresees a 70 percent decline in the available water from the Nile, while other projections estimate a 25 percent increase of Nile waters due to changing rainfall patterns in Africa's equatorial highlands and Great Lakes region.[16]

This latter possibility, if it were to come to pass, would pose a significant additional threat of erosion. The arable land of the Nile Delta makes up only 2.5 percent of Egypt's territory but houses large concentrations of the nation's population. For millennia, the annual flooding of the Nile left mud, sand, and minerals throughout the fertile delta and prevented erosion. The Aswan Dam already limits the flow of sediment from floods to replenish the delta. A significant increase in the water flow of the Nile, without compensatory sedimentation, would exacerbate erosion in this region.

According to a 2007 World Bank report, a 50-centimeter rise in sea waters could displace the inhabitants of the densely populated delta, which make up close to 10 percent of Egypt's population.[17] The 10 percent figure is more alarming considering that Egypt's population could double to about 160 million by 2050.[18] In its 1999 communication to the UNFCCC, Egypt's environmental agency highlighted the dangers of a 50-centimeter rise in sea level creating 2 million environmental refugees from the delta.[19]

Salt water from rising sea level would contaminate underground waters from the Nile that irrigate the delta. The delta produces nearly 50 percent of Egypt's agriculture, including wheat, bananas, and rice. Due to the rapid growth of the country's population, Egypt's wheat and maize production does not meet the local demand for these crops. Contamination would drastically affect the state's ability to feed its citizens. This scenario would result in a further rise of food prices, which would affect the numerous Egyptians living below the poverty line.[20]

Adaptive Capacity

Egypt is a robust and highly centralized authoritarian state with strong ties to the United States and long experience in the suppression and co-optation of political opposition.

During the past twenty years it has also accumulated a track record of reasonable success in promoting economic growth, though its achievements in this area fall short of those of the East Asian Tigers or resource-rich states like Brazil. Approximately one Egyptian in five still lives on the equivalent of less than $1 per day, a perennial benchmark for developing-world destitution, and it is reasonable to suppose that any new form of social stress is likely to manifest itself most painfully in the lives of the poorest members of Egyptian society—who are also, however, those already most dependent upon the authorities and least likely to be in a position to put organized political pressure on the state.

The most immediate climate-related peril in Egypt would be a continuation of high and unstable food prices, like those experienced in 2008. Such conditions have a demonstrated capacity to cause serious social unrest. Jacques Diouf, director-general of the United Nations Food and Agriculture Organization, said that soaring prices for commodities such as wheat, corn, and milk had the potential to cause "social tensions" and "political problems." Egypt has recently been paying record prices for food and cereal imports in a "panic buying" spree that has reduced its currency reserves.[21] The rise in food prices led to violent public clashes with security forces in 2008. One source estimates that 30 million Egyptians depend on subsidized bread, and in March 2008 six people died from exhaustion waiting in bread lines.[22] Due to these tensions, Amr Al-Shubaki of the Al-Ahram Center for Political and Strategic Studies said in a *Washington Post* interview: "The mood of the people is angry. I think it's near collapse, the state."[23] These comments are surprising, given the center's close relations with the state, but they also indicate that the possibility of state collapse is not beyond the pale of discussion among Egypt's political elite. Amin Abaza, the Egyptian agriculture minister, expressly blamed the high price of food in 2008 on "climate change."[24] Such comments indicate that the state is seeking to deflect responsibility for the crisis onto natural forces it cannot control, while suggesting that others can—the minister blamed the industrial powers for global warming. The comments reflect the state's acknowledgment that climate change is a challenge that needs to be addressed due to its detrimental effects on the public. Depending on the scope and nature of the measures that are eventually adopted, future food and water crises may strengthen the distributive role of the state while forcing the public to become even more dependent on the government.

The Egyptian state and Egyptian society certainly possess latent reserves of social ingenuity and institutional capital to address the challenges of climate change. On the institutional level, the Egyptian state has organizations to cope with the challenges of climate change. The Egyptian Environmental Affairs Agency was established in 1982 and was followed by the creation of the Ministry of State for Environmental Affairs in 1997. Egypt formed an interministerial National Climate Change Committee in the same year, and it has developed Egypt's Climate Change Action Plan, National Communication on Climate Change, and a National Energy Efficiency Strategy, "to manage its climate change activities."[25]

As of 2008, the Egyptian Ministry of Environmental Affairs was led by Maged George, a proactive minister who has sought to deal with environmental challenges on the domestic and international levels. For example, on the domestic level, one of the measures launched by the ministry was the Pollution Abatement Project, which provides financial support to Egyptian industries to assist them in complying with environmental laws.[26] On the international level, this ministry has worked closely within the UNFCCC. The ministry has hosted workshops on exchanging expertise among ninety developing countries, encouraging those present to monitor harmful emissions from the different sectors that contribute to global warming, and to implement national plans for adapting to climate change. In fact, the UNFCCC secretariat had praised the level and extent of Egyptian expertise in hosting climate change conferences.[27]

The Ministry of Water Resources and Irrigation is responsible for the management of water resources and coastal zone protection, and it supervises specialized agencies such as the Shore Protection Agency and Coastal Research Institute. The minister in 2008 was Mahmud Abu-Zayd.[28] The Ministry of Agriculture, under the leadership of Amin Abaza as of 2008, has also sought to mitigate the effects of climate change on food commodities. Within this ministry is a Desert Research Center. Further, the legislative Shura Council has also been involved in dealing with climate change.[29] How effective these institutions will be, and whether rhetoric will be met with action, remains to be seen. Egypt is notorious among Middle Eastern states for its mammoth bureaucracy, which has been slow to formulate public policy, particularly with regard to long-term challenges, of which climate change is perhaps the preeminent example.

There are a number of scientific institutions in Egypt that study and assess the affects on climate change there. These include the Egyptian Academic Society for Environmental Development and the Strategic Unit at the National Water Research Centre in Cairo. Nahla Abou El-Fotouh, of the latter, admits, however, that "until now, no specific research has been developed to determine exactly the impact of climate change on Nile water availability. All [studies] that have been published so far are only predictions."[30] She states that this problem is not only occurring in Egypt but also throughout the entire Nile Basin: "None of the Nile countries has conducted serious research on the effects of climate change. Most research and initiatives come from Western countries and institutions. But we [Egypt] are planning to initiate deep impact investigations and to find measures to reduce the threat of climate change."[31] According to the minister of environmental affairs, the government has been preparing a "national strategy study" on ways to adapt to climate change. Climate scientists staff the Egyptian Environmental Affairs Agency and have worked to obtain a vulnerability index to detect the areas that would be most affected by climate change.[32]

Two general conclusions arise from this basic picture. The first is that official Egypt, in its academic and policymaking branches, is neither ignorant nor complacent about the impact of climate change and is at least attempting to understand and prepare for the most obvious contingency. The second is that, to the extent that the climate-related social stress gives rise to traditional forms of political disruption—public protests, bread riots, underground movements bent on fomenting subversion or upheaval, and so on—there is no reason to suppose that the well-worn instruments of political repression and social control that have operated in Egypt for so long already will suddenly lose their effectiveness.

The Effects of Climate Change on Egypt's Regional Relationships

Climate change obviously does not recognize national borders, thus raising the question of how disruptions in Egypt would affect the region as a whole. Egypt has the potential to act against upstream states that disrupt the Nile's water flow, and the country might in turn be acted upon by surrounding states as an object of aggressive war to seize water or agricultural resources. Egypt could serve as a destination for environmental refugees from neighboring countries like Sudan and Chad, and from Sub-Saharan Africa, where the effects of rising temperatures and increasing aridity may prove especially severe. It may also experience significant internal population displacement if people flee the hotter south for the cooler north. Although Egypt will certainly wish under any circumstance to remain open to foreign investment and tourism, it will probably respond to climate-induced population movements from surrounding states by seeking to turn them away.

Climate change is especially likely to have an impact on Egypt's relations with its southern neighbors, with which it must share the waters of the Nile, and toward which the government in Cairo has never adopted a particularly accommodating stance. Egypt has always opposed Sudan's efforts to obtain hydroelectric power from the Nile, for instance, or to otherwise regulate or alter the Nile's flow through its territory. Such policies seem certain to become more intrusive and perhaps provocative if the Nile's water volume diminishes.

Most commentaries on Egypt's hydropolitics point toward conflict rather than cooperation. The following comments from an Ethiopian paper demonstrate how other Nile riparians view Egypt as a potential threat:

> Egypt is a country that has not abandoned its expansionist ambitions. It regards its southern neighbors as its sphere of influence. Its strategy is essentially negative: to prevent the emergence of any force that could challenge its hegemony, and to thwart any economic development along the banks of the Nile that could either divert the flow of the water, or decrease its volume. The arithmetic of the waters of the Blue Nile River is, therefore, a zero-sum game, which Egypt is determined to win. It must have a hegemonic relationship with the countries of the Nile Valley and the Horn of Africa. When, for instance, Ethiopia is weak and internally divided, Egypt can rest. But when Ethiopia is prosperous and self-confident, playing a leading role in the region, Egypt is worried.[33]

Conflicts over the Nile River have a long history, extending back to tensions between Emperor Amda Seyon of Ethiopia (r. 1314–44) and the Mamluk Sultan Al-Nasir Muhammad Qalaurn (1285–1340), with the former threatening to starve the people of Egypt by diverting the Nile if they did not end their prosecution of Coptic Christians. In the 1960s Egypt sought to take advantage of political instability in Ethiopia by supporting the Eritrean Liberation Front.[34] The case of Syria and Iraq supporting the Kurdistan Workers' Party (known as the PKK, from its name in Kurdish) as a proxy to influence Turkey's water politics could serve as an indication that Egypt, Sudan, and Ethiopia may try to use proxy powers to influence hydropolitics if the states decided to fight rather than cooperate.

In a May 10, 2007, speech at the Royal United Services Institute in London, Britain's foreign secretary, Margaret Beckett, said: "Reduce the total amount of Nile water supply so drastically and you risk exacerbating tensions between Egypt and its southern neighbors. Egypt has already warned off those countries upstream from diverting the Nile water—how much more strongly will it feel when there is far less of that water to go round in the first place."[35]

The Egyptian minister of water resources, Mahmud Abu-Zayd, objected to Beckett's speech, dismissing a war between the Nile Basin countries. He also rejected Beckett's claim that the Nile would lose 80 percent of its water, causing parts of the delta to be submerged by the sea. He has accused Beckett of having spoken from "purely political objectives," and he wondered about the timing of the statement, adding that "the majority of the countries which make such statements do not want Egypt to be stable."[36]

Although the water minister's comment may seem overdrawn, his dismissal of a potential conflict has some basis, given the recently cooperative nature of the Nile riparians. An organizational mechanism exists among the Nile riparians and meets regularly to ensure dialogue over the river's resources. An example of this cooperation was evident when Egypt offered a donation of $18 million to the Ugandan government to combat the growth of water hyacinths in the Nile and Lake Victoria.[37]

The outlook and tone of Egypt's official foreign policy has not only been cooperative along the Nile Basin but also open to engagement with the larger world and amenable to cooperation with the United Nations and regional organizations. Egypt was among the first five countries (along with Mexico, Cape Verde, Senegal, and the Netherlands) to ratify the UN Global Convention against Desertification.[38] Egypt signed the Kyoto Protocol on March 15, 1999, and the Egyptian Shura Council approved the protocol in 2004 and ratified it in 2005.[39] Egypt has coordinated meetings between Arab ministers of the environment, and the United Nations Environment Program and the United Nations Economic and Social Commission for Western Asia.[40] Egypt also cooperates with the Food and Agriculture Organization, hosting its regional office for the Mediterranean and North Africa region.[41] Egypt's Environment Ministry takes part in the Euro-Mediterranean Sustainable Development Forum.[42] And Egypt hosted an African Union Summit to discuss the 2008 food crisis.[43]

In another example of cooperation with the UN, the UNDP has worked together with the Ministry of Water and Irrigation and Egypt's National Water Research Center to develop a computer software tool, "Decision Support System for Water Resources," to produce various climate change scenarios for the Nile basin, thus improving water resource planning and management. The system is designed to be prepared in terms of extreme scenarios of climate change.[44]

Egypt has also established bilateral relations with nations to cooperate on mitigating strategies. In 2007, the Egyptian minister of state for environmental affairs signed a memorandum of understanding with the Spanish minister of the environment on initiatives dealing with climate change.[45] Egypt approved a document that forecasts a series of joint actions with the Brazilian minister of the environment.[46] The German construction bank, the World Bank, the Japan Bank for International Cooperation, the European Investment Bank, the Agence Française de Développement, and the government of Finland have all contributed to Egypt's Pollution Abatement Project.[47] A wind power generation facility under construction on the coast of the Red Sea in Egypt is financed by Japan's official development assistance.[48] This project is part of the Clean Development Mechanism to facilitate Egypt's effort in addressing climate change.[49]

Conclusion

The most pressing challenges confronting Egypt as a consequence of climate change all involve water. Even a moderate rise in sea level will directly threaten the country's major centers of population and economic productivity. Changes in continental patterns of rainfall that significantly alter the flow the Nile will also pose expensive problems of adaptation, particularly if the effect of the change is to reduce the volume of freshwater available to Egypt's burgeoning population. Such challenges are unlikely to destabilize the government in themselves, in part because of the existence of important external props to its stability, provided by the United States, Europe, and rising Asia, all of which see Egypt as an important contributor to regional stability in the Middle East. Nevertheless, water is an issue that could become a source of regional conflict in itself, and genuine, prolonged shortages are probably the straightest path linking environmental deterioration to questions of Egyptian national security.

Egypt has a dedicated political and academic infrastructure devoted to climate change, and it has taken the initiative to forge bilateral, regional, and international initiatives to deal with this challenge. It remains to be seen whether the state can follow through on these programs and achieve concrete results on the ground. Any prediction of how Egypt will fare twenty years from now depends on the success of the state's risk management and human development efforts in the face of climate change. Despite long-term plans to mitigate and adapt to climate change, it remains to be seen whether the country's decision makers are seriously committed to these programs and if the state has the capacity to protect the most vulnerable segments of the population. Its response to the food crisis of 2008 illustrates the kinds of strain such issues can put on the state. No matter how you look at it, effective climate change mitigation strategies will require coordination, patience, foresight, and efficiency, and these have not been strengths of the Egyptian bureaucracy in the past.

Notes

1. "Scientists Find Fossil Proof of Egypt's Ancient Climate," *Science Daily*, February 18, 2005, www.sciencedaily.com/releases/2005/02/050212191855.htm.

2. Mohammed Yahia, "Egypt's Looming Climate Change Nightmare," *Journal of Turkish Weekly*, January 29, 2008, www.turkishweekly.net/news/51813/egypt-s-looming-climate-change-nightmare-by-mohammed-yahia.html.

3. Ibid.

4. "Egyptian Tremors," *Jerusalem Post*, April 13, 2008, http://elliotjager.com/2008/04/egyptian-tremors.html.

5. Yahia, "Egypt's Looming Climate Change Nightmare."

6. Tarek Al-Issawi and Abdel Latif Wahba, "Egypt Sees Lower Tourism Revenue, Warns of 'Panic,'" Bloomberg.com, October 27, 2008, www.bloomberg.com/apps/news?pid=20601104&sid=ah9P_.xYZCR4.

7. Wagdy Sawahel, "Climate Change in Egypt 'to Force Millions to Migrate,'" *Science and Development Network*, June 27, 2005, www.scidev.net/en/news/climate-change-in-egypt-to-force-millions-to-migr.html.

8. "Egypt Climate Change: Background," Egyptian Environmental Affairs Agency, www.eeaa.gov.eg/ecc/ClimateBackground.htm.

9. Sheera Claire Frenkel, "Bali Parley Was Wake-Up Call, Israeli Delegates Say," *Jerusalem Post*, December 17, 2007, www.jpost.com/servlet/Satellite?cid=1196847355703&pagename=JPost%2FJPArticle%2FPrinter.

10. Yahia, "Egypt's Looming Climate Change Nightmare."

11. Mohamed El Raey, "Climate Change and Egypt," *Al-Ahram Weekly On-line* 873 (November 29–December 5, 2007, http://weekly.ahram.org.eg/2007/873/sc3.htm.

12. Ibid.

13. Ibid.

14. "Egyptian Geologist: Climate Change Will Not Affect Water Levels in Suez Canal," *Middle East News Agency Open Source Center*, document GMP20080313950018, March 13, 2008.

15. United Nations Office for the Coordination of Humanitarian Affairs, "Egypt: Scientists Uncertain about Climate Change Impact on Nile," *IRIN Humanitarian News and Analysis*, March 2, 2008, www.sciencedaily.com/releases/2005/02/050212191855.htm.

16. Ibid.

17. Yahia, "Egypt's Looming Climate Change Nightmare."

18. Anna Johnson, "Global Warming Threatens Egypt's Nile Delta," *USA Today*, August 23, 2007, www.usatoday.com/news/world/2007-08-23-egypt-nile-threat_N.htm.

19. Egyptian Environmental Affairs Agency, *Initial National Communication on Climate Change Prepared for the United Nations Framework Convention on Climate Change (UNFCCC)*, June 1999, http://unfccc.int/resource/docs/natc/egync1.pdf.

20. Yahia, "Egypt's Looming Climate Change Nightmare."

21. Javier Blas, "UN Warns of Food Price Unrest," *Financial Times* (London), September 6, 2007, www.ft.com/cms/s/0/1f0d4c6a-5ca1-11dc-9cc9-0000779fd2ac.html.

22. Quoted in "Egyptian Tremors," *Jerusalem Post*, April 12, 2008, www.jpost.com/servlet/Satellite?cid=1207649994353&pagename=JPArticle%2FshowFull.

23. Ibid.

24. "Egyptian Minister Discusses Bread Crisis at Upper House Session," Middle East News Agency Open Source Center, document GMP20080402950028, April 1, 2008.

25. "Egypt Climate Change: Background."

26. Ibid.

27. "Exchanging Expertise among Developing Countries for Reports on Climate Change," Middle East News Agency Open Source Center, document GMP20070930950046, September 30, 2007.

28. Rajab Ramadan, "Minister Abu-Zayd: Israel Threatens Us on the Issue of Water and the British Statements about the Drowning of the Delta Have Political Objectives," *Al-Misri al-Yawm*, Middle East News Agency Open Source Center, document GMP20070514950037 (14 May 2007).

29. "Egyptian Minister Discusses Bread Crisis at Upper House Session," Middle East News Agency Open Source Center, document GMP20080402950028, April 1, 2008.

30. United Nations Office for the Coordination of Humanitarian Affairs, "Egypt."

31. Ibid.

32. Johnson, "Global Warming Threatens Egypt's Nile Delta."

33. "Egypt and the Horn of Africa," *Addis Tribune* (Ethiopia), June 26, 1998, quoted by Daniel Kendie, "Egypt and the Hydro-Politics of the Blue Nile River," *Northeast African Studies* 6, nos. 1–2 (1999): 141.

34. Ibid, 115.

35. Foreign and Commonwealth Office, "The Case for Climate Security: Speech by UK Foreign Secretary Margaret Beckett at the Royal United Services Institute," May 10, 2007, www.ipb.org/Margaret%20Beckett%20-%20The%20Case%20for%20Climate%20Security.html.

36. Ramadan, "Minister Abu-Zayd."

37. "Egypt, Uganda Agree More Time Needed for Sudan's Bashir," Middle East News Agency Open Source Center, document AFP20080801558003 (30 July 2008).

38. "UN Desertification Convention May Be Delayed Till 1997," Middle East News Agency Open Source Center, document FTS19970326001685, September 11, 1995.

39. "Egypt: Shura Council's Health Committee Approves Kyoto Protocol," Middle East News Agency Open Source Center, document GMP20041026000100, October 26, 2004.

40. "Arab Meetings on Environment, Trade Start," Middle East News Agency Open Source Center, document GMP20071112950034, November 12, 2007.

41. "Egyptian President to Visit Italy 2-5 June," Middle East News Agency Open Source Center, document GMP20080601950039, June 1, 2008.

42. "George to Partake in Third Euro-Mediterranean Forum in Spain," Middle East News Agency Open Source Center, document GMP20070225950041, February 25, 2007.

43. "Egyptian Foreign Minister Says African Summit to Discuss Food Crisis," Middle East News Agency Open Source Center, document GMP20080623950040, June 23, 2008.

44. UN Office for the Coordination of Humanitarian Affairs, "Egypt."

45. "Egypt, Spain Sign MoU in Environment Affairs," Middle East News Agency Open Source Center, document GMP20070228950043, February 27, 2007.

46. Alexandre Rocha, "South Americans and Arabs Formalize Environmental Exchange," Brazil-Arab News Agency, February 7, 2007, www2.anba.com.br/noticia_diplomacia.kmf?cod=7427536&indice=340.

47. "Egypt: Minister Opens Second Phase of Pollution Abatement Project," Middle East News Agency Open Source Center, document GMP20061207950045, December 7, 2006.

48. "Emission Rights from ODA Projects; Greenhouse Gases: Japanese Company to Obtain [Emission Rights] for the First Time," Middle East News Agency Open Source Center, document JPP20070607033001, June 6, 2007.

49. "Egypt: Minister Opens Second Phase of Pollution Abatement Project."

15

The Maghreb
Algeria, Libya, Morocco, and Tunisia

Gregory W. White

Climate change will affect the Maghreb in profound ways because it is already character-ized by exceedingly fragile environmental conditions. The "ecological patrimony" of Mo-rocco, Algeria, Tunisia, and Libya is nothing if not challenging. For centuries, indeed mil-lennia, the region has been bedeviled by aridity, low and variable rainfalls, inadequate river systems, locust infestation, and encroaching desert. And now the climate change effects

Algeria

Libya

Morocco

Tunisia

anticipated in the future will have profoundly deleterious effects on the region. Given the fact that all contemporary Maghrebi states are highly authoritarian, their attempts to control and manage national economies and polities remain contingent and at the whims of *fortuna*. And the additional pressures imposed by climate change will not bode well for the region's political dynamics.

This chapter argues that the Maghreb's historical and symbiotic relationship with Europe (and North America) has long facilitated the maintenance of illiberal regimes and a poor development record. This is even more evident in the context of the "war on terror," as the region's regimes receive crucial bolstering from their North Atlantic allies. As North Africa confronts climate change, the authoritarian dimensions of the Maghrebi states will likely receive further sustenance from their North Atlantic partners in the name of maintaining stability and order. In contrast to other regions, where "state failure" is a potential outcome à la Somalia, Zimbabwe, or Sierra Leone, the Maghrebi states simply will *not* be allowed to fail, if only because of their strategic importance. Thus, their North Atlantic allies are likely to further facilitate—and perhaps even accentuate—the Maghreb's ongoing authoritarianism through programs such as the Trans-Sahara Counterterrorism Initiative and "border security" efforts designed to extend Europe's effective borders further south. In the complicated trade-off between pursuing essential reforms and ensuring stability, the Maghrebi states have opted for the latter, with crucial enabling from their North Atlantic allies. As the pressures associated with climate change mount, the opportunities for reform will dwindle, not expand.

The abiding irony, of course, is that the Maghreb emits low levels of aerosols and greenhouse gases—carbon dioxide, methane, nitrous oxide, and chlorofluorocarbons. For example, the region emits between 1.5 and 3.5 metric tons of carbon dioxide per capita annually. The advanced industrial countries, by contrast, emit far more—9.55 metric tons in the United Kingdom, for instance, and 10.24 in Germany, both relatively green societies by prevailing standards, whereas the United States, much less so, leads the pack with annual emissions of 20.14 metric tons per capita (all figures are for 2005).[1] Yet, although it is less culpable in the production of greenhouse gases, the Maghreb's geographical position and complete lack of temperate climes renders it more intensely vulnerable to climate change dynamics than its advanced industrial counterparts. In the short term, according to data provided by CIESIN (see appendix A), Morocco and Tunisia exhibit a "very serious" relative temperature vulnerability, and Algeria is "average"; and Libya is not treated. The percentage of the population that has less than 1,000 cubic meters per year per capita is in the middle to high 90 percentiles. For Morocco, 99.5 percent of the population is deemed "short of water." According to the 2007 *Fourth Assessment Report of the Intergovernmental Panel on Climate Change*, the long-term outlook is worse than dire. The Maghreb and the broader Mediterranean Basin will experience extensive drying associated with an expansion of the downward descending arm of the Hadley circulation cells at 30 degrees north and 30 degrees south latitude. The mean annual rainfall is expected to decrease by as much as 20 percent along the Mediterranean Coast between 2080 and 2099, with summertime temperatures anticipated to increase for the 2070–99 period by up to 9 degrees Celsius.[2]

What Is the Maghreb? Geographical and Spatial Considerations

It makes good sense to consider the impact of climate change on the Maghreb writ large. In a geostrategic context, regions, and not simply states, remain crucial actors.[3] At the same time the intellectual conceptualization of the Maghreb requires some explanation because what constitutes the region remains open to interpretation. As will be evident when examining the prospective impact of climate change, the Maghreb is often lost somewhere between Sub-Saharan Africa and the Middle East in area studies formulations. It almost seems as if Sub-Saharan Africanists think that the Maghreb will be considered by Middle Eastern specialists, and vice versa. This phenomenon has been evident for decades in Anglo-American academic journals and programs and in the structure of area studies associations such as the African Studies Association and Middle East Studies Association.

The Maghreb—Arabic for "where the sun sets"—is historically considered to include three central countries: Morocco, Tunisia, and Algeria. To confuse the matter a bit, however, Morocco is known in Arabic as al Mamlakah al Maghribiyah, or Kingdom of the Maghreb. If one takes into account the history of Andalusia, southern Spain has also sometimes been included in conceptualizations of the region. Mauritania and Libya are also sometimes considered part of the Greater Maghreb. This has certainly been the case since February 1989, when Algeria, Libya, Mauritania, Morocco, and Libya signed a regional pact in Marrakech to establish the Arab Maghrebi Union, known in Arabic as Ittihad al-Maghrib al-Araby and more generally by the acronym "UMA" for its name in French, l'Union du Maghreb arabe.

Despite this geographical and conceptual muddle, the Maghreb still seems to cohere as a region. Although one might also invoke historical, linguistic, religious, cultural, and economic factors in conceptualizing the Maghreb, its crucial defining characteristics, particularly in relation to environmental issues, are geographical—the Mediterranean Basin, the Atlas Mountains, and the Sahara Desert. These geographical barriers provide a natural divide between the Maghreb and regions to the north, south, and east. Thus, for the purposes of this analysis, the Maghreb includes Morocco, Tunisia, Algeria, and Libya. Admittedly, Libya is east of the Atlas Mountain range, and it is often set apart from the others because of its strikingly different historical legacy as an Italian colony, as well as its post–World War II experience before its 1969 coup.[4] Nonetheless, Libya should be included in a consideration of the Maghreb because of its central role in regional politics, as well as its Mediterranean coastline and Saharan south.

Mauritania, notwithstanding its participation in the UMA, is better considered as part of the Sahel. Thus, it joins Mali, Niger, Chad, and the Sudan, as well as parts of Senegal, Burkina Faso, and Nigeria. Proponents of the UMA might protest this recategorization, yet because the regional grouping has fallen far short of its aspirations, perhaps "removing" Mauritania from the Maghreb is acceptable.

An additional observation concerns the Sahara's decisive role in understanding the Maghreb, especially in the context of climate change. After all, both Libya and Algeria have enormous tracts of desert in their southern zones, and the Sahelian countries, by definition, have significant portions of the Sahara in their northern zones. Thus, the "border" between the Maghreb and the Sahel to the south is highly dynamic. In the end, therefore, assessing the prospective impact of climate change on the Maghreb demands a sharp focus on the Sahara, in terms of its steady encroachment on less arid terrain and its role as a source and conduit for migrants, as a site of the extraction of hydrocarbons, and as an ostensible battlefield in the global "war on terror."

Historical Backdrop and Current Context

An assessment of future prospects requires a consideration of the current context. The reality of the Maghreb today is its relatively poor development record and stressed economies. Its economic growth in recent decades has been spotty, often being held hostage to oil commodity prices and erratic rainfall. The population of the four countries combined was 62 million in 1990, had grown to 75 million in 2000, and is expected to reach 91 million in 2015. Most of this growth has occurred in urban sectors, where a lack of planning has resulted in notably anarchic urbanization. Infrastructures in Maghrebi cities are wholly inadequate in providing sanitation and other necessary services to their populations.

One of the reasons for this rapid urbanization has been the dynamic transformation in the rural sectors. Maghrebi policymakers have implemented agricultural policies over recent decades that favor perimeter irrigation. They have constructed dams to provide for the water needs of rapidly growing populations, and (in the case of Tunisia and Morocco) have pursued the development of tourism, which brings with it profound water requirements. Such efforts have largely stemmed from the precarious, asymmetrically interdependent position in which the Maghreb finds itself in the international economy.[5] The result is rural sectors that push urban migration.

With respect to agriculture, governments have neglected traditional farming practices that demanded less water—for example, durum wheat and subsistence agriculture—and have instead pushed for the development of fruits and vegetables for export. Ironically, exports face enormous constraints because the EU's Common Agricultural Policy has protected European markets from Maghrebi produce, especially since Spain and Portugal joined the European Community in 1986. Similarly, recent trade liberalization efforts have constrained agriculture production to an even greater extent. Most notably, the 2004 U.S.–Moroccan Free Trade Agreement opened Morocco to the import of U.S. commodities—grain, dairy, meat, and poultry—while allowing for the export to the American market of citrus, sardines, and olive oil. This has encouraged the production of commodities with a significant environmental impact, rather than products more suited for the Maghrebi climate. In this respect William Cline's data (characterized in appendix A) suggest that the prospective impact of climate change on agricultural productivity is "very serious" for Algeria and Morocco, with Tunisia "serious."

Chronic drought in the region has also been accompanied by periodic locust infestations. The pests were evident in the late 1980s and early 1990s. They returned vigorously in the early part of this decade and are expected to persist.[6] The relationship between climate change, rainfall amounts, and infestation has been a focus of research for entomologists; it appears that it is less the total volume of rainfall that causes locust outbreaks than its sporadic and variable nature, which is likely to become more pronounced in the future. Locust infestation has also been evident in the Sahel and the Horn of Africa. To the extent that locust infestations contribute to crop failure in these areas, migration pressures on the Maghreb from the south will likely increase further, as discussed below.

The region's focus on tourism merits special mention. In the pursuit of job creation and foreign exchange, Maghrebi governments have aggressively pursued tourism development strategies.[7] Tunisia and Morocco are the prime examples, but Libya and Algeria are also keen to further develop their sectors. Much of this development has occurred without environmental protection, and the required sewage treatment, water provision, and energy inputs present significant challenges to the existing infrastructure, and to the environment's overall carrying capacity.[8]

In the case of Morocco, for example, the "concretization" of the coastline stemming from tourist development has been a problem for years.[9] Moreover, the current king's father, Hassan II, built an extensive system of golf courses. Although some observers (well, golfers, anyway) have applauded this development, the hydrological demands of golf courses present an almost absurd challenge in a country that has severe problems providing potable water to its population.[10] Coupled with the concomitant construction of hotels, swimming pools, adventure trekking tour routes, sports fishing infrastructure, and so on, the ecological implications are vast.[11] Yet Morocco is still keen to construct more tourist attractions in the coming decade. Construction sites dot the Marrakech environs.

For all four countries, the development of respective tourism sectors is a central pillar of development strategies, made viable by crucial support from the United States, the EU, and multilateral agencies. In 2001, the World Bank provided a loan of $17 million to improve the infrastructure for Tunisia's historical patrimony sites.[12] As Hazbun points out, these efforts are actually complimentary to the Maghrebi state's authoritarianism, because it helps "the regime [to] project an outward image of stability and openness to both tourists and investors while expanding state control over domestic spaces and transnational flows."[13]

Efforts at Environmental Protection

Morocco, Algeria, and Tunisia each signed the United Nations Draft Convention on Climate Change in Rio de Janiero in 1992. In addition, at the level of policy rhetoric, each country claims to be working to fight against climate change.

Tunisia's highly authoritarian government under President Zine al-Abidine Ben Ali appears to have gone a bit further than others in offering environmental protection programs. At least it is more adept in publicizing efforts such as the construction of greenbelts and wastewater plants. Nonetheless, Tunisia's pursuit of mass beach tourism, in contrast to Morocco's pursuit of a high-end market, has presented similar challenges to Morocco's. Still, the Tunisians have stepped up potable water preservation efforts, restricted summer crop production, and controlled the opening of forestland for grazing.[14] Morocco's efforts have been less prominent. Although it has sought to form interministerial committees to combat the effects of drought and begun desalination plants in Laâyoune and Boujdour, its record on pollution control and the purification of wastewater is less positive.[15]

As for Algeria and Libya, their efforts to pursue environmental protection have been scattered and relatively ineffective. Ironically, perhaps, Algeria's minister of tourism and its minister of the environment are the same person. Libya's policies since its renunciation of weapons of mass destruction in 2003 have taken on a decidedly green tint, even if only at the level of rhetoric.[16]

The Anticipated and Estimated Scope of Climate Change for the Maghreb

Of course, specifying the extent of climate change for the region must include the usual caveats about the imprecision of the current data and the uncertainty of future projections. Many of the issues at hand are the subject of continuous investigation. In the case of desertification, for example, methodological and definitional disputes are ongoing.[17] Nonetheless, before turning to potential scenarios for the region, it is wise to offer indications of anticipated changes.

Remarkably, little attention has been paid to the Maghreb in the otherwise much-esteemed reports prepared by the IPCC, whose focus has fallen primarily on Sub-Saharan Africa. Additional insight can be gained from the Ali Agoumi study, funded by Canada's International Institute for Sustainable Development and the U.S. Agency for International Development. Agoumi projects warming of 2 to 4 degrees Celsius in this century, and more than 1 degree Celsius between 2000 and 2020.[18] More directly, he points to several areas of concern. With respect to water, for example, the water table has decreased in recent years, which has contributed to the salinization of coastal groundwater, low potability, and low freshwater volume. As for soil, the region has experienced both intensive erosion and significant degradation, and the "development" of forested areas will only exacerbate further these trends.[19] All four countries have been vulnerable to desertification. The Jifara Plain in Libya's northwest and the Ouergha watershed in Morocco are especially vulnerable to decreases in rainfall. According to the United Nations Food and Agricultural Organization (FAO), 97 percent of the Libyan population, and 80 percent of Moroccans, were already at risk from desertification in 1997.[20] Encroaching desert has been a long-standing problem in the region. Its potential deepening has prompted NATO to study its impact on Mediterranean politics generally.[21]

With specific respect to the Maghreb's precarious agriculture sector, the Twenty-Ninth FAO Regional Conference for the Near East, held in Cairo in March 2008, laid out rather dire predictions.[22] Using IPCC projections, regional specialists within the FAO argued that water runoff (the difference between rainfall and evapotranspiration) will decline, the number of dry days is expected to increase, and surface temperatures are expected to rise. The result is a decrease in the yields of key crops, the possible extinction of some species, and a silting of rivers and dams.

Additionally, the FAO's Cairo report offered sharp anxiety about the impact on rangelands for livestock, because reduced rainfall and increased evapotranspiration will cause significant changes in vegetation cover and organic carbon storage in the ecosystems. Livestock pest and disease distribution and transmission may also result in new epidemics.

Finally, coastal regions are at risk from sea-level rise and inundations, and the accompanying salinization of coastal groundwater. These are matters of an additional concern because of their impact on the tourism sectors devoted to mass beach tourism, and on the prime real estate and urban areas along the coast. The majestic Hassan II Mosque in Casablanca, for example, built by King Hassan II in the 1990s in his own honor, and with "contributions" from his subjects, juts out into the Atlantic Ocean and is often splashed with crashing waves. According to the World Bank, North Africa is at extreme risk from sea-level rise, whether one accepts the conservative estimate of a 1-meter rise, or more dire predictions of 2 to 3 meters.[23] The Maghreb is at greater risk than Latin America and the Caribbean, albeit not as jeopardized as South Asia and East Asia. The reason is simple: the large proportion of the population that lives in coastal areas. According to CIESIN data (see appendix B), Tunisia is at the greatest risk in the region (after Egypt) for a 1-meter rise in sea level; for a more substantial 3-meter rise, Tunisia and Morocco are significantly threatened.

In sum, climate change is anticipated to be quite profound for the Maghreb. Coupled with the already-thirsty hydrological demands of agricultural and industrial activities, climate change will certainly be disruptive. A detailed understanding of exactly how will require a further refinement of climate modeling. At present, despite the best efforts by scientists and modelers, much uncertainty remains. Nevertheless, enough evidence exists to support a general analysis of the likely impact of climate change.

Future Prospects

The prospects for disruptive change must be considered as "intermestic" in their blending of domestic and international dimensions. In other words, the very dichotomy between the domestic and international spheres is questionable.[24] That said, it is helpful to separate the two realms as far as possible for the purpose of analysis.

Domestic Issues

The major domestic social impact that is likely to arise from climate change will be accelerated and probably chaotic urbanization, as migrants leave stressed rural areas and move to urban centers seeking employment in industrial and service sectors. Such movements pose challenges for already-overburdened urban infrastructures and sectors unable to generate jobs. On a related note, unemployment in Mediterranean and Atlantic coastal communities is likely to increase, as already-stressed waters are unable to support the fishing sector. This has led to reports of fishermen using their underutilized fishing boats to smuggle migrants across the Atlantic passage to the Canary Islands.[25]

Within the context of accelerating urbanization, the most serious problems are likely to arise in the provision of health and social services, as shortages of potable water jeopardize nutrition and public health. Maghrebi crucial indicators such as infant maternal health, caloric intake, and the like are already at low levels. Table 15.1 shows UNDP data for 2005. Such low levels of care are likely to worsen as urban populations rise and freshwater supplies fall.

Internal conflict may also arise along lines that are more conventional. The Arabo–Berber identity schism remains central in Morocco and Algeria and often mobilizes around grievances associated with access to economic resources or political power. The urban/rural divide remains profound in all four countries, but there are also sharp regional distinctions to consider—between Fessis and Soussis in Morocco, or between Cyrenaica and Tripolitania in Libya—that are often reflective of different modes of economic activity. Differential climate effects are almost certain to heighten these disparities and to make the conflict resulting from them more salient.

The amelioration of all forms of social stress in the region is complicated by the rise of Islamism. Islamist activity is harshly controlled in all four countries, which is likely to continue in the future. Morocco's Justice and Development Party, known in Arabic as Hiza al-Adala wal-Tanmiyya and more commonly by the acronym "PJD" for its name in French, Parti de la Justice et du Développement, did well in the September 2007 elections. Yet, like its similarly named counterpart in Turkey, the PJD is willing to play by the rules of the electoral game. At the same time, the appeal of radical Islamists remains high among the population, especially given the resonance of radical Islam's critique of the West. It is important to stress that radical Islam is not ipso facto violent.[26] That said, there is no reason that Islamists in the region cannot incorporate climate change into their basket of grievances, a container that already includes criticisms of regimes as too closely allied with the West, supportive of secular modernity, and open to rapacious globalization and capitalism that benefits only a few.

Religious actors in general are likely to play a prominent role in managing the social consequences of climate change, as they already do in response to natural disasters. In the aftermath of natural calamities like earthquakes or floods, mosques are often the first and most effective responders.[27] For example, in al-Hoceima, Morocco, a powerful earthquake in 2004 caused hundreds of deaths and casualties. Given the fact that the city is located in the Rif Mountains, a primarily Berber region critical of (and distant from) the monarchy in Rabat, there were protests against the government's slow, inefficient relief efforts. Local mosques stepped into the breach left by inadequate government services. Obviously, the degradation associated with climate change is more gradual and less galvanizing than a natural disaster like an earthquake. Still, most mosques and imams in the region either are employed by governments or are carefully monitored. How the mosques' involvement in meeting the challenges caused by climate change would play out in terms of state–society relations is hard to predict.

International Issues

Climate stress in the Maghreb is also likely to create heightened external pressures, in the form of "climate refugees" from the Sahel and Sub-Saharan Africa.[28] Hein de Haas is correct

Table 15.1 Maghrebi Infant and Maternal Health and Mortality

Country	Human Development Index Rank	Infant Mortality (per 1,000 live births), 2005	Children Underweight for Age (% under age 5) 1996–2005	Births Attended by Skilled Health Professional, 1997–2005 (%)	Maternal Mortality Ratio (per 100,000 live births)
Algeria	104	34	10	96	180
Morocco	126	36	10	63	240
Tunisia	91	20	4	90	100
Libya	56	18	5	94	97
Spain	13	4	—	—	4
Germany	22	4	—	100	4
Norway	2	3	—	100	7
United States	12	6	2	99	11

Source: UNDP, *Fighting Climate Change: Solidarity in a Divided World, 2007–08* (New York: Palgrave-Macmillan, 2008).

to encourage an avoidance of invasion imagery, because it plays into anti-immigration sentiment within Europe.[29] It also reinforces a clumsy security response on the part of strategic players such as the EU, the United States, and NATO. Yet at the same time, it would be absurd to ignore the dynamic and its future impact.[30] These movements are obviously not solely the result of environment factors, but climate change likely plays a central role, particularly given the increasing evidence that it is displacing rural populations in Sahelian countries. Since the 1990s, the influx of migrants has presented an issue around the Spanish enclaves of Ceuta and Melilla on Morocco's northern coast, which is likely to worsen in years to come.[31] There will also be mixed flows from (or through) the Maghreb to Europe via the eastern passage across Libya and Tunisia to Italy.[32] In terms of trans-Mediterranean diplomacy, the Italian authorities have worked closely with the Libyan and Tunisian interior ministries in apprehending and detaining African migrants in camps on the North African coast, much to the consternation of human rights observers. The EU has likewise been urging Morocco to police its southern frontiers more effectively. Enlisting the Maghrebi governments in immigration control has been part of Europe's effort to "thicken" its southern border by extending a buffer deeper into North Africa.

The prospects for civil conflict stemming from migrant flows into the Maghreb should not be underestimated. In many cases, it has already emerged. In September 2005 *al-Shamal*, a Tangier newspaper, referred to Sub-Saharan Africans as "black locusts" invading the north. The Moroccan authorities promptly banned the paper, but the sentiment is still emblematic of perceptions in the country. Similar anti-immigrant sentiment is evident in Tunisia and Libya. Libya's long-standing openness to migrants from Sub-Saharan Africa to work in its oil fields is well documented. It is part of Qaddafi's "Africa Policy," developed in the aftermath of the 1992 UN sanctions. Although the oil economy is doing well in Libya, the country is not equipped to handle the pressures posed by its immigrant population. As in so many other contexts, immigrants often become scapegoats for poor economic performance.

Europe's efforts to put strong migrant controls in place have continued apace since the 1990s. These have taken the form of barrier fences around Ceuta and Melilla, which gained international notoriety after migrants' efforts to breach them in 2005 and 2006 resulted in death and injury.[33] Sophisticated "electronic fences" have also been placed in the ocean, initially by Spain and Italy, and increasingly with the support of Frontex, a EU agency created in 2005. These watery borders stretch across the Mediterranean and the Atlantic passage to the Canary Islands. As a result, increasing barriers to migration might prompt internal, civil conflict, as various actors contend with the challenges. This might take on the form of ethnic and racial clashes emerging from the inflow of migrants from the south. Morocco's population of Sub-Saharan African immigrants is sizable, as people trek to the northern Rif Mountains trying to gain access to Ceuta and Melilla or congregate in the southern stretches of Morocco seeking passage to the Canary Islands.

State Failure and Political Legitimacy

It is indicative of how seriously climate change is now viewed that consideration of its security implications includes the outright failure of states. Yet what is considered a failure is arguably a matter of opinion. From a Weberian perspective, one should prize legitimate authority. Yet, given the lack of democratic and liberal standards of public life in the Maghreb, the use of force by the state is anything but legitimate; it is failure of a kind, perhaps, but not synonymous with widespread social collapse. All four Maghrebi states are authoritarian. Freedom House's categories are blunt yet instructive; it lists Morocco as "partly free" and Algeria, Tunisia, and Libya as "not free."[34] Morocco has increasingly distinguished itself with liberal reforms, especially in the 1990s and this decade.[35] Yet, generally speaking, civil liberties are highly constrained throughout the region, judiciaries are not independent, press freedoms are circumscribed, and banking systems are highly dysfunctional. Moreover, each country has difficulties associated with succession. Morocco could rely on dynastic principles if its forty-something king, Mohammed VI, were to pass away. Successors to Algeria's Abdelaziz Bouteflika, Tunisia's Zine al Abidine Ben Ali, and Libya's Muammar Qaddafi remain the subject of constant debate and speculation, especially as all three are now quite elderly.

As noted at the outset, a crucial source of legitimacy for all the Maghrebi governments comes from North Atlantic actors (the EU and the United States) and not from internal, domestic constituencies. Of course, all four regimes rest upon social bases whose members benefit from their illiberality. Yet even with the pressures and challenges associated with prospective climate change, state failure is *not* going to occur because the Maghrebi state is so thoroughly imbricated with the European state, and vice versa. It is impossible to think of Italian state formation in the nineteenth and twentieth centuries without taking into account colonialism in Libya, as well as the influential Italian settler community in Tunisia.[36] Algeria, to state the obvious, is at the very heart of French history. In turn, the history of Maghrebi state formation cannot be understood without taking Europe into account, especially the legacy of European colonialism.[37]

It is also impossible to understand the Maghrebi states in isolation from their position on Europe's southern flank. Since the early 1960s, European officials in Brussels, and their national counterparts in Madrid, Paris, and Rome, have worked closely with Maghrebi officials on a wide array of technical issues: security, energy, transportation, communication, and trade.[38] Officials of interior ministries in the Maghreb routinely meet with their North Atlantic counterparts to share intelligence and support. Cooperation also extends to hydrocarbons, energy, and communication. The beginnings of an integrated energy infrastructure date from the late 1970s, featuring pipelines that transport Algerian hydrocarbons through Morocco and Tunisia to Spain and Italy, respectively.[39]

The Maghrebi states may thus (continue to) fail in terms of crucial indicators associated with liberal democratic reforms, but they will do so with the tacit support of North Atlantic powers, whose influence will help ensure that political illegitimacy does not give rise to social collapse. The Maghrebi states will simply not be allowed to fail in this latter sense, above all because of their crucial role in the "war on terror," a topic to which the chapter now turns in conclusion.

Climate Change as a Security Issue

The Maghreb, as has already been noted, is vulnerable to climate change induced by the advanced industrial countries. To this an additional incongruity must be added: the reality that assessments about the impact of climate change often devolve into concerns about security by Western actors, whose interventions may further exacerbate the social and political inequities that climate change prompts.

For millennia, ethnic groups in what are now Sudan, Chad, Niger, and Mali have transgressed official, contemporary borders with Libya, Algeria, and the Western Sahara. Today, such transgressions are seen as security threats. They have been and will likely continue to be cast as interstate conflicts. Further, Europe and the United States have also provided security resources to patrol oil fields in Algeria and Libya.[40] Such efforts will likely intensify in the near future. Assessing the likelihood of attacks against installations is difficult. If it were to occur, however, it would probably not be the result of an effort to seize energy resources. The aim would almost certainly be to disrupt energy flows, or perhaps to dramatize social grievances that have been heightened by high levels of energy consumption elsewhere. The environmental context for such actions will likely continue to be ignored, and the region will be further securitized as a consequence.

This dynamic was especially evident in the aftermath of the September 2001 terrorist attacks against the United States. The Pan-Sahel Initiative (PSI) was initiated in 2002 and was succeeded by the Trans-Sahara Counterterrorism Initiative (TSCTI) in 2005. In February 2007, U.S. secretary of defense Robert Gates also announced the creation of a new Africa Command (AFRICOM), which may have unanticipated negative consequences in terms of the perception of the local population. The Barack Obama administration has continued the development of AFRICOM, a command in which the counterterrorism mission may be presumed to loom large.[41]

The problem, simply put, is that as governments are seen to be collaborating with the United States, their legitimacy may be further undermined. In this respect Cédric Jourde's analysis of Mauritanian politics is relevant for the Maghreb and Sahel together: "By representing Mauritania almost exclusively in security terms, and consequently by implementing security policies there, the U.S. government has interfered in local Mauritanian politics, and specifically in upholding authoritarianism. In effect, as American support to the regime increased from the end of the 1990s to the current era, so too did the regime's capacity to thwart a real democratization of its institutions and practices."[42]

In other words, such securitization provides not only tangible support to an authoritarian regime in the form of military training and equipment that is central to the regime's "coercive architecture" but also offers important symbolic support. As Jourde further details: "The alliance [has] fed growing opposition to both the United States and the local regime. PSI and TSCTI have contributed to the resentment that the population feels towards a neo-patrimonial regime that favours those with close connections to the presidential clan and that uses its control over security forces to prevent popular mobilization."[43] By fueling the region's regimes in the name of security, North Atlantic actors may prompt the very circumstances they seek to avoid.

What is the solution? One thing to bear in mind is that the upside of the Maghrebi states' close ties with North Atlantic actors is that they would be in a position to change course given the right kind of signals. Even Libya has been careful to engage with European and, increasingly, U.S. diplomacy. Moreover, there is no potential actor on the horizon that might deploy anti-U.S. or anti-European rhetoric. Certainly, a state actor emerging on the order of Venezuela's Hugo Chávez or Iran's Mahmoud Ahmadinejad is hard to envision.

In the end, therefore, Maghrebi states will remain open to North Atlantic influence. This may represent an opportunity for the Obama administration. If Washington were to exhibit leadership on climate change, join in efforts to address its causes and implications, and eschew blunt instruments like the TSCTI, the Maghrebi nations just might follow. Although the glass is half empty and filled with dirty water when it comes to prospects for social justice in the region, if the international community turned to a full-hearted effort to address climate change, the Maghrebi states would likely follow and benefit. At the local level, such efforts would include greater attention to agricultural reforms emphasizing sustainable development, clean industrialization technologies, sustainable tourism projects, and alternative energy development. At the international level, global efforts to slow the demand for hydrocarbons and reduce carbon emissions would also redound positively in the region. It would remove the sources of funding from the region's rentier regimes—Algeria and Libya—and ideally provide a catalyst for needed reforms, lower fuel import costs for Morocco and Tunisia, and demonstrate that economic reform by the North Atlantic is preferable to a facile, securitized response.

Notes

1. U.S. Department of Energy, www.eia.doe.gov/environment.html.

2. See Michel Boko, Isabelle Niang, Anthony Nyong, Coleen Vogel, Andrew Githeko, Mahmoud Medany, Balgis Osman-Elasha, et al., "Africa," in IPCC, *Climate Change 2007: Impacts*, 433–67. See also L. R. Kump, J. F. Kasting, and R. Crane, *The Earth System* (Upper Saddle River, NJ: Prentice Hall, 2004).

3. See Peter Katzenstein, *A World of Regions: Asia and Europe in the American Imperium* (Ithaca, NY: Cornell University Press, 2005).

4. Lisa Anderson, *State and Social Transformation in Libya and Tunisia, 1820–1980* (Princeton, NJ: Princeton University Press, 1986).

5. Gregory White, "The Maghreb in the World's Political Economy," *Middle East Policy* 14, no. 4 (2007): 42–54.

6. The UN Food and Agricultural Organization provides a "Locust Watch" site at www.fao.org/ag/locusts/.

7. See, e.g., Waleed Hazbun, "Images of Openness, Spaces of Control: The Politics of Tourism Development in Tunisia," *Arab Studies Journal* 16, no. 1 (2008): 10–35.

8. Peter Haas, *Saving the Mediterranean: The Politics of International Environmental Cooperation* (New York: Columbia University Press, 1990).

9. Mohamed Berriane, "Environmental Impacts of Tourism along the Moroccan Coast," in *The North African Environment at Risk*, edited by W. Swearingen and A. Bencherif (Boulder, CO: Westview, 1996).

10. David Owen, "The Sporting Scene: Swinging in Morocco," *New Yorker*, May 21, 2001.

11. Deborah McLaren, *Rethinking Tourism and Ecotravel* (Bloomfield, CT: Kumarian, 1998).

12. To be sure, Tunisia has a high-end market, too. Denny Lee placed it third in an articled titled "The 53 Places to Go in 2008," *New York Times*, December 9, 2007. Lee writes: "Tunisia is undergoing a Morocco-like luxury makeover. A new wave of stylish boutique hotels, often in historic town houses, has cropped up alongside this North African country's white-sand beaches and age-old medinas, drawing increasing numbers of well heeled travelers." Libya was 10th, Algeria 43rd, and Morocco did not place. For an article that utterly disregards the political context of the country, also see Eric Lipson, "Where Europe, Africa, and the Middle East Meet in Tunisia," *New York Times*, May 24, 2008.

13. Hazbun, "Images of Openness," 10–35.

14. Ali Agoumi, *Vulnerability of North African Countries to Climatic Change: Adaptation and Implementation Strategies for Climate Change* (Winnipeg: International Institute for Sustainable Development, 2003).

15. Ibid.

16. Elisabeth Rosenthal, "A Green Resort Is Planned to Preserve Ruins and Coastal Waters," *New York Times*, October 16, 2007.

17. Mamdouh Nasr, *Assessing Desertification and Water Harvesting in the Middle East and North Africa: Policy Implications*, Discussion Papers on Development Policy (Bonn: Zentrum für Entwicklungsforschung, Universität Bonn, 1999).

18. Agoumi, *Vulnerability of North African Countries to Climatic Change*.

19. Ibid.

20. United Nations Food and Agricultural Organization, *Climate Change: Implications for Agriculture in the Near East* (Rome: United Nations Food and Agricultural Organization, 2008).

21. William Kepner, José Rubio, David Mouat, and Fausto Pedrazzini, eds., *Desertification in the Mediterranean Region: A Security Issue* (Dordrecht: Springer, 2006).

22. Ibid.

23. Susmita Dasgupta, Benoit Laplante, and Craig Meisner, *The Impact of Sea Level Rise on Developing Countries: A Comparative Analysis*, World Bank Policy Research Working Paper 4136 (Washington, DC: World Bank, 2007), www-wds.worldbank.org/external/default/WDSContentServer/IW3P/IB/2007/02/09/000016406_20070209161430/Rendered/PDF/wps4136.pdf.

24. R. B. J. Walker, *Inside/Outside: International Relations as Political Theory* (Cambridge: Cambridge University Press, 1993).

25. Hannah Godfrey, "On a Voyage of Peril to the Mirage of Europe," *Observer* (London), November 19, 2006.

26. See John Entelis, "Political Islam in the Maghreb: The Nonviolent Dimension," in *Islam, Democracy, and the State in North Africa*, edited by John Entelis (Bloomington: Indiana University Press, 1997), 43–73.

27. Earthquakes are not tied to climate change, but floods can be. Hard rainstorms, after several years of dry or partially dry years, result in sharp runoffs.

28. See Molly Conisbee and Andrew Simms, *Environmental Refugees: The Case for Recognition* (London: New Economics Foundation, 2007); and Diane C. Bates, "Environmental Refugees? Classifying Human Migrations Caused by Environmental Change," *Population and Environment* 23, no. 5 (2002): 465–77.

29. Hein de Haas, *The Myth of Invasion: Irregular Migration from West Africa to the Maghreb and the European Union* (Oxford: International Migration Institute / James Martin 21st Century School, 2007).

30. For a map illustrating the main paths of population movement, see "Key Facts: Africa to Europe Migration," BBC, July 2, 2007, http://news.bbc.co.uk/2/hi/europe/6228236.stm.

31. Gregory White, "Sovereignty and International Labor Migration: The 'Security Mentality' in Spanish–Moroccan Relations as an Assertion of Sovereignty," *Review of International Political Economy* 14, no. 4 (2007): 690–718.

32. Bruno Siragusa, "Lampedusa, European Landfall," in *The Long March to the West: Twenty-First Century Migration in Europe and the Greater Mediterranean Area*, edited by Michel Korinman and John Laughland (Portland, OR: Valentine-Mitchell, 2007).

33. There is a diagram of planned enhancements to the fences at Ceuta and Melilla at http://news.bbc.co.uk/2/shared/spl/hi/pop_ups/05/europe_enl_1128701984/html/1.stm.

34. See the Freedom House website, www.freedomhouse.org.

35. Gregory White, "'The End of the Era of Leniency' in Morocco? Mohammed VI's Halting Glasnost," in *North Africa: Politics, Religion and the Limits of Transformation*, edited by Yahia H. Zoubir and Haizam Amirah-Fernández (London: Routledge, 2008).

36. Dirk Vandewalle, *Libya since Independence: Oil and State-Building* (Ithaca, NY: Cornell University Press, 1998).

37. See, e.g., Anderson, *State and Social Transformation*; Jamil M. Abun-Nasr, *A History of the Maghrib in the Islamic Period* (Cambridge: Cambridge University Press, 1987); and Michel Le Gall, "The Historical Context," in *Polity and Society in Contemporary North Africa*, edited by I. William Zartman and Mark Habeeb (Boulder, CO: Westview Press, 1993).

38. White, "Maghreb in the World's Political Economy," 42–54.

39. Mark Hayes, *Algerian Gas to Europe: The Transmed Pipeline and Early Spanish Gas Import Projects*, Program on Energy and Sustainable Development at Stanford University and James A. Baker III Institute for Public Policy of Rice University Working Paper 27, May 2004, www.rice.edu/energy/publications/docs/GAS_TransmedPipeline.pdf. Hayes's study includes a map showing major existing and proposed pipeline routes, which illustrates the increasing integration of energy infrastructure between the Maghreb and Europe.

40. Anthony Cordesman, *A Tragedy of Arms: Military and Security Developments in the Maghreb* (Westport, CT: Praeger, 2003).

41. See "Trans-Sahara Counterterrorism Initiative [TSCTI," GlobalSecurity.org, www.globalsecurity.org/military/ops/tscti.htm, and the associated map "Pan-Sahel Initiative," at www.globalsecurity.org/military/ops/images/psi-map.jpg, which offers a counterterrorist conceptualization of the region's situation.

42. Cédric Jourde, "Constructing Representations of the 'Global War on Terror' in the Islamic Republic of Mauritania," *Journal of Contemporary African Studies* 25, no. 1 (2007): 77–100, at 93.

43. Ibid.

16

West Africa I

Côte d'Ivoire, Nigeria, and Senegal

Linda J. Beck and E. Mark Pires

Unlike the multiple countries examined in some other chapters of this volume, this chapter's three West African countries have few common denominators. They do not even share a common border. Instead, there are immense socioeconomic, political, and demographic disparities among them. A former middle-income country with 21 million inhabitants, Côte d'Ivoire has been a major recipient of foreign direct investment in Africa, but it has suffered economically and politically from a recent civil war that continues to reverberate in a fragile postconflict context. With the largest population in Africa, an estimated 150 million inhabitants, oil-producing Nigeria is an economic powerhouse whose political and economic development has been undermined by endemic corruption and a virulent form of ethnopolitics. Finally, Senegal, the smallest of the three countries in territory and population with 12 million inhabitants, has been a model of political stability despite its dire economic circumstances, which are tied to a lack of natural resources and adverse climatic conditions. However, what these three countries do have in common is that they have been regional leaders that therefore should be relatively well positioned to address the negative impacts of climate change that are predicted to affect the entire region. But we argue here that their actual capacity to do so is weakened by various political and socioeconomic factors specific to each country.

The sixteen countries in West Africa cover a territory of approximately 6.2 million square kilometers, roughly equivalent to three-quarters of the contiguous United States.

Cote d'Ivoire

Nigeria

Senegal

This extensive region is marked by diverse natural environments, including tropical rainforests in the south, savannas and the semiarid steppe of the Sahel in the central interior, and the Sahara Desert to the north. This north–south ecological gradient is represented to varying degrees in all three countries discussed in this chapter.

In terms of climate, both seasonal and annual precipitation patterns vary considerably in the region, with extreme events such as floods and droughts being common occurrences. According to the United Nations Environment Program, average annual rainfall in the region decreased significantly in the second half of the twentieth century, coincident with more frequent, severe, and widespread droughts.[1] Though the normal vagaries of the West African climate already pose considerable challenges to the region, various scientific reports on global climate change, including key analyses provided by the IPCC, point to Africa as the world region expected to bear the brunt of its negative consequences.[2]

Although none of the diverse ecological zones of the region is expected to remain unaffected, the Sahel is considered a climate change "hotspot" due to recurrent drought and increased human pressures that have contributed to considerable land degradation and desertification in recent decades. A recent report from the German Advisory Council on Global Change addressing the implications of climate change for international security concludes that climate change in this already-distressed subregion "will lead to increasing regional destabilization, including increased potential for violent conflict."[3]

All three countries examined in this chapter are expected to experience several similar consequences due to climate change in the region, although to varying degrees, including (1) the negative effects of potential sea-level rise on densely populated and economically important coastal areas; (2) a decline in agricultural production due to changes in temperature, moisture, and other atmospheric variables that affect plant growth; and (3) decreasing water availability, particularly in the northern Sahelian region of Senegal and Nigeria and urban coastal areas. To analyze their impact on regional security, the remainder of this chapter is divided into three sections that provide analyses of the political and socioeconomic risk factors that influence each country's capacity to address the negative impacts of climate change.

The Political Threat of Climate Change in Postconflict Côte d'Ivoire

By any reckoning, the Côte d'Ivoire's capacity to adapt to the climate change hypothesized by modern earth science must be judged poor, the more so given recent socioeconomic and political developments there. This assessment reflects the dramatic deterioration in the political and economic situation of Côte d'Ivoire since the 1980s, when it was categorized as a middle-income country with a high level of political stability under the one-party state of the Parti Démocratique de Côte d'Ivoire (PDCI), led by President Félix Houphouet-Boigny.[4]

Political and Socioeconomic Risk Factors in Côte d'Ivoire

Côte d'Ivoire first began experiencing an economic downturn with the fall of commodity prices, particularly cocoa, starting in the 1980s.[5] As elsewhere in Africa, economic decline led to political pressure for a return to a multiparty system that occurred during the 1990 presidential election. President Houphouet was nevertheless able to garner 80 percent of the vote against his opponent, Laurent Gbagbo, the leader of a small left-leaning party, the Front Populaire Ivoirien (FPI). Instability set in, however, with the power struggle that ensued after Houphouet's death in 1993, resulting in a political deadlock between Prime

Minister Alassane Ouattara and Henri Konan Bedié, the president of the Ivorian legislature and Houphouet's heir apparent.

Although a constitutional reform instituted by Houphouet assured that Bedié would succeed him upon his death, Bedié had to stand for election once Houphouet's term ended in 1995.[6] In an effort to sideline his opponent, President Bedié revived the concept of Ivoirité ("Ivorian-ness"), originally introduced in the 1970s to unify the country under a common national identity. Bedié transformed Ivoirité into a discourse of xenophobia, using it to justify revision of the electoral code to include a requirement that both parents of all presidential candidates be of Ivorian nationality. This effectively disqualified Ouattara, whose mother was Burkinabè, from running as the candidate of his newly created party, the Rassemblement des Républicains (RDR).[7] With the ensuing opposition boycott, the electoral mandate Bedié received in 1995 was at best tainted.

The continuing power struggle between Bedié and Ouattara ultimately gave rise to a military coup in 1999 led by Brigadier General Robert Guëi, a first in Ivorian history. The initial mission of the military regime was to level the playing field for competitive elections among civilian politicians, though once in power, Guëi decided to run in the 2000 presidential election. Although the Supreme Court disqualified his most formidable challengers, Bedié and Ouattara, Guëi nevertheless lost the election to Laurent Gbagbo. Guëi then made an unsuccessful power grab that led tens of thousands of Ivorians to take to the streets.[8]

After Guëi fled west toward the Liberian border, clashes ensued between supporters of Gbagbo and Ouattara over whether a new election should be held. This resulted in hundreds of deaths and the deepening of ethnoregional and religious tensions between the predominantly Muslim and Voltaic ethnic groups in the north that serve as the political base for Ouattara's RDR, and the predominantly Christian south, which is made up largely of two ethnic groups, the Bété (Kru) in the west, which is the base for Gbagbo's FPI, and the Akan in the central-eastern region, which provides Bedié's PDCI with political support.[9]

By early 2002, the country seemed to be regaining political stability following a national reconciliation forum and the creation of a government of national unity that included four RDR ministers. The military mutiny in September 2002 that led to Côte d'Ivoire's first civil war was therefore somewhat unexpected. In the ensuing conflict, a rebel movement led by a former student union leader, Guillaume Soro, gained control of the northern region of the country. The Mouvement Patriotique de la Côte d'Ivoire (MPCI) was initially presumed to be tied to the RDR because of the northern base of the rebel movement and its rejection of the policy of Ivoirité. The negative ramifications of Ivoirité, however, have extended far beyond elite politics, intensifying xenophobia and violence against the millions of first- and second-generation immigrants who constitute over a quarter of the population.[10]

With the assistance of four thousand French troops joined by eight thousand UN peacekeepers, progovernment forces were able to confine the MPCI to the north and also push two other western-based rebel groups there: the Mouvement Populaire Ivoirien du Grand-Ouest and the Mouvement pour la Justice et la Paix, both of which supported Guëi, who died during the initial coup attempt. The three rebel groups then joined forces under the banner of the Forces Nouvelles led by Soro, which continues to effectively control the north despite the Ouagadougou Peace Accord signed in March 2007 by President Gbagbo and Soro, who was designated prime minister by presidential decree. By sidelining both his political opponents and overriding guidelines established by the UN Security Council to resolve the crisis, the Ouagadougou Accord not only handed back uncontested executive authority to President Gbagbo but also postponed yet again the presidential elections that were initially scheduled to take place in 2005.

It remains unclear whether the Ouagadougou Accord will hold, especially given persistent problems regarding disarmament, incorporation of rebel officers into national military ranks, restoration of government administration in the north, adjudication of citizenship status, and revision of the electoral registry.[11] Critical to the peace process is successful conduct of the presidential elections, which were scheduled to be held in November 2008 but have been repeatedly postponed. In February 2010 voter registration was suspended indefinitely after violent protests against government handling of the process.[12]

Although the security situation has improved somewhat with the Ouagadougou Accord, economic hardship remains widespread, particularly among those who have been displaced by the conflict. Some of the 750,000 displaced individuals have been able to return, though many other "immigrants" have been forced to live in "host villages" far from where they had earned their livelihood, which was primarily in the cocoa-producing regions where output continues to fall.[13] Unfortunately, as long as the political crisis continues, attention to economic policy will remain secondary, with foreign direct investment also suffering. Although still relatively high in comparison with most other African countries, foreign direct investment in Côte d'Ivoire declined by more than $50 million in 2006 according to the United Nations Conference on Trade and Development.[14]

As reflected in the creation of immigrant "host villages," human rights groups accuse officials of continuing to encourage a culture of violent xenophobia.[15] Freedom House describes the government as openly favoring Christianity and targeting Muslims, who are predominant in the rebel-held north, while Amnesty International reports that progovernment supporters continue to incite violence against "Dioulas," a generic term for anyone with a Muslim family name originating from the north or other predominantly Muslim countries in the region.[16] Human Rights Watch reports that extortion at checkpoints is systematic and widespread, while sexual violence against girls and women remains prevalent.[17] Meanwhile, a number of militias and youth activists, collectively known as "the young patriots," have proliferated as a de facto militant wing of Gbagbo's hard-line supporters who engage in violent criminal acts with impunity. None of this bodes well for the capacity of Côte d'Ivoire to confront the negative impacts of climate change.

The Negative Impact of Climate Change in Côte d'Ivoire

The CIESIN analysis summarized in the appendixes to this volume indicates that Côte d'Ivoire, given its poor adaptive capacity, is particularly susceptible to the socioeconomic and political disruptions that could result from climate change. Areas of concern that may trigger disruptive change would likely stem from problems associated with declines in agricultural productivity, water availability, and coastal environmental conditions.

Although Côte d'Ivoire is not generally classified as part of the Sahel, this could conceivably change in the coming decades, at least in the drier ecological zone in the north, which border the Sahelian countries of Mali and Burkina Faso. This region remains under effective control of the Forces Nouvelles. Climate-related changes that could lead to increased drought, rainfall variability, and desertification in this ecologically and politically fragile region are likely to adversely affect agricultural production of both food and cash crops, specifically cotton. A recent IPCC report indicates that the mixed rain-fed and semiarid agropastoral systems practiced in West Africa's Sahelian zone are expected to be negatively affected by climate change.[18] Similar conclusions were also reached in a study conducted by the Nairobi-based International Livestock Research Institute, which examined alterations in the length of the growing season under similar climate change scenarios. According to this study, Côte d'Ivoire is likely to experience substantial declines of more than 20 percent in the duration of the growing season by 2050 in six different agroecological zones, including those typically found in the drier northern region.[19] In other Sahelian countries, such changes in agropastoral systems have led to land-use conflicts and violence, most recently between Peuhl (Fulani) pastoralists and farmers from other ethnic groups in Burkina Faso.

Although threats due to climate change in northern Côte d'Ivoire stem from conditions associated with increased aridity, problems that could arise in the southern part of the country, especially along the 515 kilometers of Atlantic coastline, are instead connected to the potential for a rise in sea level and severe storm events in the Gulf of Guinea. According to the IPCC, this could lead to an increased erosion of barrier beaches and possibly affect densely populated urban areas and economically important coastal fisheries and palm oil plantations.[20] In 2000, 40 percent of Côte d'Ivoire's population lived within 100 kilometers of the coast, with approximately 6 percent of the population living at elevations less than 10 meters above sea level.[21] Though the impact on the country's main foreign exchange earner, cocoa, remains uncertain, disruptions to this production system would obviously exacerbate the country's predicament as it tries to cope with climate change.

Threats from a rise in sea level may also be compounded by changes to hydrological conditions in central Côte d'Ivoire, which has been heavily deforested in recent decades. Since independence in 1960, the country has lost 40 percent of its forest cover.[22] Such land cover change in the watersheds of the country's major river systems, combined with the potential for increased precipitation in its southern region, could contribute to greater runoff and more frequent floods. Considering these potential environmental changes, attention should also be focused on the port city of Abidjan, the nation's main commercial center, which is important to national and regional economies, especially for the country's landlocked Sahelian neighbors. By 2005 Abidjan's population had reached 4.5 million (28 percent of the country's population) and was growing at an annual rate of 3 percent.[23] The pressures from increased urbanization and development in the hazard-prone coastal region will present significant land-use and urban planning challenges for future Ivorian administrations. Reports of global-warming-related losses of coastal real estate have already been published in the world press.[24]

Whether conditions become wetter or drier in various parts of Côte d'Ivoire, a combination of population growth, expanding urbanization, and the continued pursuit of economic development is expected to significantly increase the country's demand for freshwater resources by 2030. The United Nations projects that overall water availability in Côte d'Ivoire will decline significantly, from 6,000 cubic meters per capita per year in 1990 to just under 2,000 in 2025, transforming it from a state of water "abundance" to one of water "vulnerability."[25] Even more dramatic is the CIESIN forecast that the portion of the Ivorian population experiencing freshwater "scarcity" (less than 1,000 cubic meters per capita per year) will increase from none in 2000 to nearly 23 percent by 2030 (appendix B).

The Capacity of the Ivorian State to Respond to Climate Change

Perhaps the gravest concern regarding climate change in Côte d'Ivoire is that its impact is likely to vary regionally, with the northern, most politically unstable region taking the brunt of the adverse effects. Although there are unlikely to be any "winners" due to climate change, regional variation in the severity of environmental consequences may be perceived as favoring southerners and thereby create conditions that provoke migration from the north along with changes in the established agropastoral system that could intensify conflicts along ethnoregional lines.

Furthermore, the negative impact of climate change may tip the scales toward a resumption of fighting, especially given the failure of disarmament efforts and the collapse of prior peace accords. Under these conditions migration would once again be a likely strategy for many Ivorians. At the outset of the civil war in 2002, the Office of the United Nations High Commissioner for Refugees estimated that more than 25,000 Ivorian nationals fled to Liberia, inverting the flow of refugees that poured into the Côte d'Ivoire during the Liberian civil war in the 1990s.[26] Such a mass migration would only compound the adverse environmental impact of climate change and consequent socioeconomic and political tensions.

Although state failure is possible in Côte d'Ivoire, it does not yet seem likely, even if its current political and socioeconomic problems are further compounded by predicted climate changes. The resilience of the Ivorian state—despite the political stalemate, coup d'état, and civil war that have marked the post-Houphouet era—is a strong indicator of its capacity to avoid this worse-case scenario. Nonetheless, it remains a plausible outcome if fighting resumes.

One factor that distinguishes Côte d'Ivoire from other West African countries where civil wars resulted in collapsed states is that most of its human development indicators (e.g., literacy, GDP per capita, and foreign direct investment) remain relatively high for the region, although some have slipped as a result of socioeconomic strains associated with the civil war (e.g., life expectancy and infant mortality).[27] Social resilience and institutional capital in general have been severely weakened by the prolonged fighting. In particular, the sense of belonging to a unified political community has declined among Ivorians due to the politicization of ethnicity, religion, region, and nationality, which will undoubtedly impede the kind of long-term trade-offs that the management of climate change will require.

One complicating factor for Côte d'Ivoire is concern about climatic change beyond its borders, because any rise in the influx of refugees from other countries would pose a serious threat to the country's economy and political stability. Most of West Africa has a long history of migration occasioned by environmental, economic, and political stress. This is likely to increase, especially as a result of a deterioration of climate conditions in the Sahel, and the high potential for increased economic as well as political instability among neighboring states. To date, the only thing that has tempered refugee flows into Côte d'Ivoire has been its civil war. The effects of climate change, however, are likely to intensify migration pressures not only for Côte d'Ivoire but also for Nigeria.

Regional Variations of Climate Change in Nigeria

As in Côte d'Ivoire, areas for concern in Nigeria that may trigger disruptive change will likely stem from drought-related problems in the northern Sahelian region of the country where ethnoreligious tensions have been high. In addition, coastal regions with higher population densities are also expected to confront problems related to rising sea levels. Although Nigeria is somewhat more politically stable than Côte d'Ivoire, its poor track record of addressing the pollution and environmental degradation associated with oil production in the Niger Delta does not bode well for its government's capacity to cope with the environmental problems that result from climate change.

Political and Socioeconomic Risk Factors in Nigeria

Unlike in Côte d'Ivoire, in Nigeria military regimes were the rule rather than the exception following independence in 1960. Since its first coup d'état in 1966, the country has experienced seven military coups and countercoups. Nevertheless, it completed a prolonged and violent transition to civilian rule with the election of former military leader Olusegun

Obasanjo in 1999, ironically the same year that Côte d'Ivoire experienced its first military coup.[28]

Under both military and civilian rule, political corruption and virulent forms of ethno-politics have undermined the country's political stability and economic potential as a major oil producer.[29] Nigerian ethnopolitics date back to the formation of a single British colony that encompassed three major ethnoregional identities and more than three hundred other ethnolinguistic groups. Ethnoregional rivalries among civilian rulers along with rampant corruption have been used to justify Nigeria's numerous coups. Indeed, the first coup in 1966 resulted almost immediately in an ethnically motivated countercoup against the Igbo military officers who had taken power, spiraling Nigeria into the bloody Biafran civil war (1967–70).

With each successive regime, military and civilian alike, corruption became increasingly widespread and systemic. After civilian rule was reestablished in 1979, Nigeria's second republic quickly crumbled when the military once again intervened in the midst of charges of widespread electoral fraud during the reelection of President Shehu Shagari in 1983. By the time General Sani Abacha took power, following the annulment of the 1993 presidential elections that were intended to return Nigeria to civilian rule, the country had descended into a "culture of corruption," which has created nationwide fuel shortages in this oil-producing giant of Africa, and e-mail scams that have made the country synonymous with Internet fraud.[30] In some areas, powerful and violent political "godfathers" have gained control over politicians who are dependent on them to provide protection. In return, the godfathers are permitted to use government institutions to serve their own interests in a new variation on Nigerian prebendal (clientelist) politics.[31]

Although corruption and mismanagement in the oil industry have bled Nigeria of billions of dollars, they have also intensified the mobilization of ethnoregional groups in the Niger Delta, the site of both the Biafran civil war and the oil industry. In addition to a resurgence of the Biafran secessionist movement, there has been a rise in various ethno-nationalist groups as a reflection of growing frustration with the central government's failure to address environmental degradation and provide local access to oil revenues.[32] Many of these groups, such as the Movement for the Emancipation of the Niger Delta, are well armed, and they engage in unchecked violence and criminal activities such as kidnapping. Hostage taking, which was initially begun in 2006 to draw international attention to the delta crisis, has become a lucrative enterprise. According to Human Rights Watch, more than two hundred expatriate oil workers and a growing number of local officials have been held for ransom, while other political figures have been implicated in sponsoring and arming the militia groups that carry out these criminal acts.[33]

Government security forces also regularly perpetrate human rights abuses, both in the delta and elsewhere in the country, with virtual impunity and often praise. Since 2000, more than eight thousand Nigerians have been shot and killed by police, according to official statistics. Such grim statistics are often represented as an indication of effective police work rather than a scandal. For example, the inspector-general of the Nigerian police boasted that 785 suspected "armed robbers" were killed in a ninety-day period from June to September 2007.[34] Amnesty International reports that the police use armed robbery as a blanket charge to jail those who refuse to pay bribes and also to justify extrajudicial killings.[35]

Other accusations of state-sponsored human rights abuses have been associated with the extension of shari'a law from civil to criminal cases in twelve of the thirty-six Nigerian states between 2000 and 2001. These shari'a courts have failed to conform to international standards or respect due process as defined by the Islamic legal code. Although nearly all death sentences have been thrown out on appeal, or simply not carried out, there have been more than sixty amputations and numerous floggings. The introduction of shari'a in these northern states further aggravated ethnoregional divisions, forcing thousands of Christians to flee the region. High levels of poverty, aggravated by deteriorating environmental conditions in the predominantly Muslim north and growing pressures to boost fiscal allocations to oil-producing areas in the primarily Christian south, are a persistent source of political instability in Nigeria.[36]

Unfortunately, political violence and corruption have also infected the electoral process, increasing in severity since the return to civilian rule in 1999.[37] Although President Obasanjo's reelection in 2003 was marred by allegations of widespread vote rigging, political intimidation, and an estimated one hundred election-related deaths, the conditions surrounding the 2007 elections were even worse.[38] In addition to voter intimidation by gunmen who openly stole ballot boxes in front of journalists,[39] some candidates allegedly hired thugs to engage in violence and even commit political assassinations that contributed to a doubling in election-related deaths.[40] Although the election of President Umaru Musa Yar'Adua, Obasanjo's hand-picked successor, represented the first transition from one civil administration to another, the 2007 elections were described by the International Crisis Group as "the most poorly organized and massively rigged in the country's history," which is particularly significant given the nature of Nigerian elections.[41]

Nevertheless, while the Independent National Election Commission was "vigorously manipulated by the presidency . . . [and] became an accessory to active rigging," the bright spot in the electoral process was the independent stands taken by other branches of the Nigerian central government.[42] In 2006, the Senate defeated a bill that would have permitted Obasanjo to run for a third term despite a two-term constitutional limit, and the Supreme Court made a number of rulings that went against the interests of the ruling party, including the reinstatement of the candidacy of Vice President Atiku Abubakar, who had been feuding with Obasanjo over his succession.[43]

The Economic and Financial Crimes Commission has also demonstrated its independence by filing corruption charges against a number of former governors, ministers, and even the daughter of Obasanjo who served as a senator. The popular head of the commission was removed, however, in December 2007 after the arrest of a former governor, who was a major financier of Yar'Adua's campaign.[44] Such actions only serve to further undermine the legitimacy of his administration, which curtails its capacity to tackle difficult policy issues, including threats to food production due to deteriorating ecological and climatic conditions.

Any decline in Nigeria's food production poses a political threat because it is most likely to occur in the politically volatile northern region. Such declines pose an equally great economic threat despite the nation's oil wealth because it is now a large importer of foodstuffs, although it was a net exporter of agricultural products at independence. The country has a dual economy composed of an enclave oil sector with few links to the rest of the economy alongside a more typical African economy heavily dependent on agriculture. Although petroleum represents 70 to 80 percent of federal revenue and 90 percent of export earnings, it only constitutes 25 percent of GDP, as opposed to the agriculture sector, which represents 45 percent of GDP and employs the majority of the population.[45]

The Negative Impact of Climate Change in Nigeria

By 2030 considerable parts of Nigeria may confront issues related to climate change, which could seriously affect agricultural production, water availability, and coastal environmental conditions. Desertification in the north and erosion in the middle belt and south of the country are major concerns, placing 90,000 to 134,000 square kilometers of arable land at risk of degradation as a result of climate change, according to government estimates.[46]

In the semiarid steppe environment of northern Nigeria, which makes up one-quarter of the country's territory, there is a potential for a significant decline in agricultural production due to increased drought and rainfall variability associated with global warming.[47] Given the country's historical ethnoreligious conflicts, this negative impact of climate change could not only further undermine the economic base for approximately one-quarter of its population, which is located its Sahelian region, but also exacerbate preexisting tensions between its Muslim north and Christian south.

Despite the potential for drought-related disruptions to food production, there are indications that climate change in Nigeria may result in some positive outcomes for agriculture in the medium-term future. Although the majority of analyses forecast a negative impact on food production in most of Africa due to climate change, one recent study by researchers at Obafemi Awolowo University in Ile-Ife concludes that the impact of changes in physical variables such as rainfall, temperature, solar radiation, and atmospheric carbon dioxide concentrations may increase the yields of several major staple crops across Nigeria's various ecological zones during the first half of the twenty-first century.[48] This study cautions, however, that the increased yields may be followed by declines in the second half of the century, if temperatures continue to rise as projected in several climate change scenarios. Rising temperatures would induce higher rates of evapotranspiration, which would limit plant production as "the additional water need created by higher temperatures may not be met by projected increases in rainfall."[49] Such a scenario presents an interesting challenge to the development of long-term plans for adapting to climate change, demonstrating how positive effects such as increased food production in the near term could be reversed by natural feedback mechanisms that will not become apparent until a point in time well beyond the typical policy planning horizon. Unfortunately, most African countries still lack the institutional capacity to develop effective medium-term let alone long-term agrometeorological coping strategies to address the anticipated consequences of climate change on future food production.[50]

The anticipated decline in future water availability has serious implications for Nigeria, especially given the sizable population living in the semiarid north and the rapidly growing urban areas in the south. As in the case of Côte d'Ivoire, the United Nations also projects that Nigeria will see its national water situation deteriorate from one of "abundance" to one of "stress," with an anticipated decline from about 3,000 cubic meters per capita per year in 1990 to about 1,300 by 2025.[51] Also of concern is the CIESIN data (appendix B), which indicate that the portion of Nigeria's population experiencing a scarcity of freshwater will increase from 26 percent in 2000 to approximately 45 percent by 2030.

In addition to drought-related conditions that could affect food production and water availability in northern Nigeria, climate-related changes such as a rise in sea level may also pose serious risks to other demographically and economically important parts of the country. For example, considerable population settlements, industries, and other major economic activities are found along Nigeria's 835 kilometers of coastline, including the rapidly growing megacity of Lagos and the oil-producing region of the Niger Delta. In 2000, approximately one-quarter of Nigeria's population lived within 100 kilometers of the coast and an estimated 7 percent was concentrated on land at elevations less than 10 meters above sea level.[52] This coastal concentration of people and economic activity represents one of Nigeria's greatest liabilities in the context of climate change.

An early study by scientists from the University of Maryland and the Nigerian Institute for Oceanography and Marine Research that examined Nigeria's vulnerability to a 1-meter rise in sea level concluded that the potential socioeconomic costs would be significant, with flooding and erosion affecting an estimated 18,000 square kilometers, displacing 3.2 million residents, and destroying nearly $18 billion in property and infrastructure (in 1990 dollars) if no protective measures are taken.[53] Although indications are that such risks and losses could be significantly lowered through careful coastal development planning and investment in mitigation measures such as harbor upgrades, seawall construction, and beach replenishment projects, other observers have pointed out the difficulties of implementing such plans given the pressing immediate needs and growing pressures of the Nigerian population living along the coast, along with other parts of the country.[54] The potential need to relocate significant numbers of people inland from the coast, together with additional displacements if future conditions provoke further migration from a more arid north, not to mention possible migration flows from neighboring countries adversely affected by climate changes in the Sahel, are likely to pose significant challenges to Nigeria's ruling class. Unfortunately, it has proven ineffective in meeting the basic needs of its population despite the nation's immense oil wealth, and it has been tragically inept at confronting the pollution and environmental degradation associated with oil production in the Niger Delta.

The Capacity of the Nigerian State to Respond to Climate Change

The capacity of the Nigerian state to respond to the negative impact of climate change or any other public policy challenges will undoubtedly be thwarted by the limited legitimacy of the current regime and continuing high levels of corruption and ethnoregional tensions, which will only be aggravated by the anticipated environmental degradation. Nigerians are renowned for their ingenuity, but unfortunately also for the level of corruption in their country, which has thwarted the capacity of the government to address both the economic and environmental problems associated with the oil industry. The country's social resilience is further undermined by the incessant ethnic tensions, which are no longer confined to the three major ethnoregional groups. Unfortunately, the historic ethnoreligious tensions between the predominantly Muslim north and Christians in the south will most likely be intensified by regional variations in the impact of climate change, with the northern region bearing the brunt of declining agricultural production and access to water.

Political violence and human rights abuses by both state and nonstate actors are daily occurrences in Nigeria. Given the impunity with which they act, it is highly likely that the added socioeconomic stress associated with adverse climate change will increase violence, particularly in the ecologically fragile Sahel region in the north and the volatile Niger Delta in the south. In this context, mounting environmental pressures could very well lead to a coup d'état or even civil war. There are already a number of secessionist organizations with their own militias not only in the southeastern Delta region but also now among the Yoruba in the southwest. With the largest armed forces in Sub-Saharan Africa, the Nigerian military remains a major political force. Having ruled for all but four years from 1966 to 1999, the military's task of keeping soldiers in their barracks is a continual challenge.

In terms of a worst-case scenario, climatic change could be a contributing factor to state failure. Although this is a real possibility, a continuation of localized conflicts or civil war would appear more plausible, given both Nigeria's political history and vested domestic and international economic interests in maintaining at least a minimal level of political stability.

Internal migration due to environmental degradation is a likely strategy to cope with climate change, but it would undoubtedly aggravate land pressures and urban poverty. A probable alternative strategy is increased external migration via Nigeria's extensive migration channels in Europe and North America. But this is unlikely to be a sufficient release valve for the country in light of the probable influx of economic refugees from Niger and possibly other Sahelian countries due to the deteriorating environmental conditions associated with climate change in the region. Although such transnational immigration patterns may stress Nigeria as well as Côte d'Ivoire to a breaking point, fortunately for Senegal, the third country in this comparative analysis, this is less likely to be an issue.

Senegal's Capacity to Cope with Adverse Climate Change

Among the three West African cases examined here, Senegal confronts the most widespread and adverse environmental problems associated with climatic change. Nevertheless, its ability to cope with these challenges surpasses those of Côte d'Ivoire and Nigeria. The capacity of Senegal to address declining agricultural production, water availability, and other adverse environmental conditions is directly linked to its relative political stability and high rate of emigration, both within Africa and to Europe and North America, which has provided a release valve for population pressures as well as much needed added income from remittances. Although this is also true for Nigeria, and to a lesser degree for Côte d'Ivoire, much smaller percentages of their significantly larger populations have been able to emigrate, which has not provided much relief from mounting socioeconomic pressures, particularly in the face of a significant influx of economic refugees from the Sahel.

Political and Socioeconomic Risk Factors in Senegal

Senegal has long been held up as a model of political stability in Africa. Since achieving independence in 1960, Senegal has never experienced an attempted coup d'état, let alone a successful one.[55] Similar to other African countries, it did, however, have an extended period of authoritarian rule under a de facto one-party system led by the poet-statesman Léopold Sédar Senghor from 1963 to 1981. Senghor's administration was not without its critics, but it was relatively benign and distinctively inclusive of Senegal's ethnic and religious groups. This is not surprising, given that Senghor, a Serer Catholic, was a member of an ethnic and religious minority in Senegal, which is politically and economically dominated by Wolof Muslims.[56]

President Senghor orchestrated the succession of Abdou Diouf through constitutional manipulation, much as Houphouet-Boigny did in Côte d'Ivoire before his death. Nevertheless, Senghor's departure was relatively unique in Africa, in that it was voluntary and occurred more than a decade before the wave of democratization that swept most of Africa's presidents for life from office. A Wolof Muslim, Diouf nevertheless maintained the inclusive nature of the party-state of the Parti Socialiste (PS) while extending Senghor's first steps toward a multiparty system before standing for election in 1983.[57]

President Diouf was nevertheless able to maintain a dominant party system through a combination of electoral manipulation and fraud, along with the distribution of state resources through the extensive PS clientelist networks. Domestic and international pressure ultimately resulted in an electoral code that leveled the playing field by 1993, though the PS was able to retain power until the 2000 presidential elections, which longtime opposition leader Abdoulaye Wade of the Parti Démocratique Sénégalais (PDS) won by defeating Diouf in a runoff election.[58]

Although regime stability remains strong in Senegal, there has been a fair amount of political instability within the administration and the new ruling party. In addition to the formation of splinter parties from the PDS, a common practice in Senegal that was ultimately the undoing of the PS, Wade has gone through a rapid succession of prime ministers. The last two were sacked when they were perceived as posing a threat to Wade's political dominance, which is now seen as extending to his son, Karim, who is apparently being groomed to succeed his father.

There have been growing concerns about a decline in civil liberties and the concentration of power in the hands of the president, despite Wade's campaign promise to institute a parliamentary democracy in Senegal. There are numerous examples of political intimidation and violence against opposition leaders, such as the attack on a relatively minor opposition leader, Talla Sylla, who was nearly beaten to death in 2003. Freedom of the press has been seriously trounced upon, including the closure of radio stations for allegedly biased reporting. The Senegalese journalist Abdou Latif Coulibaly received death threats following his publication of a controversial book criticizing the Wade administration, and a Radio France International reporter was expelled from the country for conducting an interview with a leader of the Mouvement des Forces Démocratique de la Casamance (MFDC), a secessionist group that opposes a negotiated settlement to the long-standing conflict in the southern Casamance region.[59]

The Casamance conflict has undoubtedly tarnished Senegal's reputation by placing an international spotlight on the human rights abuses committed by both security forces and the members of the MFDC. Claims for Casamance independence that date back to the colonial period were stepped up in the 1970s by Father Augustin Diamacoune Senghor, who cofounded the MFDC in the early 1980s when a peaceful demonstration led to mass arrests and state repression. Although not an ethnonationalist movement, the MFDC is dominated by ethnic Jolas, who complain of "malign neglect" by the Wolof-dominated Senegalese state and the "invasion" of their region by northern merchants, fishermen, and farmers, who are seen as infringing on local access to and management of the region's natural resources.[60]

Despite a series of peace accords, the latest of which was signed in December 2004, sporadic fighting by the MFDC's southern front continues under the leadership of Salif Sadio, who refuses to negotiate for anything less than Casamance independence.[61] Nevertheless, it is remarkable that this conflict has not deteriorated into widespread communal violence in other areas of the country. Politicians from this and other regions continue to refrain from using divisive ethnopolitics, and Senegal's major political parties have a national base among all ethnoregional and religious groups.

Another source of socioeconomic tension related to access to natural resources has been farmer/herder conflicts, which at times have resulted in political violence. Although these conflicts are often among groups of Senegalese, the worst violence over land use took place in 1989 between Senegalese farmers and Mauritanian pastoralists and flared into massive human rights abuses in both countries.[62]

In addition to these simmering rural conflicts, social tensions are also rising in urban areas. Urbanization is clearly linked to declining agricultural productivity and land pressures in both northern and central Senegal. As has occurred elsewhere in Africa, a demonstration was held in April 2008 against rising food prices due to the increasing costs of imported foodstuffs and also Senegal's poor harvest in 2007. Although the demonstration is an indicator of mounting socioeconomic pressures, the willingness of the government to authorize a public demonstration for the first time in three years may also suggest a more accommodating approach toward opposition parties. On the other hand, efforts to abolish ten local government councils that are opposition strongholds, along with the unilateral decision to postpone local elections until 2009, have been interpreted as evidence of the Wade administration's increasing authoritarian tendencies.[63]

As a result of the country's weak economic base, emigration has been a preferred strategy for a growing number of Senegalese, resulting in large expatriate communities across the African continent, Europe, and North America. With the tightening of immigration laws, many Senegalese now risk their lives to gain illegal entry to European Union countries. According to the Spanish Red Cross, an estimated 1,000 Senegalese drowned while attempting to get to the Canary Islands during an eight-month period in 2006.[64]

Efforts to strengthen Senegal's weak economic base have focused on increased investment in mining (most recently zircon), and the refining of imported petroleum, which is reexported to other countries in the region. This diversification will become increasingly important if agricultural harvests continue to decline, despite efforts by the Senegalese government, under the new Agricultural Offensive for Food Abundance (GOANA), to double rice production during the next seven years. Most observers believe this target is unrealistic. Replacing Senegal's former reliance on peanuts, fish products are now the country's second-largest export commodity after refined oil. Unfortunately, this industry is increasingly hampered by dilapidated equipment and the depletion of stocks from overfishing, and it is likely to be adversely affected by global warming.[65]

The Negative Impact of Climate Change in Senegal

Senegal is the most Sahelian of the three countries discussed here, and it has long struggled to cope with the persistent environmental problems that afflict that region. In recent decades, the country has experienced severe agricultural and livestock losses, along with extreme human suffering and fatalities stemming from prolonged droughts, dating back to the catastrophic 1968–73 episode that devastated the entire West African Sahel.[66] In comparison with other African countries that have poor natural resource endowments, Senegal has nevertheless been able to rely on its relatively well-developed human resources and institutions to cope with the challenge of living in a drought-prone environment. However, these resources may be stretched to their limit under the conditions expected to accompany global climate change.

In addition to already-severe problems of deforestation, desertification, and salinization of soils, Senegal struggles with farmer/herder conflicts over access to land, particularly in its more marginally productive central areas. There are also problems related to the management of riparian lands and water resources in the Senegal River Basin along the country's northern border. As in the case of Côte d'Ivoire, the mixed rain-fed and semi-arid agropastoral systems practiced in these more Sahelian regions are likely to experience negative consequences due to declining agricultural productivity and water availability.[67] Estimates for changes in agricultural productivity due to global warming by the late twenty-first century indicate that Senegal may experience a "very serious" downturn in food production, possibly at a rate eight times the expected level of global decline.[68] In terms of water supply, the data summarized in appendix A of this volume indicate that in 2000, two-thirds of the Senegalese population already experienced water scarcity conditions, a figure expected to increase to almost three-quarters by 2030.

The anticipated impact of climate change along Senegal's 531-kilometer coastline is similar to that for Côte d'Ivoire and Nigeria. In 2000, nearly three-quarters of Senegal's population resided within 100 kilometers of the coast, with approximately 27 percent concentrated on land at elevations less than 10 meters above sea level.[69] The region around the Cap Vert Peninsula, including the capital city of Dakar, has been a focal point of population growth and economic activity, representing approximately 30 percent of the country's population and 90 percent of its industrial installations.[70] Given that Dakar's population of 2.5 million is estimated to grow at an annual rate of 4 percent, demographic and economic pressures in this densely populated urban area are particularly acute.[71] Rapid urbanization translates into increased pressure on already-scarce resources, such as wood for urban charcoal markets, and the diversion of water to supply the burgeoning urban population and industrial facilities.

Several significant tourist centers along the Senegalese coast, particularly to the south of the Cap Vert Peninsula, as well as biologically productive and economically important mangrove estuaries in the Sine-Saloum and Casamance River deltas, are also at risk from the coastal flooding and inundation that would accompany a rise in sea level. In the mid-1990s, a study by researchers from the University of Maryland and Dakar's Cheikh Anta Diop University concluded that a 1-meter rise in sea level poses a considerable risk of damage to coastal infrastructure due to erosion and inundation, especially around economically important tourist facilities and the port of Dakar.[72] The study estimated that more than 6,000 square kilometers of land could be flooded, threatening up to $700 million worth of property, and forcing the relocation of up to 180,000 coastal dwellers if no protective measures are taken. Other negative impacts associated with this scenario include saltwater intrusion into coastal aquifers and surface waters that are essential to local agricultural production and domestic water supply, including that of Dakar's rapidly growing population.

The country's important artisanal fishing industry is also considered at risk due to the effect of rising ocean temperatures on nutrient availability in the marine food chain.[73] The inexorable demographic and economic pressures bearing down on the coastal environment pose challenges for urban development and hazard mitigation similar to what can be anticipated for Côte d'Ivoire and Nigeria, though perhaps to an even greater extent in Senegal, given its relatively smaller geographic size and comparatively higher proportion of land area in the Sahelian zone.

The Capacity of the Senegalese State to Respond to Climate Change

Emigration to urban areas, and to other parts of Africa, Europe, North America, and now even Asia, is the most likely response to environmental degradation in Senegal. Even when this involves risking their lives, young Senegalese men are likely to see emigration as their best option for escaping the high rate of unemployment and underemployment, especially given the well-established migration channels and high levels of remittances upon which many Senegalese families have come to depend for their day-to-day survival.

Although the Casamance region has been adversely affected by irregular rains and the salinization of soil, the northern regions have arguably been harder hit. Nevertheless, internal north–south migration as found in Côte d'Ivoire and Nigeria is less likely in Senegal due to the simmering Casamance conflict. With the level of fighting in Casamance likely to fluctuate based in part on political conditions in neighboring Guinea-Bissau, where the MFDC continues to have bases, a resolution of the conflict seems unlikely in the near future.[74]

Although peace and stability in Casamance will probably remain elusive, it is unlikely that this will spread to northern Senegal, even if environmental conditions continue to deteriorate throughout the country. Because Senegal has never experienced a military coup, neither this nor state failure is probable during the period under review, although the regime appears to be destined to return to a form of semidemocracy under a dominant-party system, which arguably has already occurred given the ruling party's landslide in the 2007 presidential and legislative elections.

Despite the prolonged civil conflict in Casamance, Senegalese have a strong sense of national identity that embraces sociopolitical principles of ethnoreligious tolerance, generosity, hospitality, and charity. The level of tolerance is evident, for example, in the lack of violence against ethnic Jolas outside Casamance. Moreover, compared with other African countries that have similarly poor natural resource endowments, Senegal has traditionally been able to rely on its relatively well-developed human resources and institutional capital (e.g., Sufi Brotherhoods) to face the challenges of living in a drought-prone environment. However, these usually successful coping mechanisms may be put to the test under worsening climatic conditions. The Wade administration and its successors will need to avoid populist policies such as the GOANA plan, which raises expectations with its unrealistic promises. Given the mediocre (if not poor) record of development programs designed to enhance agricultural production, such as the massive irrigation project in the Senegal River Valley, it would be wise for the government to continue to make significant investments in the agricultural sector, while being more realistic about their projected impact.

Finally, climate change in the Sahel will probably increase the likelihood of more skirmishes between farmers and herders, particularly along the Senegal River. This could result in larger border conflicts and a massive influx of political refugees, as occurred in 1989. An increase in economic refugees from neighboring Sahelian countries that may be harder hit by climate change, or have a lower adaptive capacity, is also plausible. Unlike Côte d'Ivoire and Nigeria, however, Senegal is an improbable destination for refugees from most of its Sahelian neighbors, who are more likely to migrate to other West African countries or to destinations outside the region.

Conclusion

There are two primary sets of factors to consider when assessing the capacity of any country to address the negative impact of climatic change. Obviously, the anticipated level of adverse effects is critical, including the three examined here: declines in agricultural production, a reduction in water availability, and rising sea levels. But in addition to the consequences of climate change, it is necessary to also consider the political and socioeconomic factors that influence the capacity of individual countries to respond to the challenges they face.

Among the three West African countries discussed here, the country facing the most adverse climatic changes, Senegal, is ironically and perhaps fortunately the one with the greatest capacity to cope with the negative impacts of global warming, though not necessarily to master them. Given its limited economic base, Senegal is unlikely to be able to do this on its own; but its government's demonstrated ability to parlay its international reputation and prominent role in African affairs into enhanced access to foreign aid is likely to be an invaluable resource in addressing the deteriorating environmental conditions that will only be exacerbated by global warming.

For the other two countries, Côte d'Ivoire and Nigeria, the predicted impact of climate change is at least marginally less severe, though demographic and economic impacts are likely to be dramatically higher, due to their larger populations and economic infrastructures. Indeed, given their significantly stronger economic bases, these two countries should be better positioned to address the problems associated with climatic change. But their efforts will undoubtedly be hindered by their higher levels of instability, associated with widespread political violence, virulent forms of ethnopolitics, and debilitating levels of corruption.

Notes

1. United Nations Environment Program, *Africa Environment Outlook: Past, Present, and Future Perspectives* (Nairobi: United Nations Environment Program, 2002).

2. Michel Boko et al., "Africa," in IPCC, *Climate Change 2007: Impacts,* 433–67.

3. Schubert, H., J. Schellnhuber, N. Buchmann, A. Epiney, R. Grießhammer, M. Kulessa, D. Messner, S. Rahmstorf, and J. Schmid, *World in Transition: Climate Change as a Security Risk*, Report from German Advisory Council on Climate Change (Earthscan: London, 2008), 144, www.wbgu.de/wbgu_jg2007_engl.html, 138.

4. Aristide R. Zolberg, *One-Party Government in the Ivory Coast* (Princeton, NJ: Princeton University Press, 1969); Souga Jacob Niemba, *Politique Agricole Vivri ère en Afrique: Base du Miracle Economique en Côte d'Ivoire* (Paris: L'Harmattan, 2000); Catherine Boone, *Political Topographies of the African State: Territorial Authority and Institutional Choice* (Cambridge: Cambridge University Press, 2003).

5. The agroforestry sector constitutes 28 percent of the GDP and employs 49 percent of the labor force, with the Côte d'Ivoire producing 40 percent of world's cocoa. *Economist* Intelligence Unit, *Côte d'Ivoire Monthly Report* (London: *Economist* Intelligence Unit, 2008).

6. Alice Ellenbogen, *La Succession d'Houphouët-Boigny* (Paris: L'Harmattan, 2002); Amadou Koné, *Houphouët-Boigny et la Crise Ivoirienne* (Paris: Karthala, 2003).

7. Ramsès L. Boa Thiemele, *L'Ivoirité: Entre Culture et Politique* (Paris: L'Harmattan, 2003).

8. Francis Akindés, *Les Racines de la Crise Militaro-politique en Côte d'Ivoire* (Dakar: Codesria, 2004); Théo Doh-Djanhoundy, *Autopsie de la Crise ivoirienne* (Paris: L'Harmattan, 2006).

9. Théophile Koui, *Multipartisme et Idéologie en Côte d'Ivoire* (Paris: L'Harmattan, 2006).

10. Dennis Cordell, Joel W. Gregory, and Victor Piché, *Hoe and Wage: A Social History of a Circular Migration System in West Africa* (Boulder, CO: Westview Press, 1996).

11. International Crisis Group, "Côte d'Ivoire: Can the Ouagadougou Agreement Bring Peace?" Africa Report 127, June 27, 2007, www.crisisgroup.org/home/index.cfm?id=4916.

12. Amnesty International, "Côte d'Ivoire," in *Amnesty International Report 2008:State of the World's Human Rights*, http://archive.amnesty.org/air2008/eng/regions/africa/c%f4te-d%27ivoire.html; *Economist* Intelligence Unit, *Côte d'Ivoire Monthly Report* (London: *Economist* Intelligence Unit, 2008); International Crisis Group, "Côte d'Ivoire: Ensuring Credible Elections," Africa Report 139, April 22, 2008, www.crisis group.org/home/index.cfm?id=5400&l=1; *Economist* Intelligence Unit, *Côte d'Ivoire Country Report* (London: *Economist* Intelligence Unit, 2010).

13. Human Rights Watch, "Côte d'Ivoire," in *World Report 2007* (New York: Human Rights Watch, 2007).

14. *Economist* Intelligence Unit, *Country Profile: Côte d'Ivoire* (London, 2008).

15. Ibid.

16. Amnesty International, "Côte d'Ivoire"; Freedom House, *Freedom in the World: Côte d'Ivoire* (Washington, DC: Amnesty International, 2007).

17. Human Rights Watch, "Côte d'Ivoire."

18. "Technical Summary," in IPCC, *Climate Change 2007: Impacts,* 23–78.

19. P. K. Thornton, P. G. Jones, T. Owiyo, R. L. Kruska, M. Herrero, P. Kristjanson, A. Notenbaert, et al., *Mapping Climate Vulnerability and Poverty in Africa* (Nairobi: International Livestock Research Institute, 2006).

20. Boko et al., "Africa."

21. CIESIN, *National Aggregates of Geospatial Data: Population, Landscape and Climate Estimates,* version 2 (Palisades, NY: CIESIN / Columbia University, 2007), http://sedac.ciesin.columbia.edu/plue/nagd/place.html.

22. United Nations Environment Program, *Africa: Atlas of Our Changing Environment* (Nairobi: United Nations Environment Program, 2008), www.unep.org/dewa/Africa/AfricaAtlas/.

23. United Nations Human Settlements Program, "Côte d'Ivoire Country Page," www.unchs.org/categories.asp?catid=189; United Nations Department of Economic and Social Affairs, *World Urbanization*

Prospects: The 2001 Revision (New York: United Nations, 2002); www.un.org/esa/population/publications/wup2001/wup2001dh.pdf.

24. James Copnall, "The Ivorian Town Sinking under Waves," BBC, October 8, 2007, http://news.bbc.co.uk/go/pr/fr/-/2/hi/africa/7030291.stm.

25. United Nations Environment Program, "Water Availability in Africa" [graphic], in *Vital Climate Graphics Africa* (Nairobi: United Nations Environment Program, 2002), www.grida.no/climate/vitalafrica/english/15.htm.

26. "Côte d'Ivoire: Recent History," in *Europa Encyclopedia* (London: Routledge, 2007).

27. United Nations Development Program, *Human Development Index* (New York: United Nations Development Program, 2008), http://hdr.undp.org/en/statistics/indices/hdi/.

28. Eghosa E. Osaghae, *The Crippled Giant* (Bloomington: Indiana University Press, 1998); Stephen Wright, *Nigeria: Struggle for Stability and Status* (Boulder, CO: Westview Press, 1998).

29. Larry Diamond, *Class, Ethnicity and Democracy in Nigeria: The Failure of the First Republic* (Basingstoke, U.K.: Macmillan, 1988); A. H. M. Kirk-Greene, *The Genesis of the Nigerian Civil War and the Theory of Fear* (Uppsala: NAI, 1975); Richard Joseph, *Democracy and Prebendal Politics in Nigeria* (Cambridge: Cambridge University Press, 1987).

30. Daniel J. Smith, *A Culture of Corruption: Everyday Deception and Popular Discontent in Nigeria* (Princeton, NJ: Princeton University Press, 2007).

31. Human Rights Watch, *Criminal Politics: Violence 'Godfathers,' and Corruption in Nigeria*, October 2007, www.hrw.org/en/reports/2007/10/08/criminal-politics.

32. Charles Ukeje and Wale Adebanwi, "Ethno-nationalist Claims in Southern Nigeria," *Ethnic and Racial Studies*, March 2008, 563–91.

33. International Crisis Group, "Nigeria: Ending Unrest in the Niger Delta," Africa Report 135, December 5, 2007, www.crisisgroup.org/home/index.cfm?l=1&id=5186.

34. Human Rights Watch, "Nigeria: Investigate Widespread Killings by Police," November 18, 2007, www.hrw.org/en/news/2007/11/17/nigeria-investigate-widespread-killings-police.

35. Amnesty International, "Nigeria," *Amnesty International Report 2008*, http://archive.amnesty.org/air2008/eng/regions/africa/nigeria.html.

36. Human Rights Watch, "Nigeria," in *World Report 2009*, www.hrw.org/en/node/79250; Human Rights Watch, "Political Shari'a? Human Rights and Islamic Law in Northern Nigeria," September 2004, www.hrw.org/reports/2004/nigeria0904/.

37. Jibrin Ibrahim, *Nigeria's 2007 Elections: The Fitful Path to Democratic Citizenship*, Washington, DC: U.S. Institute of Peace, 2007; Ben Rawlence and Chris Albin-Lackey, "Nigeria's 2007 General Elections: Democracy in Retreat," *African Affairs* 106, no. 424 (July 2007): 497–506; Rotimi T. Suberu, "Nigeria's Muddled Elections," *Journal of Democracy* 18, no. 4 (October 2007): 95–110.

38. Daren Kew, "The 2003 Elections: Hardly Credible, but Acceptable," in *Crafting the New Nigeria,* ed. R. Rotberg (Boulder: Lynne Rienner, 2004), 202–46.

39. *Economist* Intelligence Unit, "Nigeria Country Profile" (London: *Economist* Intelligence Unit, 2008).

40. Human Rights Watch, "Nigeria: Presidential Election Marred by Fraud, Violence," April 25, 2007, www.hrw.org/en/news/2007/04/24/nigeria-presidential-election-marred-fraud-violence.

41. International Crisis Group, "Nigeria: Failed Elections, Failing State?" Africa Report 126, May 30, 2007, www.crisisgroup.org/home/index.cfm?id=4876.

42. Ibid.

43. Amnesty International, "Nigeria"; Human Rights Watch, "Nigerian Elections Marred."

44. *Economist* Intelligence Unit, "Nigeria Monthly Report," May 2008.

45. *Economist* Intelligence Unit, "Nigeria: Country Profile," 2008.

46. Ibid.

47. CIESIN, *National Aggregates.*

48. James O. Adejuwon, "Food Crop Production in Nigeria," *Climate Research* 32 (2006): 229–45.

49. Ibid, 232.

50. Andrew Challinor, Tim Wheeler, Chris Garforth, Peter Craufurd, and Amir Kassam, "Assessing the Vulnerability of Food Crop Systems in Africa to Climate Change," *Climate Change* 83 (2007): 381–99.

51. United Nations Environment Program, "Water Availability in Africa."

52. CIESIN, *National Aggregates.*

53. Gregory T. French, Larry F. Awosika, and C.E. Ibe, "Sea-Level Rise in Nigeria: Potential Impacts and Consequences," *Journal of Coastal Research* 14 (1995): 224–42.

54. Robert J. Nicholls and Nobuo Mimura, "Regional Issues Raised by Sea-Level Rise and Their Policy Implications," *Climate Research* 11 (1998): 5–18.

55. Sheldon Gellar, *Senegal: An African Nation between Islam and the West* (Boulder, CO: Westview Press, 1995).

56. Irving Leonard Markovitz, *Léopold Sédar Senghor and the Politics of Negritude* (New York: Antheneum, 1969).

57. Momar Coumba Dioup and Mamadou Diouf, *Le Sénégal Sous Abdou Diouf* (Paris: Karthala, 1990); and Robert Fatton, *The Making of a Liberal Democracy: Senegal's Passive Revolution, 1975–1985* (Boulder, CO: Lynne Rienner, 1987).

58. Linda Beck, *Brokering Democracy in Africa: The Rise of Clientelist Democracy in Senegal* (New York: Palgrave, 2008).

59. Abdou Latif Coulibaly, *Wade: Un Opposant au Pouvoir* (Paris: Harmattan, 2003); Amnesty International, "Senegal Country Report," 2007; Freedom House, "Freedom in the World: Senegal," 2007, www .freedomhouse.org/template.cfm?page=22&year=2007&country=7266.

60. Dominique Darbon, "La voix de la Casamance . . . une parole Diola," *Politique Africaine* 18 (1985): 127–89; Ferndinand de Jong, "Politicians of the Sacred Grove: Citizenship and Ethnicity in Southern Senegal," *Africa* 72, no. 2 (2002): 203–20; Vincent Foucher, "Cheated Pilgrims: Migration, Education and the Birth of Casamançais Nationalism" (London: SOAS, 2002).

61. *Economist* Intelligence Unit, "Senegal Monthly Report," May 2008, 3.

62. Charles Becker and A. Lericollais, "Le probleme frontalier dans le conflit sénégalo-mauritanien," *Politique africaine* 35 (1989): 149–55; R. Parker, "The Senegal-Mauritania Conflict of 1989," *Journal of Modern African Studies* 29, no. 1 (1991): 155–71.

63. *Economist* Intelligence Unit, Senegal Report.

64. Amnesty International, Senegal Summary.

65. *Economist* Intelligence Unit, "Senegal Monthly Report," December 2007.

66. Richard W. Franke and Barbara H. Chasin, *Seeds of Famine: Ecological Destruction and the Development Dilemma in the West African Sahel* (Montclair, NJ: Allanheld, 1980).

67. "Technical Summary," IPCC, *Climate Change 2007: Impacts.*

68. See appendix A.

69. CIESIN, *National Aggregates.*

70. United Nations Environment Program, *Africa Environment Outlook,* www.grida.no/publications/ other/aeo/; United Nations Department of Economic and Social Affairs, *World Urbanization Prospects.*

71. United Nations Human Settlements Program, "Senegal Country Page," www.unchs.org/categories .asp?catid=229.

72. Karen Clemens Dennis, Isabelle Niang-Diop, and Robert James Nicholls, "Sea-Level Rise and Senegal: Potential Impacts and Consequences," *Journal of Coastal Research* 14 (1995): 242–61.

73. Isabelle Niang-Diop, M. Dnsokho, A. T. Diaw, S. Faye, A. Guisse, I. Ly, F. Matty, et al., "Senegal," in *Climate Change in Developing Countries: An Overview of Study Results from the Netherlands Climate Change Studies Assistance Programme,* edited by M. A. van Drunen, R. Lasage, and C. Dorland (Wallingford, U.K.: CABI, 2006), 82–88.

74. The bases in the Gambia have basically disappeared as a result of efforts by the Gambian government and the 2004 peace accord signed by the northern front of the MFDC.

17

West Africa II: The Mano River Union
Guinea, Liberia, and Sierra Leone

Dennis Galvan and Brian Guy

Climate change in the Mano River Union (MRU) countries, though scarcely insignificant in itself, must be set against a recent history of weak and collapsed states in the region. Temperature increases and associated changes in rainfall, water availability, and sea-level rise will have only modest effects on the living conditions of the majority of the region's population.[1] The extant political and social atmosphere in the MRU countries, however, is central to understanding the underlying conditions against which environmental shifts must be considered. This chapter explores the effects of climate change within the context of fragile state–society relations and suggests that state structural limitations and a vulnerable, dependent global position are the primary issues in the MRU zone. The negative effects upon the natural environment projected by earth scientists will only serve to exacerbate this fundamental reality.

Limited state capacity coupled with social and political crisis define the recent past in the MRU zone.[2] Liberia experienced civil conflict and multiple, effective armed challenges to state sovereignty from 1989 to 2003, with the peak of political disorder concentrated in the period before Charles Taylor's 1997 election to the presidency, and the years surrounding his removal from office (2002–3). Taylor and other factions in Liberia's civil war relied in part on locally available diamond resources to fund their wars.[3] Diamonds are even

Guinea

Liberia

Sierra Leone

more abundant in neighboring Sierra Leone, which sent Taylor and others to poach and eventually occupy parts of that country.

Thus, Liberia's conflict easily spread into Sierra Leone. Taylor backed a Sierra Leonean former corporal, Foday Sankoh, whose Revolutionary United Front (RUF) challenged the Sierra Leonean government beginning in 1991, seizing major diamond-producing regions in that same year.[4] Failure to deal with the RUF challenge led to Sierra Leone's own spiral of crisis and gradual reconsolidation, the core elements of which included a brutal civil war that killed as many as fifty thousand and maimed a half million, by UN estimates. There have been, in addition, two coups (1991 and 1997); the intervention of the Economic Community of West African States (ECOWAS) and later a UN peacekeeping force; and a power-sharing agreement with Sankoh, culminating eventually in the military defeat of RUF forces, Sankoh's arrest, and his death in 2000 while awaiting trial for war crimes.[5]

Guinea, the geographical linchpin of the region, does not have significant deposits of diamonds, and although threatened by the spillover effects of the meltdown of its southern neighbors, remained more or less stable until the death of its longtime leader, Lansana Conté. During the late 1990s, Guinea was home to as many as 1 million refugees from Liberia and Sierra Leone, more than half of whom have now been repatriated.

Although the lack of "conflict diamonds" and Guinea's wealth in bauxite (a mineral that requires significant infrastructure and machinery to extract and process) produces a different political economy of conflict, Guinea is by no means immune to disorder. In early 2007 civil strife nearly broke out when unions and opposition critics of Conté took to the streets in protest over recent arbitrary interventions in the courts.[6] The violent repression of these strikes produced an escalation of the protest. The crisis was averted when Conté agreed to appoint a prime minister from a list of candidates approved by the opposition.[7]

More recently, however, Guinea has experienced an increase in unrest, protest, repression, and instability, all of which followed Conté's death in late 2008 and the seizure of power by the military in a coup led by a young officer, Captain Moussa Dadis Camara. High hopes faded and tensions increased as it became clear that the initial promises to transfer power through elections were not materializing. In the most explosive event, on September 28, 2009, Guinean soldiers (notably Camara's personal bodyguard, the "Red Berets") brutally attacked an opposition political rally, injuring more than 1,000 people and killing at least 150. The incident prompted a UN investigation due to its extremely repressive nature, especially with regard to the many instances of public violence against women. Camara, who was widely presumed to have ordered the massacre, survived an assassination attempt just two months later, and he has been in voluntary exile since January 2010. New leadership under Jean-Marie Doré has followed concerted efforts to help restore order and move to civilian rule via elections by 2011.[8]

Given the history of instability in the region, one would assume that climate change effects might push already weak states and fractious societies over the brink into crisis. Although this is not an implausible scenario, on closer examination climate change adds only a few drops of fuel to this tinderbox region. This is because (1) in one of the few strokes of good fortune experienced by the MRU region in the last five centuries, projected 2030 climate changes are modest; and (2) the regional context that precipitated nearly fifteen years of tragic disorder in this part of West Africa has shifted.

As the data summarized in appendix A to this volume make clear, the temperature increase projected for the MRU countries by 2030 is not extreme, at below 1 degree Celsius. In Sahelian environments to the north, this will have a more serious effect on production of rain-fed cereals such as sorghum and millet, the backbone of the subsistence economy there. With the exception of northern reaches of Guinea, Sahelian ecosystem cereal crops are not major food sources in the MRU countries. Tropical food crops such as rice, as well as cash crops such as oil palm, coffee, and cocoa, are more significant parts of the overall economy.[9] Available projections suggest that temperature increases will affect these crops in such a way that coffee producers may need to shift elevation in response to a rise in temperature. This is more problematic in high-population-density locales such as Rwanda-Burundi and some parts of Côte d'Ivoire than in the much less densely populated MRU countries. Even a severe sea-level rise by 2030 will have a moderate impact on the countries in the region for a few reasons: (1) The core areas of coastal population centers are centered on relatively high ground—the historic locations of European settlement, which were situated on the highest, driest spots, to avoid malaria; (2) the population is not yet heavily concentrated in coastal cities but is dispersed in interior town and rural areas; and (3) given the importance of mining and plantation crops in the region, economic activity also tends to be distributed more evenly between coastal cities and the interior.[10] One important exception is Liberia's capital, Monrovia. The city and its environs contain roughly half the country's 3.5 million people. Although the sea level would need to rise significantly more than 1 meter to affect the area directly, some models suggest that the higher cyclonic surges associated with relatively near-term climate changes may put Monrovia in a potentially devastating position.[11]

This relatively sanguine outlook does not, however, take into account the possibility that the MRU countries might follow the path of their slightly more developed neighbors (Senegal, Côte d'Ivoire), where many more people have flocked to coastal cities, a pattern that may replicate itself in the MRU. This includes settling in areas outside the high-ground colonial core zones of these cities, and significantly increasing the population at risk from sea-level rise, along with corresponding economic activity. Clearly, this is of particular concern in Liberia. Given the ineffectiveness of the states in this region, it is unreasonable to expect that the next two decades will see a robust increase in public capacity of the kind required to avoid an increased concentration of migrants in coastal cities.

Although freshwater may become more scarce in the West African region, the MRU zone is water rich and is graced with the headwaters of most of West Africa's major rivers. Studies suggest that rainfall will actually increase as a result of global climate change and resultant sea temperature increases.[12] Therefore, the challenge for these countries is less a drop-off in available water resources in the coming decades and more the long-standing problem of building the infrastructure to get safe, clean water to the majority of the population, combined with periodic flood risks.

Thus, many of the challenges associated with global climate change in the MRU zone have less to do with the severity of environmental effects than with the already very limited capacity of state and society to work together to provide public order, disaster relief, and other basic human services. A brief discussion of this context and its historical underpinnings follows.

Capacity and Crisis in the MRU Zone

The effectiveness of the state in mitigating climate change in the MRU zone depends on its capacity to deliver basic public goods—a capacity that has varied widely during the past half century. In this regard it is worth noting that the MRU countries, now widely regarded as the birthplace of the notion of a "failed state," were once exemplars of the older notion of a "juridical" state.[13] The juridical state, widespread in Sub-Saharan Africa but also found elsewhere, is a postcolonial entity whose main claim to authoritative legitimacy was the moral, symbolic, and financial support it received from the international community, independent of its ability to maintain the legitimate monopoly of violence over its entire territory, let alone extract resources and provide services there.

The MRU juridical states historically depended on their former colonial masters (the United States, the United Kingdom, and France), the hegemonic Cold War actors (the United States and the Soviet Union), and multilateral organizations of various stripes for financial and administrative assistance as well as moral and symbolic support. These props helped ensure both some degree of order and also a modest flow of international private investment into the MRU area in the period from late 1950s until the end of the Cold War. By the late 1980s and early 1990s, international attention had shifted dramatically to the situation in Eastern Europe and to the democratizing states of Latin America. Aid and other financial supports also shifted, leaving the region more vulnerable to global economic fluctuations, particularly with regard to agricultural and mining commodity prices.[14]

The countries in the MRU region faced a trio of coincident crises: (1) a secular decline in aid and investment; (2) pressure from the International Monetary Fund and World Bank to structurally adjust their economies, reducing or eliminating costly public subsidies for foodstuffs, fuel, and housing, along with privatizing or closing parastatals and reducing the size of the civil service; and (3) after 1991, shifting from single-party authoritarian states to multiparty electoral democratic regimes. The removal of financial props, combined with the social costs of structural adjustment and the prospect of a fairly sudden shift to free-for-all political pluralism, proved highly destabilizing in this and other similarly positioned regions.

Since September 2001, U.S. interest in the region and willingness to invest attention and resources has revived. In the last decade the U.S. Agency for International Development has pumped nearly $1 billion for refugee relief and reconstruction into Guinea alone. Many of the world's nongovernmental organizations and aid organizations can be found in the two conflict-torn countries, functioning in recent years as de facto state authorities, especially in terms of the provision of human services. Guinea's recent violent eruptions attracted similar interest on the world stage. Fear of social chaos as a breeding ground for extremism and terrorism has contributed to renewed attention toward the region. Even the threat of Islamic extremism is on the radar for the MRU zone, despite the fact that tolerant and syncretist Sufism in the region has proved a major bulwark against Wahhabism and other Islamist orthodoxies.

The restoration of Western props for the juridical states of the MRU helps explain the shift toward stability and the restoration of order since about 2002. One need only consider the massive nongovernmental organization presence, the influx of aid and relief resources, and the prosecution of Taylor and Sankoh by an international war crimes tribunal for Sierra Leone to realize that the MRU has ceased to be a forgotten and hopeless backwater and has become a quasi-ward of the international community.[15]

The MRU countries, faced with future climate change challenges, despite the modesty of the environmental impact, will need this degree of international support for the foreseeable future. It is also worth highlighting, in terms of capacity, the relationship between a rentier political economy based on natural resource extraction and the particular type of natural resource extracted. The "resource curse" associated with oil production results from highly state-centered processes of antiproductive, anti-investment class formation, in which elites position themselves to control a state flooded with oil rents. Sierra Leone's reasonably accessible diamonds suggest a different political economy, in which small, easily smuggled resources of great value provide the ideal context for predation and warlordism.[16] Guinea and Liberia, with international investments in commodities less lucrative than oil (bauxite and rubber), seem to experience a kind of clientelistic dependency distinct from the oil economies.[17]

Envisioning Worst-Case Scenarios

If, however, international support wanes, as in the 1990s, three scenarios are reasonably likely, with or without climate change effects. First, economic weakness and the particular forms of the resource curse in the MRU countries make intra-elite conflict over control over the machinery of the state a constant threat. In Guinea and Liberia control of the state means control of relations with important international investors (Alcoa and Russia's Rusal in the bauxite sector in Guinea; Firestone and international-aid-focused nongovernmental organizations in Liberia).[18] In Sierra Leone control of the state is also quite important, but control of the eastern region diamond resources represents a countervailing power base. As a result, in all three MRU countries we can expect conflict for control of the state to emanate from (1) dissatisfied would-be elites in the military, especially younger officers and those with international experience and training (e.g., Guinea's Camara); (2) educated elites who have not fled the country and have not been co-opted by the state in the form of meaningful public-sector employment; and (3) especially in Guinea, leaders of the country's two well-organized and effective trade unions.[19] In all cases, the modest internal population shifts resulting from sea-level rise and the decline of ecologically Sahelian cereals (as well as peanuts) will exacerbate the tendency for would-be elites to mobilize the discontented against whoever happens to control the levers of state power.

The second scenario—climate change in the Sahel, to the north of the MRU—will be more severe and could result in population shifts southward. Population will be more likely to flow to locales with robust economic activity. It is conceivable that Guinea could be positioned to grow in future years, especially if governance reforms permit a more efficient state use of mineral resources, jump-starting investment in economic diversification. If Guinea achieves even modest growth rates in the next two and a half decades (e.g., 5 percent per year), it will become a magnet for migrants from the more devastated MRU countries and also from the Sahel, which is likely to bear the brunt of climate change effects in the region. Regional migration flows, one should remember, were once typical in West Africa and might become so again.[20] West Africa's micro states were, before colonialism, linked by long-standing, regionwide migrations of people and goods that resulted in a great deal of fluidity in ethnic, religious, and other forms of identity. These age-old habits and patterns could resume in the face of new forms of environmental pressure.

Regionalization, long a pipe dream of pan-Africanists, is making something of a comeback, especially in the forms of the ECOWAS efforts at peacekeeping in the 1990s and the efforts to expand the Communauté Financière Africaine franc currency zone to include the major anglophone economies of the region (Ghana and Nigeria), producing a truly region-wide common currency—a project that is notably dependent on a considerable increase in regional cooperation.[21] If this is an initial step toward regional integration, then West Africa circa 2015, with a nascent common currency and an institutional infrastructure at least attempting common peacekeeping, would be well beyond where the nascent European Union was in, say, 1970. If regionalization were to advance to produce a more meaningful customs union, and later the harmonization of social service policies and the removal of migration barriers, then the climate-change-induced crises of the coming two and half decades (especially in the Sahel), and the associated migration effects, could be more effectively managed.

Finally, the third scenario—migration outside the region, toward Western Europe—will likely increase as a result of crises in the region and the MRU zone. Sahelian agricultural crisis to the north of the MRU will almost certainly increase flows to Europe. Economic stagnation in the MRU will probably add to the migrants seeking to cross the Sahara or risk the Atlantic passage to the Canaries in dugout canoes. In the 1990s refugee flows from Sierra Leone and Liberia came in the context of sudden crisis, the abandonment of homes, and the murder of family members. People struggled simply to walk to the nearest safety they could find. Unlike contemporary Senegalese (currently the largest source of migrants to Europe from the region), the MRU's migrants in the last twenty years have not had the time and resources to plan complex international journeys. Increasing stability in the MRU zone may alter that dynamic and increase international out-migration from the region.

Climate Change and a Return to State Collapse?

As noted above, the risk of state failure is never remote in these juridical regimes. There are a few key elements to keep in mind on this issue. To begin, as discussed above, international props are the sine qua non of a juridical state, so anything that enhances these supports, or translates support into self-sustaining bases for state consolidation, is welcome. As climate change adds one more set of pressures to these fragile regimes, continued international support is vital. Because climate change will be more severe elsewhere, international attention to the MRU zone may decline in the coming decades, possibly resulting in a shift of international attention and investment to other countries in the subregion, to the detriment of the MRU states, especially in the case of hazard-prevention work in Monrovia.

Further, it is worth noting that the elections in Liberia in 2005 and in Sierra Leone in 2007 were largely free and fair. The losers did not contest the results, and the winners so far appear positioned to surmount previous factionalism and lead their countries in reconstruction. Ellen Johnson Sirleaf, in particular, has managed to parlay her position in Liberia as Africa's first elected female president into considerable international support.[22] In both countries the elections have produced a meaningful increase in popular legitimacy, though it remains to be seen how this will translate into the government's ability to mobilize support and manage dissent.

A final key element to consider is that, though quiet, structural conditions in Guinea continue to be worrisome. The death of the decrepit autocrat Lansana Conté precipitated a succession crisis. Guinea, with its legacy of socialist-nationalist mobilization, is equipped with strong, effective unions. They did not tolerate handing power to another authoritarian military strongman, Captain Camara. As was seen, the repression of protest led to significant social unrest. Widespread corruption and unbalanced contracts with international investors have kept most Guineans from benefiting from the country's natural resource endowment, and the resentment was hard to contain. If the political crisis persists, there may be many, especially among the almost 50 percent of the population under eighteen years of age, ready to be mobilized for one or another promise of a better life. This degree of youthful hopelessness, linked with economic crisis and chaos at the state level, takes Guinea down a path not unlike that of its neighbors to the south. Many hope that the newest Guinean leadership, with an experienced leader from an opposition party who was hand-chosen by the junta, may ease the unstable standoff between a fragile regime and an embittered society.

The Domestic Political Impact of Climate Change

There are no obvious regional, ethnic, religious, or sociopolitical cleavages in any of the MRU countries that map neatly onto climate change in a way that would produce recognizable winners and losers. Guinea's Fulani population is relatively heavily engaged in the production of cereals and peanuts, which are likely to be affected by changes in average temperature and rainfall. But the Fulani are also pastoralists, and they are dispersed as a group throughout the country. Even if climate change disproportionately hurts the Fulani who farm cereals and peanuts, their portfolio, especially when seen in terms of extended kin, is diverse enough that they are unlikely to suffer any serious overall disadvantage compared with other ethnic groups.

In spite of the legacy of intense conflict in the region, there are important reserves of social and institutional capital that, though sometimes latent, could point toward greater resiliency and adaptive capacity than one might expect. First, many farmers and herders in this region are mainly subsistence producers, who have long been disengaged from the large-scale sale of cash crops, let alone the use of cash to buy agricultural inputs (fertilizer, pesticides, better tools, improved seeds) on the market. Subsistence producers and herders have long been on their own in adapting to shifting rainfall, unreliable grazing lands, locusts, and other ecologically induced stressors. There is considerable evidence of household-level ingenuity in switching crop varieties, introducing fallow, using animal manure for fertilizer, or shifting to new locales. Ingenuity and a gift for adaptation have allowed people to survive in what is an already rough and unreliable environment.[23] As forms of human and social capital, these reserves still exist, and they will be deployed by ordinary people on the ground in their struggles to make a living.

Although they cannot count on the state to lend them much or any help, they do need the state to not impede their efforts to adapt. This is typically not a problem, because the state presence is minimal in the lives of subsistence producers. But with climate change, pressure for migration will increase. In this region, migration has historically been a crucial adaptive response to stress; as Bayart describes, the Braudelian *longue durée* pattern of dealing with political conflict has been social cleavage, migration to a new locale, and the creation of new settlements and polities.[24] This was historically facilitated by low population densities and plenty of available land. In the MRU area and much of the Sahel to the north, this basic geodemographic structure is still in place. But postcolonial borders make migration extremely difficult. Regionalization is thus crucial to more fully unleashing the inherent adaptive capacity and resilience of local populations.

As another important social capital resource, patron–clientelism itself, is an important local resource for state building. We usually think of patronage politics as a pathology that needs to be replaced with impersonalism and legal-rationalism. Although this may represent an ideal direction for political change, it is not realistically foreseeable in this region during the next few decades. Given the weakness of the state itself, its irrelevance outside narrow urban elite circles, building patron–client networks that link rural localities to urban power centers in relations of mutual support and resource distribution is an important step in establishing order, and in furthering state formation. One need only look north to Senegal, where hegemonic political parties and the Sufi brotherhoods have built extensive networks of this sort, to see more advanced state formation of a kind that makes sense in historical and cultural terms.[25] Patronage links of course exist in the MRU zone, and if they expand they may provide a means for the center to reach the periphery, distribute resources, and, to a limited but measurable degree, hear the opinions of the people who live there.[26]

Finally, conflict in the MRU zone has often been misunderstood as ethnic. Ethnicity in turn is often mislabeled as a detriment to building political community and sociopolitical order. The colonial construct of "tribe," a reification of existing, fluid identity categories, is at the center of the problem here.[27] Ethnicity itself can be a resource in this region. Although warlords in Sierra Leone and Liberia tried to play upon the politics of ethnicity, this was not the real basis of their appeal or their power; nor is it a durable approach. This is in part because ethnic boundaries and affiliations have historically been fluid across the region. Ethnic categorization and reimagination has long followed migration patterns, particularly as elites and colonizers from the central dynastic states in the region (the precolonial empires of Ghana, Mali, and Songhai) traveled out to and conquered coastal, peripheral peoples. These processes resulted in the intermarriage of conquering elites and conquered peoples, the absorption of subject ethnicities into dominant ones, or vice versa, to the point where recognizably new ethnolinguistic groups have emerged.

International Relations and Climate Change

The impact of climate change on international relations in the MRU states will in all probability center on two issues. First, as was noted above, the more profound ecological and economic effects of climate change that can be anticipated in the Sahel may produce migration flows south to the MRU countries, especially the economically most viable one, Guinea. This will put pressure on the already-fragile social services, economy, and state capacity in that society, pointing toward the kinds of crises discussed above. As noted, international props will be crucial for a country like Guinea to weather the effects of increased migration.

If Guinea falls into a period of internal conflict, antigovernment rebels in Côte d'Ivoire might be in a position to use Guinean territory as a base to restart their conflict with the government in Abidjan. Although RUF remnants in Sierra Leone and fragments of Liberia's various factions remain weak, the "ungluing" of sociopolitical order in those countries in the 1990s leaves open the possibility for the rapid reconstitution of militias, especially if chaos in a neighboring country causes conflict to spill over. Moreover, at an earlier point in the 1990s, Guinea's neighbor to the north, Guinea Bissau, underwent its own political crises, instigated in part by its role as a base of operations for the rebels in Senegal's southern region, the Casamance. At one point in the 1990s, it was possible to imagine disintegration in both Guinea and Guinea Bissau, which could have seriously inflamed the Casamance's low-grade civil war, thus threatening Senegal as a whole—the most stable, prosperous, and democratic country in West Africa. A crisis in Guinea in the coming years could restart the region's fires and cause them to spread in new directions.

Given that Guinea is the most important and promising economy in the MRU region and is the geographical linchpin of the subzone, its reestablished if tenuous stability is crucial to that of its many neighbors. As quasi-wards of the international-aid-and-development community, the MRU countries are quite positively disposed toward the outside world. The Guineans rely on American, Canadian, and Russian investment in the vital bauxite mining sector.[28] They are thus very far removed from the socialist militancy of founding president Sekou Touré. The Sierra Leonean newspapers were filled with appreciation of the United Kingdom's role in ending the civil war and pushing ECOWAS and later the UN to use force to defeat the RUF rebels. Liberians, especially under their new president Sirleaf, a U.S.-trained technocrat, look to renew and deepen their historic ties to the United States, and they deeply appreciate U.S. help. Given the likelihood that these countries will remain dependent on international props and supports, there is every reason to consider their international stance as quite stable. Future Chinese investment and technical assistance in the region, may, however, diversify their portfolio of partners and friends.[29]

Islamic militancy is not a looming threat in the region. Islam is very important in Guinea, 85 percent of whose population is Muslim. It is significant in Sierra Leone (60 percent Muslim) but a minority religion in Liberia (20 percent Muslim). The Islam that is practiced in Guinea and Sierra Leone especially is syncretist and Sufi, open to interpretations and practices that permit blending with local animist traditions, to the point of sometimes rendering the faith unrecognizable to more orthodox believers. Moreover, the sociology of syncretist religion in the region is such that almost every family has some members who are more or less Muslim, other who are more or less animist, and still others who hedge their bets, worshipping the ancestral spirits while saving to make the hajj to Mecca one day. As a consequence, religiously based extremism has had a hard time gaining traction in a region where orthodoxy requires a kind of cognitive, and sometimes social, rupture with one's brother or uncle or cousin. Although the threat of militant Islam may be crucial to keeping the United States interested in the MRU region, the truth is that the threat lies in isolated cells in disparate locales, which so far have not been able to mobilize mass followers because their message tends to be out of touch with the social realities of syncretist Islam.[30]

In conclusion, it is worth noting how far two small superpower investments will go in this part of the world. First, American efforts to accelerate the consolidation of the African Contingency Operations Training and Assistance (ACOTA) program to train and equip African peacekeepers to respond to local crises would constitute a major form of international support for these weak juridical states.[31] An effective African peacekeeping force of the kind envisioned, equipped, and transported by the United States under ACOTA would have dealt with the threat of Taylor in Liberia or Sankoh in Sierra Leone in short order. It would also send a signal about the degree of consolidation of the African state to would-be rebels across the continent, pushing them to express their frustrations not by raising militias in the bush but by crafting appealing messages in the name of forming actual political organizations. Promoting a revitalized ACOTA under the rubric of the Africa Command would also do much to mitigate African suspicion of U.S. military intentions in the region.

Second, as the discussion above underscores, state capacity is crucial to crafting effective responses to climate change. In this connection it is worth recalling that, demographically and territorially, the states in this region make no sense. They are artificial postcolonial anomalies, and with the passage of time they will likely become less important relative to deeper processes of social and cultural change. Regionalization—as usefully modeled, albeit with challenges, by the EU, and as undertaken, albeit incompletely and imperfectly, by ECOWAS—is the future. U.S. support for the fiscal and currency dimension of regionalization, as well as the political dimensions—customs union, harmonization of social services to facilitate the opening of capital and labor markets, and finally of borders—is perhaps the single most important step the United States can take to avoid further costly crises in this region, whether they are induced by the structural weakness of states or by the added pressure that will come from even the modest climate change envisioned by 2030.

Notes

1. The susceptibility of Liberia's capital, Monrovia, to sea-level change is the major exception, and is discussed below.

2. "Fears Guinea Could Be Africa's 'Next Failed State,'" *BBC Monitoring Africa*, January 25, 2007.

3. Diamonds in this region are easy to access because deposits lie near the surface, as opposed to Southern Africa's deep mines.

4. John L. Hirsch, *Sierra Leone: Diamonds and the Struggle for Democracy* (Boulder, CO: Lynne Rienner, 2001).

5. "ECOWAS—Burden of Keeping Peace in the Region," *This Day* (Liberia), November 15, 2007; "A New Democracy Spinning Its Wheels (Sierra Leone)," *The Star* (Johannesburg), October 3, 2007.

6. International Crisis Group, "Guinea: Change or Chaos," *Africa News*, February 14, 2007, www.crisis group.org/en/regions/africa/west-africa/guinea/121-guinea-change-or-chaos.aspx.

7. "Guinea: Army Violence against Civilians Escalates," *AfricaFocus*, February 18, 2007, www.africa focus.org/docs07/guin0702.php; "Guinean Military Gets Pay Rise: Is Conté Buying Loyalty?" *Africa News*, March 11, 2007.

8. The *New York Times* provided ample coverage of these events; see Adam Nossiter, "Guinea Seethes as a Captain Rules at Gunpoint," October 2, 2009; Adam Nossiter, "In a Guinea Seized by Violence, Women Are Prey," October 9, 2009; and Neil MacFarquhar, "U.N. Panel Calls for Court in Guinea Massacre," December 21, 2009. See also "Coup Bid after Death of Lansana Conté," *Gabonews* (Libreville), December 23, 2009, http://allafrica.com/stories/200812230638.html.

9. U.S. Department of State, "Background Notes: Guinea," March 2010; "Background Notes: Liberia," April 2010; and "Background Notes: Sierra Leone," April 2010, all at www.state.gov/r/pa/ei/bgn/.

10. Ibid.

11. Susmita Dasgupta, Benoit Laplante, Siobhan Murray, and David Wheeler, *Climate Change and the Future Impacts of Storm-Surge Disasters in Developing Countries*, Center for Global Development Working Paper 182, September 2009, http://ideas.repec.org/p/cgd/wpaper/182.html.

12. Heiko Paeth and Hans-Peter Thamm, "Regional Modeling of Future African Climate North of 15°S including Greenhouse Warming and Land Degradation," *Climatic Change* 83 (2007): 401–27.

13. Robert Jackson and Carl Rosberg. *Personal Rule in Black Africa* (Berkeley: University of California Press, 1982).

14. For a discussion of the global positioning of the region and its lasting effects, both environmental and otherwise, see Gabriela Kütting, "Globalization, Poverty and the Environment in West Africa: Too Poor to Pollute?" *Global Environmental Politics* 3 (November 2003): 4.

15. As a case in point, reports suggest that President Conté chose to back down in early 2007 in his confrontation with unions and civil society after his own military was contacted by U.S. and U.K. military colleagues, who pressured the doddering and sometimes incoherent Conté to be reasonable. This is an example of the kind of support that was lacking in the crisis years, which can make a tremendous difference in ensuring that juridical states of this sort do not become failed or crisis states. Yet, as the latest episodes in Guinea have shown, the need for such support is often difficult to predict.

16. William Reno, *Warlord Politics and African States* (Boulder, CO: Lynne Rienner, 1998).

17. This may be because resource rents are not large enough or the technical requirements of extraction are high enough to prevent Guinean and Liberian elites from becoming truly autonomous from the interests of mineral investors. See "Statement of Haskell Sears Ward," House Committee on Foreign Affairs, Subcommittee on Africa and Global Health, March 22, 2007, www.internationalrelations.house.gov/110/war032207 .htm.

18. Note that rubber represents more than 90 percent of Liberian export dollars; U.S. Department of State, "Background Notes."

19. I.e., the Guinean Workers Union (USTG) and the National Confederation of Guinean Workers (CNTG).

20. See Jean François Bayart, *The State in Africa: Politics of the Belly* (New York: Longman, 1993). For an example of regional migration patterns, see Philippe David, *Les navétanes: Histoire des migrants saisonniers de l'arachide en Sénégambie des origines à nos jours* (Dakar: Nouvelles Editions africaines, 1980).

21. Moin Siddioi, "Is a Common Currency for West Africa Feasible?" *African Business*, July 1, 2003, www.allbusiness.com/africa/971524-1.html.

22. "Ellen Speech to Swedish Parliament," *The Analyst* (Liberia), November 28, 2007; National Public Radio, "A Conversation with Ellen Johnson-Sirleaf," May 17, 2007, www.npr.org/templates/story/story.php?storyId=10227733.

23. See Sara Berry, *No Condition Is Permanent: The Social Dynamics of Agrarian Change in Sub-Saharan Africa* (Madison: University of Wisconsin Press, 1993); and Jane Guyer, "Traditions of Invention in Equatorial Africa," *African Studies Review* 39, no. 3 (December 1996): 1–28.

24. Bayart, *State in Africa*.

25. See, e.g., Leonardo Villalón, *Isamic Society and State Power in Senegal: Disciples and Citizens in Fatick* (Cambridge: Cambridge University Press, 1995).

26. Mohamed Saliou Camara, *His Master's Voice: Mass Communication and Single-Party Politics in Guinea under Sékou Touré* (Trenton, NJ: Africa World Press, 2005).

27. For a complete dismantling of the trope of tribe and associated myths about primordial ethnicity, see Africa Policy Information Center, "Talking about 'Tribe': Moving from Stereotypes to Analysis," November 1997, updated February 2008, http://community.learnnc.org/weblogs/partners/phe/3_Talking%20about%20'Tribe'.pdf.

28. Guinea is the world's largest exporter of bauxite (15 million tons per year), and industry experts claim that Guinea's high-grade reserves constitute more than one-third of the world's known recoverable bauxite reserves. See "Statement of Haskell Sears Ward."

29. This is at present a hot topic. See Adam Nossiter, "Guinea Boasts of Deal with Chinese Company," *New York Times*, October 13, 2009; Chris Alden, Daniel Large, and Ricardo Soares de Oliviera, eds., *China Returns to Africa: A Rising Power and a Continent Embrace* (New York: Columbia University Press, 2008); and Firoze Manji and Stephen Marks, eds., *African Perspectives on China in Africa* (Oxford, Fahamu, 2007).

30. See U.S. Department of State, Guinea International Religious Freedom Report 2003, www.state.gov/g/drl/rls/irf/2003/23712.htm.

31. This program originated in 1996 as the African Crisis Reaction Initiative. See Major Barthelemy Diouf [Senegalese Army], "Supporting the African Crisis Reaction Initiative," www.almc.army.mil/alog/issues/MayJun02/MS749.htm; and "Africa Crisis Response Initiative (ACRI), African Contingency Operations Training and Assistance," www.globalsecurity.org/military/agency/dod/acri.htm.

18

Southern Africa

Lesotho, South Africa, Swaziland, and Zimbabwe

Ngonidzashe Munemo

Slowly and belatedly, we are acknowledging the reality of global climate change. As denial gives way to recognition, we are also beginning to imagine what these changes mean not just for the environment but also for the viability of human society. Climatologists and environmental scientists predict significant changes to the climate. These changes range from the melting of the ice caps, associated sea-level rise, a general increase in global

Lesotho

South Africa

Swaziland

Zimbabwe

temperature, and subsequent declines in the availability of freshwater to more powerful and devastating severe weather, such as droughts, storms, tornadoes, and hurricanes. On the basis of these predictions, a number of studies suggest intermediate effects of climate change that include a reduction in agricultural production and economic decline in the more severely affected parts of the world.

Less developed and contested are subsequent questions about what each of these changes to the climate might mean for the viability of society—specifically, how climate change might transform governance and security. The emerging consensus is that when the narrow traditional conception of security is expanded, climate change, through its intermediate social transformations, can be viewed as a legitimate threat to security.[1] These scholars suggest that the existing dynamics of intra- and interstate relations will be stressed by sea-level rise, temperature rise, more severe weather, a decline in access to freshwater and the intermediate social outcomes associated with poor agricultural performance, increased competition for freshwater, economic decline, and increased migration, producing new lines of conflict.[2] Thus for some scholars, the processes already set in motion are going to bring about changes to the climate that will transform domestic governance and economies in a number of regions in the world.

Of course, not everyone is convinced about the link between environmental degradation and security. For instance, Deudney contends that environmental concerns should not be considered matters of national security.[3] Generally, he argues that the likelihood that environmental changes will cause a decline in human well-being is too broad a criterion for incorporating such concerns into the realm of national security. He insists that security should be limited to threats posed by intentional, organized groups. This is quite different from how he describes environmental threats: as largely unintentionally harmful, produced by uncoordinated businesses, public agencies, or individuals.[4] Thus, for some, disquiet remains over claims that environmental issues can be considered security issues because of their capacity to spur international conflict.

Although Deudney's insistence that environmental degradation does not constitute a national or regional security problem is spirited, the analysis that follows suggests that there are reasonable grounds on which to reject this idea. As is well summarized by Gleick, levels of "agricultural productivity, the availability and quality of freshwater resources, and access to strategic minerals" will be the environmental factors most likely to significantly influence international relations.[5] Consistent with Gleick and others who see climate as a real threat to security, this chapter sketches out what the anticipated climate changes in individual countries (Lesotho, Swaziland, South Africa, and Zimbabwe) portend for governance and regional security in the region of Southern Africa. It does so by (1) summarizing the expected climate changes looking ahead to 2030, (2) stressing the social transformations likely to result from these changes, and (3) suggesting how each of these intermediate social transformations will alter regional security.

There are a number of reasons that it makes sense to consider the impact of climate change on Southern Africa. First, though the region is divided into several different countries, those countries' current climates and the future changes likely to occur are largely similar. Second, as the discussion of the economies of Lesotho, Swaziland, and Zimbabwe below illustrates, the region as a whole is tied to its powerhouse: South Africa. Thus, a political and economic crisis in one country in the region is likely to be transferred to the others through their interaction with South Africa. Finally, Southern Africa has a regional body, the Southern Africa Development Community (SADC), that has a mandate to provide socioeconomic cooperation and integration and, recently, also political and security cooperation among its fifteen members.[6] In this context, changes to the security status of one member are likely to draw in the other member states due to commitments enshrined in the SADC protocols.

How the Climate Is Changing

During the past decade and half, various studies have explored and summarized the anticipated global climate changes.[7] The emerging consensus from this work is that the planet's biological and physical systems have exhibited changes consistent with a much warmer climate.[8] For instance, Arctic temperatures are rising twice as fast as the global average, the frequency of heavy rain over land has increased, so too have the frequency and severity of tropical cyclones in the North Atlantic, heat waves are increasing in intensity and frequency, and some parts of the globe have seen an increase in the severity of droughts.

These changes offer a sobering picture of what we can expect in the coming decades, when many of these processes are expected to accelerate. Though not all changes to climate will emerge rapidly,[9] those changes that are already apparent are poised to transform ecosystems and human systems in complex ways. In varying degrees of severity, we have reason to expect that a warmer planet will result in an increase in agricultural degradation due to inadequate water or the increase in pests and diseases. Furthermore, low-lying coastal areas and major river mouths are at risk of sea-level rise and destruction due to storms surges.

For reasons having to do with chronic poverty, weak governance, and high dependence on agriculture, Africa is the region estimated to be most vulnerable to the effects of climate change. For instance, in his widely cited book on the likely impact of global warming on agricultural performance by 2080, Cline's models suggest that the impact of climate change on African agriculture is going to be severe. According to Cline, "there is a predominant pattern of large negative changes from business as usual warming (excluding carbon fertilization) in dryland African agriculture."[10] Although initial models of irrigated agriculture (including Egypt) produced positive gains, subsequent models excluding Egypt suggested that even irrigated agriculture across Africa was going to be negatively affected by global warming.

These continent-wide changes are illustrative of the range of the first-order effects of climate change in Africa. However, the aggregate picture masks some regional differences and underestimates the effect of climate change in a number of countries. As Cline notes in his study, aggregate results may appear better than the reality on the ground. The positive gain in agricultural productivity estimated for irrigated agriculture, for instance, is largely a function of the inclusion of Egypt, which has the largest base of irrigated agriculture and the second-largest agricultural output on the continent, along with the relatively small net decline anticipated for Nigeria, the country with the largest agricultural output in Africa.[11]

What this suggests is that while a global and continent-wide picture is a useful place to start, a regional and country-specific focus is needed to more accurately understand climate change and its political and human dimensions. Looking at Southern Africa, Magadza argues that the region is particularly sensitive to changes in its water resources. On this front, the picture is quite bleak for a number of countries in the region. For instance: the Okovango Delta is Botswana's only source of surface water; Zimbabwe is expected to reach its water resource development capacity by 2035; and South Africa's industrial and mining heartland, Gueteng, reached the limits of its water resources in the 1980s, such that it became necessary to import water from Lesotho.[12] Magadza also uses the El Niño–produced 1991–92 regional drought as a proxy for some of the major effects of climate change and finds that Southern Africa can expect a significant decrease in moisture during the growing season, leading to a decline in food production and a reduction in the flow of the Zambezi, Limpopo, and Save rivers. With reduced flow in these important rivers, the downstream effects of low hydroelectric generation and impaired economic activity are not hard to imagine.

Of course, each country is likely to experience different magnitudes of some of these regional climate changes. For instance, according to CIESIN data, it is estimated that by 2030, Lesotho's mean temperature will increase by 1.03 degrees Celsius. As a consequence, the higher rates of evaporation associated with the temperature increase, notwithstanding potential gains that some scientists anticipate from the process of carbon fertilization, are expected to produce a very serious decline in agricultural productivity in Lesotho. The situation is slightly different in Swaziland, where the average temperature is expected to rise by 0.86 degree Celsius and agricultural production will likely be lower.

A similar difference in the specific climate changes countries will endure can be seen if we compare the estimates for South Africa and Zimbabwe. Current models suggest that South Africa's average annual temperature will increase by about 0.97 degree Celsius. In addition to potentially lowering agricultural production, it is estimated that climate changes will increase the proportion of South Africans living on less than 1,000 cubic meters of freshwater per year from 55.4 percent of the population to 58.2 percent—an increase of 2.7 percent—by 2030. Estimates for Zimbabwe show smaller increases in the average temperature (an increase of about 0.81 degree Celsius), but more dramatic agricultural and freshwater availability effects. The combination of an increase in temperature and higher levels of evaporation are expected to have a more serious impact on Zimbabwe's agricultural performance. More significantly, the proportion of Zimbabweans living on less than 1,000 cubic meters of freshwater will increase from 44.9 percent in 2000 to 56.3 percent by 2030. All these figures are summarized in appendix A.

It must be noted that the overall impact of these climate changes will be conditioned by preexisting environmental stresses and by each country's institutional capacity for adaptation. For instance, Lesotho is already struggling to cope with overgrazing, soil erosion and exhaustion, desertification resulting from demographic pressures, and periodic droughts. Furthermore, the country is already institutionally struggling with mounting demographic pressures, a recent decline in economic performance, and a progressive deterioration of the government's ability to provide public services. Similarly, the projected climate changes for Swaziland will interact in complex ways with its current environmental problems of overgrazing, soil depletion and degradation, droughts, and limited supplies of potable water. Because the Swazi government is already confronting extreme poverty, mounting demographic pressure, while itself displaying signs of criminalization and delegitimization in the eyes of society, its capacity or willingness to respond to new environmental challenges is not obvious.

In South Africa's case, the changes to its climate are going to complicate its unique challenges stemming from the absence of significant arterial rivers or lakes. The majority of South Africa's population (55.4 percent) already has limited access to freshwater. Soil erosion is widespread, and the pollution of rivers from agricultural runoff, urban discharge, and acid mine drainage are well-recognized problems. In addition, South Africa's uneven economic development is only likely to widen under the pressure of rising temperatures and reduced rainfall. For Zimbabwe, any additional stresses produced by climate change may well complicate the challenges of the unity government that took office in February 2009. As was noted in the introduction to this volume, the worst effects of climate change can be anticipated in countries with weak states, located in the high northern or southern latitudes. These conditions are prevalent in Southern Africa.

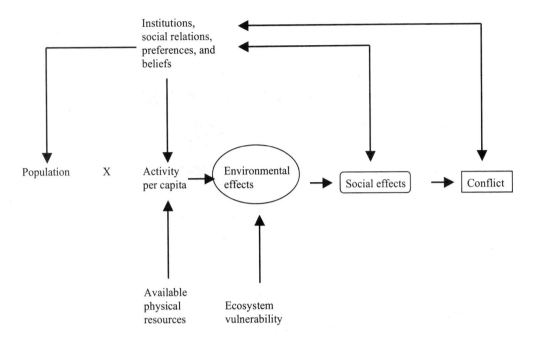

Figure 18.1 The Connection between Climate Change and Security
Source: This figure is from "On the Threshold: Environmental Changes as Causes of Acute Conflict," by Thomas Homer-Dixon, *International Security* 16, no. 2 (Autumn 1991): 86. © 1991 by the President and Fellows of Harvard College and the Massachusetts Institute of Technology. Reprinted by permission.

Climate Change and Security in Southern Africa

In general terms, how might changes to the climate affect domestic and/or regional security? In his 1991 *International Security* article, Thomas Homer-Dixon developed an analytical framework (figure 18.1) through which the causal pathways between environmental change and conflict could be understood.[13] Describing the implications of his framework, Homer-Dixon begins by stressing that the starting point of any analysis is predicated on the understanding that the aggregate effect of human activity on a given environmental system is a function of two variables: "the product of total population in the region and physical activity per capita, and second, the vulnerability of the ecosystem"[14] Specific to questions about security, Homer-Dixon's framework suggests that environmental effects are causally connected to conflict primarily through four intervening social effects: "decreased agricultural production, economic decline, population displacement, and disruption of legitimized and authoritative institutions and social relations."[15] Thus, by paying particular attention to these social effects of environmental change, Homer-Dixon also summarized a number of the pathways by which climate change will likely transform security.[16] In the next section I describe these four social effects and consider how climate change will likely transform regional security in Southern Africa.

Access to Freshwater

Historically, rivers have figured quite prominently in the strategic policy and action of nations. Two factors help explain why rivers have held such an important position in the

Table 18.1 Per Capita Water Availability in Southern Africa (1,000 cubic meters per capita per year)

Country	Available Water		Water Status
	1995	2025	
Angola	17.0	7.2	Sufficient
Botswana	10.1	5.7	Sufficient
Lesotho	2.6	1.3	Stress
Malawi	1.9	0.9	Scarce
Mauritius	2.0	1.5	Stress
Mozambique	12.1	5.9	Sufficient
Namibia	29.6	15.2	Sufficient
South Africa	1.2	0.7	Scarce
Swaziland	5.3	2.7	Sufficient
Tanzania	3.0	1.4	Scarce
Zambia	14.4	7.2	Sufficient
Zimbabwe	1.8	1.0	Stress

Note: Water is "sufficient" if greater than 1,700 cubic meters per capita per year; "stress" implies availability between 1,000 and 1,700 cubic meters. Water is "scarce" below 1,000 cubic meters per capita per year.
Source: Tom Gardner-Outlaw and Robert Engelman, *Sustaining Water, Easing Scarcity: A Second Update*, (Washington, DC: Population Action International, 1997), cited by Herbert H.G. Savenije and Pieter van der Zaag, "Conceptual Framework for the Management of Shared River Basins, with Special Reference to SADC and EU," *Water Policy* 2 (2000): 10.

business of state formation and consolidation. First, as noted by Gleick, "freshwater is a fundamental resources, integral to all ecological and social activities, including food and energy production, transportation, waste disposal, industrial development, and human health."[17] However, the ability of most states to benefit from the exploitation of rivers is constrained and threatened by the fact that "freshwater resources are often shared by two or more nations."[18] Owing to the interaction of these two factors, it is not surprising that interstate conflict over access to and the control of rivers and their freshwater resources has been common.[19] Because climate change is anticipated to result in a noticeable decline in the availability of freshwater in most regions of the world, there is good reason to expect that interstate conflict over access to and the control of freshwater will become more acute, and potentially more destabilizing.

The literature on water and conflict has tended to focus on the Middle East.[20] However, Southern Africa is not insulated from the potential security threats posed by competition over access to freshwater. If anything, the region has a number of the elements that make conflict over access to freshwater likely: preexisting water scarcity (table 18.1), a high proportion of people living within river basins, many rivers whose watersheds are shared among several countries (table 18.2), and a history of international tension and competition over water resources. Adding to these common challenges, van Wyk contends that conflict over access to freshwater is made all the more possible because the largest and most powerful state in the region, South Africa, "accounts for eighty percent of Southern Africa's water use, [but] only ten percent of the total water resource [of the region] is available in South Africa."[21]

The likely intensification of water scarcity as a result of the climate changes projected for Southern Africa will make water conflicts more likely and harder to resolve. A brief look at the Zambezi River and its basin, the fourth-longest river in Africa, illustrates the many

Table 18.2 Selected Shared Water Sources in Southern Africa

River	Total Area (square kilometers)	River Basin Countries
Cunene	117,000	Angola and Namibia
Limpopo	423,000	Botswana, Mozambique, South Africa, and Zimbabwe
Okavango	586,000	Angola, Botswana, Namibia, and Zimbabwe
Orange	973,000	Botswana, Lesotho, Namibia, and South Africa
Rovuma	155,000	Mozambique and Tanzania
Save	104,000	Mozambique and Zimbabwe
Zambezi	1,234,000	Angola, Botswana, Malawi, Mozambique, Namibia, Tanzania, Zambia, and Zimbabwe

Source: Jo-Ansie van Wyk, "Towards Water Security in Southern Africa," *African Security Review*, 7 no. 2 (1998): 1, www.iss.co.za/PUBS/ASR/7No2/VanWyk.html.

dangers that lie ahead in a warmer, dryer world. Originating in northwestern Zambia, the 2,574-kilometer-long Zambezi snakes its way east through Angola, back into western Zambia, then along the Namibia/Zambia border, where it becomes the meeting point of Namibia, Botswana, Zambia, and Zimbabwe. It then forms the entire stretch of the Zimbabwe/Zambia border, before flowing through Mozambique on its way to the Indian Ocean. As illustrated in the map at the start of this chapter, the basin of the Zambezi River covers a total of eight countries (Angola, Botswana, Malawi, Mozambique, Namibia, Tanzania, Zambia, and Zimbabwe). According to the Southern Africa Research and Documentation Centre, the Zambezi Basin amounts to about 25 percent of the combined total area of these eight countries.[22] This makes it the largest river basin wholly within the SADC region and the most shared natural resource in the region.

With so many interested parties, tension over the exploitation of the Zambezi River has not been uncommon. For instance, Zimbabwe's proposal in the early 1990s, as part of the Matabeleland Zambezi Water Project, to build a pipeline from the Zambezi River to supply its drought-prone city of Bulawayo produced some tension among riparian countries.[23] In addition to the proposed pipeline, Zimbabwe's existing water withdrawal from the Zambezi, to supply its coal-based Hungwe thermal power station and its irrigated agriculture, threatens the downstream water supply for Mozambique.

Zimbabwe is not the only country to propose expanding its exploitation of the Zambezi River. With more than 94 percent of its freshwater originating from outside its borders, Botswana also proposed to draw water from the Zambezi under the North–South Carrier Extension.[24] South Africa was also keenly interested in building on Botswana's plan by developing its own water diversion scheme, the Zambezi Aqueduct, to meet its own water shortfall. Predictably, Botswana and South Africa's musings over "diverting the Zambezi River at Kazungula" generated significant friction among SADC members.[25]

Although these earlier claims on the resources of the Zambezi River were resolved amicably by the adoption of the SADC Protocol on Shared Watercourse Systems (1995) and by the creation of the Zambezi River Authority,[26] van Wyk is justified in his concern that Angola and Zambia did not sign the protocol and that only Zambia and Zimbabwe are part of the authority. Thus, under conditions of a weak commitment to the institutions governing the exploitation of the Zambezi River, it is likely that the additional water stresses produced by climate change will test the limits of these agreement.

Even in cases where there is a strong institutional arrangement for sharing water re-sources, as with the Lesotho Highlands Water Project (LHWP) on the Orange River, the potential for future conflict is not entirely absent. Under the terms of the LHWP Treaty (1986), Lesotho "transfer[s] water from the water-rich highlands of Lesotho to the dry Guuteng of South Africa . . . and to supply electricity to Lesotho."[27] As noted by Matete, the LHWP is one of the largest water transfer schemes in the world and is vitally import for water-scarce South Africa. The scheme supplies water to a region of South Africa that ac-counts for "40% of the country's GDP, more than 50% of its industrial output, and supports more than 30% of the total population."[28] Because this project supplies freshwater to South Africa's core industrial and population center, it is possible that as South Africa struggles to meet its own domestic needs for water, it might pursue more aggressive methods to secure the amount of water delivered through the LHWP.

Competition for Food and Migration-Related Disputes

Like rivers, threats to food supplies and agricultural systems also tend to generate con-flict. As discussed above, climate change is expected to result in significant negative effects on agricultural production and national food supplies. In developing countries, declin-ing food production will most likely lead to two types of security threats. At the domestic level, as food production declines, internal strife produced by a combination of rural–urban conflict over access to limited food or nomadic–sedentary tension over competing land uses can be expected. At the interstate level, as food becomes scarcer, it is also more likely that surplus-food-producing countries might be tempted to use this new "resource" against neighbors and or rivals.[29]

In addition, failing economies, less-than-adequate water supplies, and collapsing ag-ricultural and food systems will increase the local, regional, and national movement of people. Whether this migration involves existing networks or the creation of new destina-tions, this movement of people in search of a better livelihood will generate its own set of conflicts.[30] The competition for resources, food, land, water, or jobs created by migration will likely increase the incidence of ethnic conflict, xenophobic clashes, and possibly con-flict between states. Because climate change can be expected to result in increased migra-tion, we can reasonably foresee an intensification of migration-related conflicts at both the national and regional levels.[31]

South Africa recently experienced a wave of xenophobic attacks on migrants that signals the kind of conflicts that are likely to increase in frequency and intensity as the effects of cli-mate change take hold. Within Southern Africa, South Africa is the economic and political giant. Because of its dominance and promise of greater opportunity, South Africa has histor-ically been and continues to be the prime destination of migration flows from other coun-tries in the region. For instance, the existing data suggest that close to 35 percent of male wage earners in Lesotho work in South Africa. South Africa also attracts a large number of migrants from the other countries in the region. Most notably, as the political and eco-nomic crisis in Zimbabwe worsened, scores of Zimbabweans risked crossing the crocodile-infested Limpopo River, not to mention the razor wire erected on the South African side to hold them back, in search of opportunity in South Africa. The numbers were staggering. Although the South African authorities were said to have arrested and returned about a thousand illegal immigrants from Zimbabwe every day, about three times that many were estimated to cross the border successfully each day at the peak of the flows in mid-2007.[32]

Regional migration to South Africa from the surrounding countries is only likely to in-tensify as other countries suffer the agricultural decline and economic stresses anticipated from the effects of climate change. As the number of economic and "environmental refugees" increases in South Africa, we can expect the level of migrant-based conflicts to increase. With high domestic unemployment levels and continuing high hopes for economic opportunity and social improvement, the situation in South Africa will be easily combustible.[33]

Over the weekend of May 10–11, 2008, Alexandra Township, north of Johannesburg, exploded into a frenzy of violence directed against immigrants from Malawi, Mozambique, and Zimbabwe.[34] Alexandra was the starting point of a wave of attacks on Makwerekwere that spread from Gauteng to Kwa-Zulu Natal and Durban over a four-week period.[35] With a population of about six hundred thousand and an unemployment rate of about 40 percent at the time, Alexandra contained within it many of the aspirations, tensions, and cleavages of urban South Africa. According to reports in the South African media, gangs of Alexandra township residents "whipped, threw stones, and shot at" suspected illegal immigrants whom they accused of robbing them.[36] Foreigners were not only attacked in the streets of Alexandra but also in their homes. Makeshift houses belonging to foreigners were hacked down by angry mobs, property was looted, and occupants were at risk of being burnt with what was left of their homes and possessions. These initial attacks left two people dead and more than forty injured.[37] The police deployed close to five hundred officers in and around Alexandra, and many immigrants took refuge at local police stations. In the following few weeks, the attacks would subside during the day only to flare up again at night. Fairly quickly, makeshift barricades made of smoldering tires and debris from ransacked shops transformed parts of Johannesburg's central business district into a war zone.[38]

When the violence ended, more than fifty people were dead, hundreds had been injured, and thousands were left displaced.[39] Immigrants were accused of coming into South Africa "to get our jobs."[40] Some South Africans also placed the blame on South African companies for giving jobs to foreigners because employers believed they would "work harder than South Africans."[41] The spontaneity, intensity, and national reach of this episode of xenophobia in South Africa offers a disquieting glimpse of the form of tension likely to intensify as the stresses of climate change in Southern Africa force more and more people to look for opportunity in the regional economic powerhouse.

Economic Decline and Security

Economic decline is another potential outcome of a warming climate and the subsequent impaired agricultural systems. As I have summarized elsewhere, the impact of the 1991–92 Southern African drought on Zimbabwe's economy is a good analogue of the connection between the environment and the economy.[42] In 1991–92, Zimbabwe suffered its most severe drought in a century.[43] According to government documents, poor rains, measuring more than 300 millimeters below the country average of 662 millimeters per year, resulted in communal farmers harvesting an abysmal 10 percent of their normal grain output. At the aggregate level, cereal production in 1992 was only 22 percent of the five-year average, with crop failure estimated at 92 percent for communal farmers. The total cereal crop of 44.6 metric tons per 1,000 capita was more than 410 metric tons per 1,000 capita below the 1981 harvest. As a result of this once-in-a-hundred-years drought, close to 5.6 million people (nearly half the population) were in need of government assistance.

Notably, the 1991–92 drought had significant downstream effect on the overall economy of Zimbabwe. A combination of water shortages, electricity shortages, input supply difficulties, and reductions in demand for both agricultural inputs and other basic consumer goods severely impaired the operation of Zimbabwe's light-industrial sector.[44] For instance,

Benson estimates that the 1991–92 drought was associated with "year-on-year declines of 5.8 percent in GDP at factor cost, . . . and 8.5 percent in industrial GDP."[45] In all, the drought was estimated to be responsible for, at a minimum, a 25 percent decline in the volume of manufacturing output, a 6 percent reduction in foreign currency receipts from manufactured exports, and an overall decline in manufacturing output of about 9.5 percent in 1992.

Of course, as noted by Benson and Clay, the overall effect of the 1991–92 drought was amplified by the intermediate structure of the Zimbabwean economy, which transmitted the effects of the water shortage to all the sectors of the economy that were integrated with the agricultural sector. According to Benson and Clay, an intermediate economy is one in which a more "diversified structure exists with economic growth occurring via the development of a labor-intensive, low-technology manufacturing sector, that is typically dependent on domestically produced, renewable natural resources and imported inputs and capital equipment, but with natural resources still representing a relatively important part of export earnings."[46] As a consequence, the increasing complexity in an intermediate economy means that the effects of drought and other shocks are more likely to be widely transmitted to other sectors.

Although the other countries in the region are not intermediate economies, there is nevertheless good reason to be concerned about the effect of climate-change-induced natural disasters on these economies. For instance, with respectively 86 percent and 70 percent of their populations dependent on rain-fed subsistence agriculture for their livelihood, the simple economies of Lesotho and Swaziland are going to be devastated by the combination of increases in annual temperatures, higher rates of water evaporation, and reductions in the rainfall. Even South Africa's somewhat complex economy will not be insulated from the economic effects of climate change, as evinced by its recent electricity supply crisis.[47]

Thus, to the extent that climate change will lower the production of crops that are key raw materials in many light industries, or completely destroy subsistence agriculture, there is good reason to expect a significant negative effect on economic systems. In traditional security studies, economic decline and social impoverishment have long been argued to produce and intensify group conflict over scarce resources in ways likely to result in insecurity in Southern Africa. As Gurr noted a little over two decades ago, "in circumstances of increasing scarcity and persisting economic decline, . . . there is every reason to think that group conflict over distribution [will] intensify.[48]

The effect of Zimbabwe's economic decline on relations with those of its neighbors is suggestive of how economic decline in one country could easily affect regional security. For instance, the increased migration caused by Zimbabwe's political and economic meltdown initially led Botswana to erect an electric fence along its border with Zimbabwe. Botswana later decided to reinforce its fence by deploying troops along the border,[49] as a result of the worsening situation in Zimbabwe, which saw two rounds of elections in March and June 2008, an increase in political violence and economic collapse, an official inflation rate that reached more than 232 million percent, and the introduction of a Z$100 billion note.[50] For its part, South Africa has taken to a combination of razor wire barriers along its border with Zimbabwe and more frequent military patrols of the boundary. The danger, of course, is that once borders are militarized in this way, there is a real risk of skirmishes between the forces on both sides, which may provide a pretext for open conflict between states.

Although it is unlikely that climate change will engender such rapid and deep economic decline elsewhere in the region as we have witnessed in Zimbabwe, its effect on subsistence agriculture, commercial agriculture, and electricity generation will intensify the general-ized economic stresses and social impoverishment that are already prevalent in Southern Africa and will thus deepen social strife and make scarcity disputes more likely.

Conclusion

Much of what has been outlined here as the probable climatic conditions looking ahead twenty or thirty years from now describes environmental changes that have already begun in many parts of Southern Africa. Temperatures have been increasing steadily, droughts have been more frequent and severe, and the region is sensitive to freshwater availability. As these climatic changes have taken hold, we have already witnessed evidence of some of the intermediate effects of global warming in the form of declining agricultural produc-tion, freshwater scarcity, economic decline, and increased migration. This chapter suggests that the intensification of climate change and its intermediate effects will likely translate into conflict over food, access to water, and economic opportunity.

Notes

1. See, among others, Jessica T. Mathews, "Redefining Security," *Foreign Affairs* 68, no. 2 (Spring 1989): 162–76; Arthur H. Westing, "An Expanded Concept of International Security," in *Global Resources and Inter-national Conflict: Environmental Factors in Strategic Policy and Action*, edited by Arthur H. Westing (Oxford: Oxford University Press, 1986), 183–200; and Richard H. Ullman, "Redefining Security," *International Secu-rity* 8, no.1 (Summer 1983): 129–53.

2. See Neville Brown, "Climate, Ecology and International Security," *Survival* 31, no. 6 (November 1989): 519–32.

3. Daniel Deudney, "The Case against Linking Environmental Degradation and National Security," *Jour-nal of International Studies* 19, no. 3 (1990): 461–76.

4. Ibid.

5. Peter H. Gleick, "The Implications of Global Climatic Changes for International Security," *Climate Change* 15 (1989): 309–25, at 310.

6. Current member states of the SADC, with dates of membership, are Angola (1980), Botswana (1980), Democratic Republic of Congo (1997), Lesotho (1980), Malawi (1980), Madagascar (2005), Mauritius (1995), Mozambique and Namibia (1990), Seychelles (1997–2004 and 2007–), South Africa (1994), Swaziland (1980), Tanzania (1980), Zambia (1980), and Zimbabwe (1980).

7. IPCC, *Climate Change 2007: Impacts*, 23–78.

8. Most general circulation models of climate change suggest that the mean temperature of the Earth has risen at a rate of about 0.13 degree Celsius per decade over the past five decades.

9. E.g., the full extent of global warming is yet to be felt due to the cooling associated with high-levels of aerosols (from the burning of coal and biomass) in the atmosphere. However, because aerosols have a much shorter atmospheric lifetime compared with greenhouse gases, there is reason to expect higher rates of warming in the future.

10. William R. Cline, *Global Warming and Agriculture: Impact Estimates by Country* (Washington, DC: Peterson Institute for International Economics, 2007), 53.

11. Ibid., 56. Cline's estimate seeks to account for the effects of anticipated changes in temperature and freshwater availability but does not consider the potential inundation or salinization of large areas of the Nile delta as a consequence of sea level rise.

12. C. H. D. Magadza, "Climate Change: Some Likely Multiple Impacts in Southern Africa," *Food Policy* 19, no. 2 (1994): 165–91.

13. See Thomas Homer-Dixon, "On the Threshold: Environmental Changes as Causes of Acute Conflict," *International Security* 16, no. 2 (Autumn 1991): 85–88.

14. Ibid., 85.

15. Ibid., 91.

16. Ibid, 91–98.

17. Peter H. Gleick, "Water and Conflict: Fresh Water Resources and International Security," *International Security* 18, no. 1 (Summer 1993): 79.

18. Peter H. Gleick, "Climate Change and International Politics: Problems Facing Developing Countries," *Ambio* 18, no. 6 (1989): 336.

19. See Malin Falkenmark, "Fresh Waters as a Factor in Strategic Policy and Action," in *Global Resources and International Conflict*, ed. Westing, 85–113.

20. See Franklin M. Fisher and Annette Huber-Lee, "Economics, Water Management and Conflict Resolution in the Middle East and Beyond," *Environment* 48, no. 3 (April 2006): 2–41.

21. Jo-Ansie van Wyk, "Towards Water Security in Southern Africa," *African Security Review* 7, no. 2 (1998): 1, www.iss.co.za/PUBS/ASR/7No2/VanWyk.html.

22. SARDC, www.sardc.net/imercsa/zambezi/Cep/map.htm. The Nature Conservancy provides a good map of the Zambezi River Project Area at www.nature.org/wherewework/greatrivers/files/zam_basin_map .pdf.

23. Ashok Swain, "Water Wars: Fact or Fiction?" *Futures* 33 (2001): 769–81.

24. van Wyk, "Water Security." Had it been implemented, the North–South Carrier Extension scheme would have also supplied water to South Africa.

25. Magadza, "Climate Change," 172.

26. For a discussion of the some of the provisions of the protocol and other agreements governing the exploitation of the Zambezi River, see Herbert H. G. Savenije and Pieter van der Zaag, "Conceptual Framework for the Management of Shared River Basins, with Special Reference to SADC and EU," *Water Policy* 2 (2000): 9–46; Osborne N. Shela, "Management of Shared River Basins: The Case of the Zambezi River," *Water Policy* 2 (2000): 65–81; M. J. Tumbare, "Equitable Sharing of the Water Resources of the Zambezi River Basin," *Physical Chemistry of the Earth*, part B, vol. 24, no. 6 (1999): 571–78; and Mikiyasu Nakayama, "Politics behind Zambezi Action Plan," *Water Policy* 1 (1998): 397–409.

27. World Bank, *Implementation Completion and Results Report (IBRD-43390): On a Loan in the Amount of US$45 Million to the Lesotho Highlands Development Authority for Lesotho Highlands Water Project— Phase1B*, Report ICR168, June 14, 2007, 1, www-wds.worldbank.org/external/default/WDSContentServer/ WDSP/IB/2007/08/22/000020953_20070822101922/Rendered/PDF/ICR168.pdf.

28. Mampati E. Matete, *The Ecological Economics of Inter-Basin Water Transfers: The Case of the Lesotho Highlands Water Project* (PhD diss., Faculty of Natural and Agricultural Sciences, Department of Agricultural Economics, Extension, and Rural Development, University of Pretoria, 2004), 4.

29. See Peter Wallensten, "Food Crops as a Factor in Strategic Policy and Action," in *Global Resources and International Conflict*, ed. Westing, 143–58; and Gleick, "Climate Change," 337.

30. For a discussion of the migration and security, see Myron Weiner, "Security, Stability and International Migration," *International Security* 17, no. 3 (Winter 1992–93): 91–126.

31. Homer-Dixon, "On the Threshold," 76–116.

32. Peter Biles, "The Lure of Plentiful South Africa," BBC, July 21, 2007, http://news.bbc.co.uk/2/hi/pro grammes/from_our_own_correspondent/6908182.stm. Estimates suggest that there are close to 3 million Zimbabweans in South Africa.

33. For a discussion of unemployment and inequality in South Africa, see Michael MacDonald, *Why Race Matters in South Africa* (Cambridge, MA: Harvard University Press, 2006); and Jeremy Seekings and Nicoli Nattrass, *Class, Race, and Inequality in South Africa* (New Haven, CT: Yale University Press, 2005).

34. "South African Mob Kills Migrants," BBC, May 12, 2008, www.news.bbc.co.uk/go/pr/fr/-2/hi/africa/ 7396868.stm.

35. "Makwerekwere" is a derogatory term used to describe immigrants in South Africa.

36. "Mob Kills Two in Suspected Xenophobic Attack," *Mail & Guardian Online*, May 12, 2008, www.mg .co.za/printPage.aspx?area=/breaking_news/breaking_news__national/&articleId=338997; and "Death Toll Climbs in SA Violence," BBC, May 27, 2008, http://news.bbc.co.uk/2/hi/africa/7420708.stm.

37. "Cops Monitoring Alexandra after Attacks," *Mail & Guardian Online*, May 12, 2008, www.mg.co.za/ printPage.aspx?area=/breaking_news/breaking_news_national/&articleId=3389038.

38. Nicole Johnston and Percy Zvomuya, "Mob Violence Turns Jo'burg CBD into War Zone," *Mail & Guardian Online*, May 18, 2008, www.mg.co.za/printPageaspx?area=/breaking_news/breaking_news_national/&articleId=339522.

39. See Paul Simao, "Mbeki's Rule in Limbo as Townships Burn," *Mail & Guardian Online*, May 27, 2008, www.mg.co.za/printPageaspx?area=/xenophobia_home/xenophobia_news/&articleId=340223.

40. Dickson Jere, "Zim Exiles Face New Fear and Loathing in SA," *Mail and Guardian Online*, May 14, 2008, www.mg.co.za/printPageaspx?area=/breaking_news/breaking_news_national/&articleId=339207.

41. Ibid.

42. See Ngonidzashe Munemo, *Politics during Dry Times: Incumbent Insecurity and the Provision of Drought Relief in Contemporary Africa* (PhD diss., Columbia University, 2008), appendix G, 249–55.

43. C. Benson, "Drought and the Zimbabwe Economy, 1980–1993," in *A World without Famine? New Approaches to Aid and Development*, edited by Helen O'Neil and John Toye (London: Macmillan, 1998), 241–78; Carol B. Thompson, *Drought Management Strategies in Southern Africa: From Relief through Rehabilitation to Vulnerability Reduction* (Gaborone: Gaborone Food Security Unit, Southern African Development Community, 1993); and *The Drought Relief and Recovery Programme, 1992/93* (Harare: Zimbabwe Government Publications, 1993).

44. Charlotte Benson and Edward Clay, *The Impact of Drought on Southern-Saharan African Economies: A Preliminary Examination*, World Bank Technical Paper 401 (Washington, DC: World Bank, 1998).

45. Charlotte Benson, "Drought and the Zimbabwe Economy, 1980–1993," in *World without Famine?* ed. O'Neil and Toye, 248.

46. Benson and Clay, *Impact of Drought*, 17.

47. Benson and Clay, ibid., characterize simple economies as "predominantly rain-fed agriculture and livestock semi-subsistence economies with a limited infrastructure, low levels of per capita and high levels of self-provisioning in the rural population." A complex economy, in contrast, is marked by the presence of a highly developed industrial sector "with a relatively small agricultural sector and proportionately small forward and backward linkages between agriculture and the other water-intensive activities and the rest of the economy." Though not a perfect fit, the South African economy closely approximates the latter structure, while those of other Southern African states resemble the former.

48. Ted Gurr, "On the Political Consequences of Scarcity and Economic Decline," *International Studies Quarterly* 29, no. 1 (March 1985): 60.

49. Reuben Pitse, "Botswana Prepares for War?" *Sunday Standard* (Gaborone), July 2, 2008, http://sundaystandard.info/print.php?NewsID=3380.

50. "Zimbabwe Introduces Z$100bn note," BBC, July 19, 2008, http://news.bbc.co.uk/2/hi/africa/7515823.stm.

19

The Northern Andes

Bolivia, Colombia, Ecuador, and Peru

Kent Eaton

Bolivia, Colombia, Ecuador, and Peru are commonly referred to as "Andean" countries, and the Andes Mountains undoubtedly provide the visual imagery that is most commonly associated with each of these four countries. However, in addition to the Andes, the national territories of Bolivia, Colombia, Ecuador, and Peru all contain substantial non-Andean geographies. Colombia, Ecuador, and Peru, for example, can be divided into three

Bolivia

Colombia

Ecuador

Peru

starkly different subnational regions: Andean, Amazonian, and coastal (with Colombia having two different coasts, Atlantic and Pacific). Even in Bolivia, which lost its coastal access in a nineteenth-century war with Chile, the mountain ranges of the Andes account for less than half the country's territory. Much of the Bolivian land mass lies in the valleys and plains of the Amazon River Basin east of the Andes. Thus, though these four countries certainly have the Andes in common, they also share significant non-Andean spaces and much geographic diversity.

This geographic diversity provides a critical point of entry into the consideration of climate change and regional security in the northern Andes. The significant geographic variation within each of these four countries is important, for two reasons. First, it draws our attention to the reality that changes in temperature, agricultural productivity, and sea level will generate effects that vary substantially *within* these countries. Coastal areas will experience the immediate effects of sea-level changes, Andean regions will suffer the direct impact of glacier loss, and Amazonian regions will undergo deforestation and the loss of animal and plant species. Second, the reality that climate change will have different effects in different subnational regions is important because political and socioeconomic conflicts between these regions have worsened dramatically in recent years. Indeed, after the decades-old armed conflict in Colombia, one of the most important security challenges facing these countries today is the worsening of tensions between coastal, Andean, and Amazonian regions, some of which are articulating quite radical demands for autonomy.[1] Territorial conflict between subnational regions is particularly important because it is taking place at a time of widespread crisis in national political institutions and in the context of weak or weakening national political parties. Although climate change is by no means the root issue behind these intrastate territorial conflicts, it stands to worsen these conflicts still further. Here I argue that over the course of the next two decades, the exacerbation of already-emerging territorial conflict represents the chief security issue posed by climate change in the region.

After presenting summary data on how the climate is likely to change in the northern Andes as of 2030, I briefly discuss the (unlikely) prospects for state failure and interstate conflict as a result of these changes. Turning to what are more plausible security threats, the next section focuses in greater detail on the ways that climate change is likely to compound already-worsening tensions between subnational regions. I argue that glacier loss, the decline of agricultural productivity, and sea-level rises will each have the effect of heightening territorial grievances, and that the successful mediation of these grievances will probably exceed the capacity of national political institutions. The subsequent section steps back from these territorial conflicts to discuss the implications of climate change for foreign policy in the region. Given the ongoing regional trend toward more decentralized forms of governance, the chapter concludes with a discussion of the importance of governmental responses to climate change at the subnational level.

Climate Change in the Northern Andes

According to data collected by the IPCC, by 2030 the countries of the northern Andes region will face a number of challenges as a result of climate change.[2] Summarizing the available data, Bolivia in 2030 will be in the worst position of the region's countries in agricultural productivity and in temperature changes (where its aggregate vulnerability score is much higher than that of the other three countries). In contrast, with respect to sea-level changes, Ecuador will be in the worst position in the event of a 1-meter rise in the sea level, and Ecuador and Peru will be in equally disadvantageous positions with a 3-meter rise in the sea level, unlikely in our time frame, but a reasonable proxy for extreme weather events, whose incidence may well increase along with average global temperature. In Latin America as a region, according to the IPCC, by 2030 the number of people under water stress will increase from 7 to 77 million. The IPCC also determined that glacier retreat is a critical issue in all four countries, with negative effects on water consumption and hydroelectric power generation. Constraints on hydroelectric power are likely to increase the construction of oil-burning electricity plants, further contributing to global warming. In terms of agricultural productivity, rainfall is expected to increase in Bolivia and decrease in southern Peru. The latter development is liable to harm food crop yields generally and to inhibit the development of higher-value agricultural exports at a time when many Peruvians have come to see these exports as an attractive alternative relative to more environmentally costly activities in the mining sector. Finally, again according to the IPCC, another key issue for these countries is the low quality of weather forecasting in a region that has seen an increasing incidence of extreme weather events.

State Failure and Interstate Conflict in the Northern Andes

Even in a region that has been known for high levels of political instability and political violence, the four countries included in this chapter stand out. Each country is currently experiencing high or rising levels of social conflict and political turbulence, making the northern Andes the most volatile subregion in all of Latin America. Although other countries in the region—such as Brazil, Chile, and Mexico—have recently begun to experience the consolidation of democratic politics, and the increasingly routine alternation in power of rival political parties, the countries of the northern Andes remain quite unstable politically.[3]

For example, in the last seven years alone, Bolivia has witnessed the collapse of its party system, the overthrow of two presidents, and the emergence of profound political conflicts within an assembly that was elected in 2006 to write a new Constitution for the country but that divided along territorial lines (pitting the western Andean half against the eastern Amazonian half). Ecuador has suffered from even greater political instability, having been governed by eight different presidents in the last fourteen years, several of whom were forced from office as a result of opposition by newly mobilized indigenous groups or the withdrawal of support by the armed forces.[4] Though Peru has experienced fewer "interrupted presidencies" in recent years, its political system has been forced to deal with the legacies of a brutal internal armed conflict, the breakdown of democracy under President Alberto Fujimori (1990–2000), and the disappearance of most of its traditional political parties. Alone in the region, Colombia continues to suffer from the destabilizing effects of an ongoing and deadly internal armed conflict, one that has begun to seriously affect the neighboring countries of Ecuador and Venezuela.[5] Looking at the region as a whole, political instability, economic stagnation, widening inequality, and decreasing physical security have led analysts, including those at the Council on Foreign Relations, to conclude that "regional collapse is a possibility, something that would constitute a major setback, not simply for U.S. interests in the hemisphere but for the world."[6]

Despite these challenges, complete state failure, if interpreted to mean the collapse of public order, is unlikely as a result of disruptive climate change. State failure is most likely in Colombia, but not for reasons that have much to do with climate change. Indeed, based on projected changes in temperature, agricultural productivity, and sea level, Colombia is the best-positioned country in the subregion. The loss of *páramo* (high-elevation grass-lands) could increase migration from Bogotá to Colombia's three other urban centers (Ba-ranquilla, Cali, and Medellín) or to rural destinations, but this pales in importance relative to the other challenges facing the Colombian state.[7] According to the Office of the United Nations High Commissioner for Refugees, Colombia already has the highest number of in-ternally displaced persons in the world, with an estimated 2 million displaced individuals, or nearly 5 percent of the population.[8] Because so many Colombians have been brutalized by internal armed conflict, largely at the hands of paramilitary forces that seek to displace peasants from their lands, additional migrants produced by climate change are unlikely to have the decisive effect that might be expected elsewhere. It is also important to note that, in contrast to the obvious limitations on the ability of Colombia's national government to exert control over national territory, the country's four largest cities have produced some of the most effective and innovative mayors in Latin America.[9] These well-led municipal governments may be better positioned to mitigate some of the negative effects of climate change than the national government, particularly given the degree to which its attention continues to be consumed by the ongoing conflict with guerrilla insurgencies.

As discussed in greater detail in the subsequent section, Bolivia, Ecuador, and Peru are likely to experience growing challenges to the authority of the central state from sub-national regions. Nevertheless, the negative impact of climate change is merely an addi-tional factor relative to the deep-seated socioeconomic, ethnic, and political divisions that are fueling these challenges, and even their combined effects are unlikely to produce state failure.

The prospects for serious interstate conflicts as a result of climate change are only slightly more likely than state failure. On the one hand, the countries of the northern Andes have been something of an exception to the general pattern of limited interstate conflict in Latin America.[10] Though Latin America has been described as a zone of peace, the northern Andean subregion has indeed witnessed significant interstate conflicts, in-cluding Bolivian and Peruvian involvement in the nineteenth-century War of the Pacific, the deadly Chaco War between Bolivia and Paraguay in the 1930s, and chronic border con-flicts between Ecuador and Peru that were resolved only in the mid-1990s.[11] More recently, the 2008 raid by the Colombian armed forces against a top operative of Fuerzas Armadas Revolucionarias de Colombia (FARC; Revolutionary Armed Forces of Colombia) on Ecua-dorian territory led to the temporary rupturing of diplomatic relations between Colombia and both Ecuador and Venezuela.

On the other hand, despite the incidence of interstate conflicts in the past, aggression among these four countries is unlikely in the future, still less for reasons that would be at-tributed to climate change. The 2008 conflict is a case in point; despite the seriousness of the incursion by Colombian forces into Ecuador, the dispute was quickly resolved through high-level diplomacy at the Rio de Janeiro Summit. A number of factors reduce the likeli-hood that serious interstate conflict would take place as a result of climate change. First, the border regions between these countries are not heavily populated, particularly where in-ternational borders cut through the relatively sparsely populated territory of the Amazon. This reduces the likelihood that disruptions caused by climate change would result in large flows of refugees across borders. Second, to the extent that climate change will increase ex-ternal migration in the future, current patterns of out-migration suggest that the United States and European Union are more likely destinations than neighboring countries.

Third, interstate conflicts between these four countries over rivers are not significant, in contrast to many of the other regions that are highlighted elsewhere in this volume (i.e., the conflicts between Egypt and Sudan, Turkey and Syria, India and Pakistan, and Kyr-gyzstan and Uzbekistan). Because the Andean mountain chain mostly follows a north–south course through these countries, rivers for the most part run west to the Pacific or east toward the Amazon, and few cross international borders in significant ways. There are some exceptions to this generally reassuring pattern, however, particularly with respect to potential disputes over water resources in the southern Ecuadorian and northern Peru-vian highlands. As we approach 2030 the negative impact of mining activities on access to water may well be perceived as a more important problem than glacier loss due to climate change.[12] Another possible exception is conflict between Bolivia and Peru over rising levels of pollution in Lake Titicaca, an important source of freshwater that lies on the border between these two countries.

Intrastate Territorial Conflicts in the Northern Andes

Although state failure and significant interstate conflict are both improbable as a result of climate change in the northern Andes, it is very likely that climate change will accentu-ate important intrastate conflicts as of 2030. The growing salience of intrastate conflicts between subnational regions is one of the defining features of politics in each country, with the possible exception of Colombia, where political life continues to be dominated by the conflict between the armed forces and guerrilla insurgencies. Intrastate conflict in the con-temporary period reflects the resurgence of earlier and quite deep political patterns in the region. As was the case elsewhere in postindependence Latin America, the nineteenth cen-tury in the northern Andes was a period of protracted and often bloody conflict between subnational regions under the control of local leaders or caudillos, none of whom had suf-ficient military power to hold onto the center for very long. According to the literature on state formation in Latin America, strong regional identities and interests delayed the con-solidation of the state and, to varying degrees, limited the development of its monopoly on the use of force.[13] Though most of the twentieth century was consumed by national-level struggles between groups with national identities (i.e., unions, political parties, and the armed forces), latent subnational cleavages remained in place.

At the dawn of the twenty-first century these subnational cleavages began to take on once again the significance they claimed in the nineteenth. This was the result chiefly of political decentralization and economic liberalization. With respect to decentralization, for example, all four of our Andean states have shifted from the appointment of subnational political authorities by the central government to their direct election by local constituents. Colombia introduced elections for mayors and governors in the late 1980s and early 1990s, Ecuador incorporated elections for parish-level authorities in 2003, and Peru and Bolivia introduced elections for regional presidents/prefects in 2002 and 2005, respectively. Sub-national elections have given subnational politicians greater incentives to promote and de-fend the interests of subnational units.

At the same time, the ability of national politicians to mediate interregional conflicts has been reduced, in part because they no longer control the careers of subnational officials. With respect to economic liberalization, market reforms have eliminated the type of statist interventions that previously sought to redress the often considerable developmental gaps between subnational regions within the same country. For important periods in the twen-tieth century, the state subsidized agricultural and industrial ventures in marginal areas in order to promote a more regionally balanced process of economic development (often unsuccessfully). Throughout the northern Andes, the shift to market-oriented policies and export-oriented strategies in the last two decades has had a differential impact on subnational regions. Some have benefited from more liberal foreign investment codes and more competitive exchange rates while others have been disadvantaged and, consequently, oppose the turn toward neoliberalism.

Against this backdrop, the intrastate territorial struggles that will be generated by climate change will tend to reinforce the already-fraught relationships that have developed between subnational regions in each country. In Bolivia, the critical regional conflict pits the poorer, less developed, and less productive Andean departments in the west (i.e., Chuquisaca, Cochabamba, La Paz, Oruro, and Potosí) against the relatively more developed departments in the east (i.e., Beni, Pando, Santa Cruz, and Tarija), where the country's most significant natural gas deposits are located and where most of its productive agricultural activities are concentrated.[14] Ecuador's most important regional conflict has unfolded between the vibrant and export-oriented coastal region, centered on the port city of Guayaquil, and the less dynamic, inward-oriented Andean region, which has benefited disproportionately from the largesse of the national government in Quito.[15] In addition to this historic "Guayaquil versus Quito" conflict, groups in Ecuador's Amazonian region, the location of its significant oil wealth, have also begun to mobilize to secure a greater share of the revenues that are generated by oil exports, at the expense of the central government. Unlike in both Bolivia and Ecuador, in Peru the sites of political and economic power coincide. As a result, Peru's most important territorial cleavage reflects the resentment of the interior provinces against the overwhelming political and economic dominance of Lima.[16] Finally, Colombia's significant territorial cleavages—which produced some of the continent's worst conflicts in the nineteenth century—have become an increasingly important factor in the country's lengthy armed conflict, particularly since decentralization in the 1990s has encouraged armed combatants on the both the left and right to appropriate the greater resources that are now controlled by subnational governments.[17]

Because a full discussion of climate-induced territorial conflicts in these four countries is beyond the scope of this chapter, the following subsections focus on those possible climate changes that are likely to be most important in each country, and they illustrate the nature of each country's most salient interregional conflicts. These include the loss of glaciers in Andean Bolivia, the decline of agricultural productivity in southern Peru, and sea-level rise in Ecuador. The following paragraphs do not touch in depth on the Colombian case due to the probability that the effects of climate change will be less pronounced there, and also because of the dominance of its ongoing internal armed conflict, which makes it especially difficult, if not impossible, to weigh the relative significance of anticipated climate change there.

The Loss of Glaciers in Bolivia

The climate impacts hypothesized for 2030 are more likely to be an appreciable additional factor in Bolivia than in the other three Andean countries we are considering. Glacier loss is one of the most important challenges, and the impact of glacier loss in the Andean western half of Bolivia may lead to particularly disruptive consequences. Currently, the residents of La Paz and El Alto rely on glaciers for a third of their water consumption. Glacier loss is likely to encourage significant migration from these Andean cities to Santa Cruz, which is the center of economic activity in the eastern, Amazonian half of the country. This eastward migration pattern was dramatically accelerated by the privatization of inefficient state-owned mines in the western highlands in the 1980s. Glacier loss and water stress are likely to further accelerate and prolong it in the future.

Political actors in Santa Cruz are already using anti-migrant rhetoric in their campaign for autonomy from the La Paz–based national government, which culminated in the successful referendum on autonomy in May 2008. On the one hand, internal migration due to climate change could worsen the interregional split that has given rise to this autonomy movement. The leaders of the autonomy movement, grouped together in the Pro–Santa Cruz Civic Committee (Comité Cívico Pro–Santa Cruz), argue that Santa Cruz cannot continue to absorb migrants from highland regions that have failed to generate jobs for their residents and that have been unable to promote competitive economic activities. Migration induced by climate change is also likely to worsen the malapportionment that has so angered the residents of Santa Cruz. During the past decade, Bolivia's National Congress failed to use updated census figures to recalibrate the number of representatives sent from each of the country's nine regions, a failure that harmed Santa Cruz, considering that its population growth has exceeded the national average. On the other hand, though most signs point toward worsening regional conflict as a result of climate-induced migration, it is possible that additional flows of Quechua and Aymara migrants from the highlands in the west to the lowlands in the east could challenge the hegemony of the white and mestizo groups that have led the drive for autonomy in Santa Cruz. In this sense, migration could dilute the starkness of the ethnic divisions that are currently fueling this interregional conflict between the "indigenous highlands" and the "white/mestizo lowlands."

Although it is common to distinguish between "fight" and "flight" as dominant responses to environmental stresses like water scarcity, the Bolivian case illustrates how flight can in fact lead to fight. In the first instance, water stress due to glacier loss in the Andean highlands is likely to produce migration to Amazonian regions, particularly due to deeply held but misleading beliefs that the Amazon is essentially an empty space (*tierras baldías*). As a result of this view, Santa Cruz did not experience the radical agrarian reform measures that were generated by Bolivia's 1952 Revolution, which effectively eliminated the landowning class in highland regions. As a result, to this day Santa Cruz has the most concentrated patterns of land tenure in the country. In recent years, growing land pressure in Santa Cruz has induced both indigenous communities and migrants from the highlands to conduct land invasions of large agricultural holdings. In addition to worsening conflicts over land tenure, deforestation in Bolivia's lowland regions has also produced conflicts between indigenous communities and loggers, which are likely to worsen as greater numbers of migrants appear in response to water scarcity in the Andes.

Given that Bolivia is the poorest country in the group, its national government is also currently least able to afford the mitigating effects—such as the construction of water reservoirs and the repair of leaking water pipes—that could help reduce the impact of glacier loss. However, despite Bolivia's disadvantaged position, it is important to note the possible effects of the significant gas deposits that were discovered in Bolivia in the last decade (and that are second in the region only to Venezuela). Not only might revenues from future gas exports finance the construction of reservoirs and other mitigating infrastructure (provided a gas pipeline for these exports is built), but these revenues might also finance food imports and/or compensate for the loss of agricultural exports if increased rainfall does in fact limit crop yields. The prospect of using natural resource rents to help mitigate the effects of climate change is particularly important in Bolivia because the emergence of water nationalism in recent years appears to have eliminated privatization as a means of securing additional financing for investments in water infrastructure. In this respect, it is also important to note that water scarcity and struggles over water have already produced very important conflicts in Bolivia. Specifically, residents of the country's third largest city, Cochabamba, mobilized in the so-called Water War of 2000 against the sale of the municipal water company to the Bechtel Corporation. Ironically, Bechtel's contract would have obliged it to invest in infrastructure that was intended to expand the supply of water to the city.

Glacier loss is also an issue in Ecuador, whose Andean capital city also depends heavily on glaciers for water, and whose national government has proposed to build a tunnel through the mountains that ring Quito to the east in order to capture Amazon Basin runoff for the city. In Colombia, the loss of *páramo* due to increasing temperatures will also have a negative impact on cloud formation and on the availability of water in Bogotá. Unlike in Bolivia, however, there are more options for migrants seeking to leave Quito and Bogotá, including both coastal and Amazonian destinations. For this reason, the negative effects of migration due to water scarcity are likely to be more regionally diffuse than in Bolivia.

The Decline of Agricultural Productivity in Peru

One of the chief effects of climate change in Peru is the expected decline of rainfall in the country's southern departments of Arequipa, Moquegua, Puno, and Tacna. Less rainfall would have a negative impact on agricultural productivity in this already-arid region. More limited rainfall would thus make the construction of irrigation projects that much more critical. In recent years, however, several conflicts have taken place between regional departments—now governed by regional presidents who are elected in their own right—over the design and financing of irrigation projects. One case in point is the strong resistance in the department of Cuzco to Arequipa's attempt to build irrigation canals that would divert water from the Colca River.

Beyond interdepartmental feuds over water rights, a decline in rainfall and agricultural productivity in southern Peru would exacerbate intrastate territorial conflicts in two additional ways. First, obstacles to agricultural production would serve to reinforce Peru's problematic dependence on mineral exports (i.e., silver, tin, copper, gold, zinc, and lead), which currently account for more than half the country's export earnings. Due to their more positive impact on employment generation and generally less negative impact on the environment, agricultural activities represent an appealing alternative to Peru's mining-based economy, as demonstrated by the new agricultural products now grown in the northern department of Lambayeque. Inadequate rainfall in the south, however, would decrease the viability of this "Chilean model" of high-value agricultural exports. In addition to the negative impact on water quality and other environmental costs, Peru's continued pursuit of a mineral-based export model is problematic because the uneven distribution of mining revenues and royalties has created significant tensions between have and have-not regions.

Second, any climate-induced constraint on the economic development of Peru's southern departments, which constitute some of the country's poorest areas, would likely accentuate already high levels of anti-Lima sentiment. Consider the department of Arequipa. Before the onset of economic liberalization in the 1990s, industrial firms in Arequipa benefited heavily from state subsidies and the department housed Peru's second-most-important industrial base after Lima. Liberalization led either to the bankruptcy of most of these firms or to their relocation to Lima. In 2002, Arequipa exploded in violence over attempts by the national government to privatize the region's electricity plant, foreshadowing the type of territorial conflicts that will likely become more common with climate change.

The Increase in the Sea Level in Ecuador

Despite the difficulty of forecasting the magnitude of probable sea-level rise, it is clear that Ecuador is the country in the northern Andes that is likely to be most adversely affected by this process. Bolivia has no coast, only one of Colombia's four major cities is on the coast

(Barranquilla), and Peru's coastal capital, Lima, is built on cliffs high above the Pacific, which would reduce the direct impact of an increase in the sea level.[18] As is the case with glacier loss in Bolivia and reduced agricultural productivity in Peru, a 1- to 3-meter change in the sea level would worsen territorial conflicts within Ecuador. Specifically, a higher sea level could prove to be devastating for Ecuador's largest city, Guayaquil, and could complicate the already-tense relationship between Guayaquil and Quito.

Ever since the cacao boom of the middle to late nineteenth century, many residents of the city of Guayaquil and the province of Guayas have resented the neglect they feel they have suffered from the Quito-based national government, whose revenues have disproportionately been generated by coastal economic activities. Guayas represents just one-tenth of Ecuador's territory but one-quarter of its economic product. In recent years high levels of political volatility in Quito and, after 2006, a dramatic shift to the left in the orientation of the national government under President Rafael Correa, have fueled a growing autonomy movement among business elites and political leaders in Guayaquil. In January 2000 the residents of Guayas voted overwhelmingly for greater autonomy (more than 80 percent in favor) in a nonbinding referendum called by the provincial government, and three other coastal provinces followed suit later that year (El Oro, Los Rios, and Manabí). In January 2008 more than 200,000 residents of Guayaquil demonstrated on behalf of greater independence from Quito in the attempt to influence the terms of the debate over autonomy that took place within the country's constitutional convention. As in Santa Cruz, Bolivia, the demand for autonomy on the part of Guayaquil is fueled by substantial anger about the underrepresentation of Guayas in national legislative institutions.

Unless the national government responds aggressively to changes in sea level that would threaten residents of Guayas, these changes may substantially worsen Ecuador's most important interregional conflict. If the lackluster response of the national government to the devastation caused by the El Niño phenomenon in 1997–98 is any indication, Guayaquil's fears about how Quito is likely to respond to sea-level rise, or the increasingly frequent severe weather events associated with it, are well founded. In 1997 and 1998, flooding caused by El Niño triggered the collapse of several major Guayaquil financial institutions, leading to a profound economic crisis on the coast.

Foreign Policy Implications

As the local effects of externally generated changes in climate become more acute in the northern Andes, governments are likely to adopt foreign policies that seek redress from more developed countries, chiefly including the United States. For example, in an Associated Press interview in November 2007, Bolivian president Evo Morales argued that he would seek legal recourse against wealthy countries for damages attributable to climate change. Left-leaning presidents like Morales in Bolivia and Correa in Ecuador are perhaps more likely to criticize the United States on climate change, but antagonism toward the United States for its role in altering the climate is likely to outlast the continent's current tilt to the left. Only in Colombia is climate change not likely to assume a more prominent place in relations with the United States. For the foreseeable future, Colombia's foreign policy will be dominated not by climate change but by such issues as extradition requests, human rights abuses by military actors, and the drug trade more generally. Thus, with the exception of Colombia, climate change is likely to join the list of grievances that currently dominate the relationship between this subregion and the United States, including the impact of American agricultural subsidies on Latin American exports and the steadfast pursuit by the United States of a supply-side approach to the war on drugs.

Though climate change is likely to heighten the sense of grievance that many Latin Americans feel with respect to the United States, two caveats are in order. First, though the worsening effects of climate change may complicate relations with the United States, they may also make countries more dependent on U.S. financial and technological assistance to mitigate some of these negative effects, shifting the dynamic from one of confrontation to one of cooperation. Second, antagonism toward external actors may be directed not against foreign governments but more pointedly against transnational companies, particularly in extractive industries. In the last decade, increases in world market prices for primary commodities (including minerals, natural gas, and petroleum) have triggered a boom in investment in these areas in the northern Andes and record profits for extractive industries. Relative to faraway foreign governments, transnational corporations in this sector are an easier target for aggrieved groups, given that they have an immediate and often visibly negative impact on access to water and other aspects of local livelihoods.[19] At the same time, to the extent that glacier loss threatens hydroelectric power and forces the construction of gas- and oil-burning electricity plants, dependence on foreign companies for the technical know-how and requisite financing to build these plants might increase.

Finally, as is the case in many of the other regions discussed in this volume (e.g., see chapters 6, 10, and 12 on, respectively, India, Central Asia, and Turkey), the effects of climate change in Bolivia, Colombia, Ecuador, and Peru suggest the desirability of using supranational regional mechanisms to respond to environmental stress and to share ideas about possible responses to climate change. In the northern Andes, however, there is little history of effective regional coordination between these countries. The lackluster experience with regional trade integration—as the Andean Pact countries have failed to successfully coordinate tariff reductions, despite clear efficiency gains—is a case in point. Despite some limited movement toward regionalization in the early 1990s, domestic instability, political volatility, and economic crises within these countries have all derailed attempts to construct more effective regional mechanisms via the delegation of authority upward to supranational institutions. As a result, the foreign policy effects of climate change are more likely to shape the critical bilateral relationship with the United States than to lead to new multilateral approaches within the region.

Conclusion

Unfortunately, the nature of political and civil society in these four countries leaves little room for optimism about effective responses to the challenges of climate change. With respect to political society, opinion polls routinely show dangerously low approval ratings for executive, legislative, and judicial officials, which will limit their capacity to make some of the difficult but necessary reforms that climate change demands. With respect to civil society, new social movements have emerged to challenge entrenched inequalities, but in the process they have adopted aggressively oppositional tactics that limit governability (as seen most powerfully in the "street coups" that have recently ended quite a few democratically elected presidencies in Bolivia and Ecuador). All four countries also suffer from low levels of institutional capacity (as seen in the poor weather forecasting systems described by the IPCC),[20] legal systems that tend not to work for non-elite citizens, and high levels of institutional volatility (since 1990, each country has rewritten its constitution, and Bolivia and Ecuador recently emerged from quite divisive processes of constitutional revision).

In contrast to some countries in other regions, none of the four countries discussed in this chapter has any prospect of improved agricultural productivity as a consequence of climate change, and no subnational regions are likely to consider themselves winners as a result of whatever measures may be taken to adapt to it. In the short run, at least, the socially disruptive effects of climate change are certain to be perceived as "lose–lose" propositions. Thus, for instance, both the highland regions that lose residents due to water scarcity and the lowland regions that must bear the cost of providing social services to the new arrivals are likely to feel equally aggrieved and equally restive toward those in authority.

Despite generally low levels of institutional capacity at the national level and the absence of clear winners among subnational regions, it is critical not to overlook the substantial (and growing) variation in the capacity of subnational governments in each of these four countries. In other words, not only are these countries divided between more- and less-developed regions, but the governments of these subnational regions differ in their ability to govern, with potentially important effects on the response to climate change. Each country has embraced programs of decentralization in recent years, and decentralization has exposed the heterogeneous quality of subnational governments, with improvements in service delivery subsequent to decentralization in some countries and declines in the quality of services in others.[21] The importance of this variation should not be overstated— subnational governments can only do so much, and climate change demands concerted national (and supranational) responses. Nonetheless, given the crises of legitimacy that have plagued national governments in the northern Andes, and considering the absence of effective supranational institutions in the region, the most innovative responses to the effects of climate change are likely to come from mayors, governors, and other officials at the subnational level.

Notes

1. Kent Eaton, "Conservative Autonomy Movements: Territorial Dimensions of Ideological Conflict in Bolivia and Ecuador," *Comparative Politics*, in press.

2. "Technical Summary," in IPCC, *Climate Change 2007: Impacts*, 23–78.

3. Paul Drake and Erik Hershberg, eds., *State and Society in Conflict: Comparative Perspectives on Andean Crises* (Pittsburgh: University of Pittsburgh Press, 2006); Scott Mainwaring, Ana María Bejarano, and Eduardo Pizarro, eds., *The Crisis of Democratic Representation in the Andes* (Stanford, CA: Stanford University Press, 2006); Michael Shifter, "Breakdown in the Andes," *Foreign Affairs* 83, no. 5 (September–October 2004): 126–38.

4. Arturo Valenzuela, "Latin American Presidencies Interrupted," *Journal of Democracy* 15, no. 4 (October 2004): 5–19, www.journalofdemocracy.org/articles/gratis/Valenzuela-15-4.pdf.

5. Francisco Gutiérrez and Luisa Ramírez, "The Tense Relationship between Democracy and Violence in Colombia, 1974–2001," in *Politics in the Andes: Identity, Conflict, Reform*, edited by Jo-Marie Burt and Philip Mauceri (Pittsburgh: University of Pittsburgh Press, 2004).

6. Daniel Christman, John Heinmann, and Julia Sweig, *Andes 2020: A New Strategy for the Challenges of Colombia and the Region* (New York: Council on Foreign Relations, 2004), v.

7. Located in high elevations between the upper forest line and the permanent snow line, the *páramo* is a neotropical ecosystem that combines shrublands and forests with a variety of lakes, wet grasslands, and peat bogs.

8. Office of the United Nations High Commissioner for Refugees, *Refugees by Numbers* (New York: Office of the United Nations High Commissioner for Refugees, 2006).

9. Alan Angell, Pamela Lowden, and Rosemary Thorp, *Decentralizing Development: The Political Economy of Institutional Change in Colombia and Chile* (Oxford: Oxford University Press, 2001).

10. Miguel Angel Centeno, *Blood and Debt: War and the Nation-State in Latin America* (University Park: Pennsylvania State University Press, 2003).

11. Jorge Dominguez and David Mares, *Boundary Disputes in Latin America* (Washington, DC: U.S. Institute of Peace Press, 2003).

12. Anthony Bebbington, Denise Humphreys Bebbington, Jeffrey Bury, Jeannet Lingan, Juan Pablo Muñoz, and Martin Scurrah, "Mining and Social Movements: Struggles over Livelihood and Rural Territorial Development in the Andes," *World Development* 36, no. 12 (2008): 2888–905.

13. See, e.g., Fernando López Alvez, *State Formation and Democracy in Latin America* (Durham, NC: Duke University Press, 2000); and Guillermo O'Donnell, "On the State, Democratization and Some Conceptual Problems," *World Development* 21, no. 3 (1993): 1355–69.

14. Kent Eaton, "Backlash in Bolivia: Regional Autonomy as a Reaction against Indigenous Mobilization," *Politics & Society* 35, no. 1 (March 2007): 71–102.

15. Francisco Muñoz, *Descentralización* (Quito: Tramasocial Editorial, 1999).

16. Efraín Gonzales de Olarte, *Neocentralismo y neoliberalismo en el Perú* (Lima: Instituto de Estudios Peruanos, 2000).

17. Kent Eaton, "The Downside of Decentralization: Armed Clientelism in Colombia" *Security Studies* 15, no. 4 (October–December 2006): 1–30.

18. In Peru, tensions between Lima and the interior are also sharp and worsening, but sea-level change is not likely to be a contributing issue because coastal Lima is the capital and thus more likely than Guayaquil to receive the national government's attention as the sea level rises.

19. On the impact of mining activities on local livelihoods, see Jeffrey Bury, "Transnational Corporations and Livelihood Transformations in the Peruvian Andes : An Actor-Oriented Political Ecology," *Human Ecology* 67, no. 3 (2008): 307–21, http://people.ucsc.edu/~jbury/Publications/BuryFinalHumanOrganiza tion2008.pdf.

20. "Technical Summary," in IPCC, *Climate Change 2007: Impacts.* On institutional instability in the Andes, see Paul Drake and Eric Hershberg, eds., *State and Society in Conflict: Comparative Perspectives on Andean Crises* (Pittsburgh: University of Pittsburgh Press, 2006).

21. Donna Lee Van Cott, *Radical Democracy in the Andes* (New York: Cambridge University Press, 2008); John Cameron, *Struggles for Local Democracy in the Andes* (Boulder, CO: Lynne Rienner, 2010).

20
Brazil

Jeffrey Cason

Brazil is an important player in climate change discussions, if for no other reason than its size and location. The country has long been an active agent in negotiations on the environment, and it hosted the 1992 United Nations environmental summit in Rio de Janeiro, one of the landmark events in international relations focusing on the environment. Understanding Brazil's role in the context of global climate change can shed a great deal of light on how we should understand the overall political processes—both international and domestic—associated with environmental change. It can also illustrate how dealing with climate change will require that we pay attention to inequality in the international system.

In keeping with the themes of this volume, this chapter considers the potential impacts of climate change on Brazil and some of its Southern Cone neighbors. At the same time, I consider how Brazil has been involved in international environmental politics, both as an agent defending what it perceives as Brazilian interests and as a country where institutional constraints make responding to domestic and international challenges especially complicated. Because Brazil is a country with the potential to *affect* the global climate as well as to be affected by it, it is crucial to consider its role in international environmental issues from both directions.

Brazil is especially conscious of how it is perceived as a country crucial for the environment. For several decades there has been concern about the consequences of the

Brazil

destruction of the rain forest in the Amazon River Basin, and Brazil has made the environment an important domestic political issue.[1] Environmental issues have also led to significant concerns in the country about its sovereignty; there is widespread anxiety in Brazil that because of the ecological importance of the Amazon, the region will be "internationalized," so that the country's own control over its own territory will be reduced. Although this concern might be described as, at the very least, exaggerated, it does point to the complications associated with environmental politics in the region. It also focuses on legitimate issues of sovereignty, and the role of international institutions in regulating what states may do internally when the consequences of their actions go beyond their territorial boundaries. Although we might consider the environment the quintessential "global" issue, coming to grips with it reminds us that we still deal with a world of nation-states.

This chapter has two foci. The first is on the role of Brazil in international climate change debates. The second is on the potential effects of significant climate change on Brazil itself, given its institutional stability—or lack thereof. The second focus will look, in particular, at how long-term climate change—given current projections about its effects—will play out in the Brazilian context. In contrast with some of the other regions and countries discussed in this volume, the effect of climate change on Brazil, though potentially problematic (especially when it comes to warming and drying in the Amazon and in the Northeast) is not likely to be as severe as in other places, and the security concerns are less acute. At the same time, what Brazil does in terms of its own policies, particularly vis-à-vis the Amazon rain forest, is likely to have an outsized effect on the rest of the world. Thus Brazil faces a situation in which its actions will likely have a greater effect outside than inside Brazil. As a consequence, the country will need to be given significant incentives to change its domestic policies in a meaningful way, and these incentives will need to come from wealthy countries.

Brazil has begun to take steps toward assuming a leadership position in global climate change negotiations, especially in recent years. At the same time, the Brazilian government is not quite ready to embrace a position that puts climate change at the top of the agenda, especially because of concerns about sovereignty. In part, this is related to the fact that many in Brazil think that the country is primarily a *contributor* to the environmental health of the planet, because of the Amazon rain forest and its energy policies, with their emphasis on hydroelectric power and sugar-based ethanol.

In addition, Brazil does face risks when it comes to the effect of climate change on the country. In particular, there is likely to be increasing internal migration because of drought, and this could lead to an increased level of political instability. The internal migration could also exacerbate internal political conflict, especially as it relates to the rural landless movement, one of the country's better-organized social movements, and one that will certainly not go away, given overall inequality, particularly when it comes to land distribution.

This chapter is organized in three sections. The first looks at the historical evolution of Brazil's environmental policies, and in particular at the recent evolution of its policies in a more environmentally friendly direction. The second looks at the potential effects of global climate change on Brazil, in line with the general themes around which this volume is organized. This section also discusses the potential effects on some of Brazil's Southern Cone neighbors, because environmental problems do not recognize political boundaries. The third section particularly focuses on how Brazilian political institutions might deal with the consequences of global climate change. In this final section, I consider how Brazil's actions could interact with those of its neighbors and the potential effects of such interactions.

Brazil and the Environment

The natural environment of Brazil has been a subject of intense interest since well before it achieved independence from Portugal. Starting with its "discovery" by the Portuguese, Brazil was viewed as an especially fertile place, one where anything could grow and anything was possible.[2] Brazilians have a proud relationship with their environment, and in recent years, as global environmentalism has become more important, its government has engaged more with the international community on environmental issues. As Viola has pointed out, Brazil went from having a relatively nationalist view of the environment to one where it has participated actively in international environmental negotiations.[3] Indeed, it plays an important role in environmental negotiations, if for no other reason than the Amazon Basin (most of which is in Brazil) contains 45 percent of the world's tropical forests.[4]

As was the case in most developing countries in the middle and late twentieth century, Brazil's drive toward development did not include a large dose of environmentalism. Rather, the focus was on increasing industrialization, attaining other attributes of modernization, and (at least rhetorically) significantly reducing the level of poverty. At the groundbreaking UN conference on the environment in Stockholm in 1972, Brazil took a leadership role among developing countries in insisting that environmental protection could not come at the expense of economic development.[5] This idea was particularly prevalent in the military regime that controlled Brazil from 1964 to 1985.[6] Drawing on a long tradition of Brazilian interest in expanding its sphere of influence in the South American continent, the government made a point of assuring Brazil's physical control over as much of its territory as possible.[7] In the domestic sphere the focus was conclusively on the side of populating and controlling the vast territory of the Amazon and other relatively uninhabited parts of the country and on exploiting the natural resources of these areas.

Brazil's foreign policy has traditionally been one that defended its territorial rights as well as its sovereignty. Brazil's Foreign Affairs Ministry, nicknamed Itamaraty, has been an effective defender of Brazilian interests, and Brazilian diplomats are viewed as among the most competent and experienced in Latin America.[8] And Itamaraty was not especially enamored of environmental issues when they first became part of international political discourse in the 1970s. Hochstetler and Keck cite an interview with Paulo Nogueiro Neto, who was Brazil's first environmental secretary in the 1970s. Nogueiro Neto described the prevailing view: "'Brazil was an island under siege by the rest of the world . . . that . . . had to defend itself.' Itamaraty . . . believed that developed countries would make attention to the environment into an instrument of imperialist domination. The economic agencies in government were similarly hostile, and [Economics] Minister Delfim Neto continuously blocked resources for the agency."[9] Simply put, the environment during the military regime was not a high priority for policymakers, either in the domestic arena or in foreign affairs.

In recent years, Brazil's rhetoric has changed. In part, this has been a consequence of the increased importance of environmentalism in world politics. Keck and Sikkink note the effect of transnational advocacy networks on local environmental organizations, which saw opportunities in the attention of activists from wealthier countries.[10] That said, while the general thrust of Brazil's approach to international environmental negotiations has changed, there is still a strong emphasis on the historical and present inequality of the international system in the way many Brazilian scientists and activists perceive climate change and environmental issues.[11] Many Brazilians clearly expect that the burden of change (and the cost associated with it) should be borne by wealthier countries. They recognize the country's important role in climate change discussions, and expect those in the global North, the United States and Europe especially, to foot most of the bill in making the needed adjustments to put the Earth on a more sustainable track.

The Potential Effects of Climate Change on Brazil and the Southern Cone

Brazilians are accustomed to thinking about how they affect the international environment because of their vast territory and abundance of tropical forests. At the same time, climate change clearly has the potential to affect Brazil. In this section I outline some of these potential effects and selectively consider potential effects on several other countries in the Southern Cone.

There are a number of ways in which Brazil could be vulnerable to the effects of climate change. The largest would be an acceleration of internal migration from the Northeast to the Southeast, and also from rural areas to the major cities.[12] This process has been going on for decades, but climate change—particularly the increasingly likelihood of drought in the Northeast—would exacerbate it. Internal migration is also fed by the extraordinary inequality of Brazilian society, which is among the most extreme to be found anywhere.[13]

Given that migration has historically happened in Brazil because of economic opportunities and constraints (as is the case elsewhere), the fact that the region most vulnerable to climate change—the Northeast—is also the poorest in Brazil is alarming, in terms of the potential for dislocation from the region. Per capita GDP in the Southeast is nearly three times the per capita GDP in the poor Northeast. The poverty rate in the Northeast in the late 1990s was 53 percent, while in São Paulo, the wealthiest state (and a common destination for migrants), it was just over 7 percent.[14] Poverty in the region has been reduced somewhat in recent years through government cash transfer programs (the so-called Bolsa Familia in particular), but it is still quite severe.[15]

An increase in internal migration might also lead to greater political instability. In part, this would be the case because of Brazil's already high degree of inequality. It is one of the most unequal countries in the world, to the point where one may well wonder why there has not been more political unrest already (more on that below). Additional significant internal migration, with its concomitant political demands, could lead to serious instability.

That said, Brazilian political institutions do appear relatively resilient when compared with those of many of its South American neighbors. Since the transition to democracy in 1985, there have been no real threats of military coups, and even when there were political crises (in one case, leading to the impeachment of a sitting president, Fernando Collor), they have handled things in a rather orderly way. This contrasts sharply with both Argentina and Venezuela in recent years, for example. In this sense Brazil can be seen to have relatively robust political institutions, if by robust one means institutions that have staying power, even if they are not always effective in the delivery of good government. Comparatively speaking, both Chile and Uruguay have stronger and better-developed political institutions, but Brazil has had remarkably little political instability, given its rather less disciplined politics. I discuss political institutions in Brazil in more detail below. In general, however, it is fair to say that Brazil is likely to experience some instability as a result of global climate change scenarios, but that unless these are of a massively disruptive nature, the Brazilian political system will likely absorb them without serious consequences.

This is not necessarily the case throughout the region and for some of Brazil's neighbors. In Paraguay, for example, there is more to be concerned about, because of the relative poverty of the country, its more fragile politics, and its heavier dependence on agriculture,

Table 20.1 The Weight of Agriculture in the Economy, Brazil, Chile, and Paraguay

Measure of Economic Weight	Brazil	Chile	Paraguay
Agricultural exports as a percentage of total exports	4	6	3
Agriculture as a percentage of GDP	7	4	20

Source: World Bank, World Development Indicators. The most recent figures available for each country are from 2008.

which, according to prevailing climate change estimates, is likely to be affected very negatively. Some comparative data should help in thinking through the implications for Paraguay, relative to Brazil and Chile, as can be seen in table 20.1.

Thus, though both Brazil and Chile are significant agricultural exporters, they do not rely as heavily on agriculture for their economic well-being. As a consequence, climate change that affects agricultural production will not have an outsized impact on their economies. In the case of Paraguay, this impact would be more significant, and it would likely introduce more potential for political unrest. Not only is Paraguay more reliant on agriculture as a percentage of its GDP and in its exports—32 percent of Paraguay's population is employed in agriculture (compared with 21 percent in Brazil and 13 percent in Chile—but serious climate change could also wreak havoc on this large segment of the population.[16]

Instability is especially likely, given Paraguay's relatively weak political institutions. Its government, despite having transitioned away from a fairly harsh authoritarian regime in 1989, remains among the least democratic in Latin America. Having had a thirty-five-year dictatorship before the democratic transition did not help matters much when it came to establishing democracy. In 1996 Paraguay suffered a military coup attempt, and it faces recurring problems with corruption, including a significant trade in drugs and other contraband to Brazil and Argentina.

Although there is a potential for greater political instability in the region because of Paraguay's fragility, Brazil's armed forces, like those of any other country, can be expected to defend the nation's frontier against hostile action or encroachment by outsiders, whether driven by environmental pressures or other motives. Barring some large and disruptive change, beyond what is reflected in current climate estimates, Brazil's response to climate change is highly unlikely to include military aggression. Brazil is quite skilled at diplomacy, and it is recognized among its Latin American counterparts as having the most professional and competent diplomatic corps in the region. It is also part of a region that has seen relative peace for an impressive period of time.[17] It is difficult to imagine a circumstance in which Brazil would strike out at its neighbors, particularly because it already has such a large store of resources within its borders.

Because of the inequality noted above, internal conflict is somewhat more likely, but this is probably not a large threat. The most likely potential conflict here would come as a result of the activities of rural landless workers, whose interests are now represented by an increasingly active social movement called Movimento dos Trabalhadores sem Terra (MST, Landless Workers' Movement).[18] MST's constituents have clear (and justified) grievances, and they will continue to demand land and resources for poor and dispossessed people. Climate change scenarios that make it increasingly difficult for more marginal populations to make ends meet will increase the ranks of those supporting the MST, even while perhaps making it more difficult to satisfy the movement's demands.

Climate-induced emigration is likely to become a familiar feature of the developing world in years to come. In Brazil, however, this option is likely to be difficult for those most directly affected by climate change, at least when it comes to international migration. There is no obviously better place to go outside Brazil for those with the fewest resources. As noted above, however, increasing internal migration is likely, from poor regions to rich ones, and from the countryside to the cities.

As for Paraguay, one of the more vulnerable countries in the region when it comes to climate change, it does not have a strong enough military to contemplate significant external aggression. Environmentally induced migration to neighboring Brazil or Argentina is likely, however. Economic opportunities in those two countries are superior to what is available in Paraguay. Again, the likely trend will be toward increasing urbanization within and beyond Paraguay, a process that often brings with it rising rates of unemployment, crime, food and energy shortages, and so on.

When it comes to civil conflict, there is also a potential for such conflict in Paraguay. The potential effects of climate change on agricultural production, from which a large share of Paraguay's population makes its living, make such conflict more likely. Mitigating against it would be what could be considered a deferential response to authority in the country, which has been conditioned by many decades of authoritarian rule. Unlike some of its neighbors, Paraguay has not experienced a significant and deep incorporation of the masses into politics. There has never been a Paraguayan equivalent of Juan Perón, Getúlio Vargas, or Salvador Allende, whose leadership might inspire the kind of mass political awakening that these other countries have had. On the contrary, Paraguay is much more accustomed to repression.[19]

Repressive governments are not necessarily robust in the face of change, however. Paraguay is thus the country in the Southern Cone in which state failure in the face of climate-induced social and political unrest is at least a possibility. It is unlikely elsewhere. Brazil, Chile, and Uruguay have governments that are institutionalized and well established. They are not in danger of becoming failed states. Argentina is regularly unstable, but not in a way that might lead to state collapse. Chile is probably the strongest state in South America. It is both orderly and well respected, to the point that one might call politics in Chile boring in the current context. Presidents and governments are not flashy, there is no overt instability, and politics tends to revolve around the center. After the transition from military rule in the late 1980s, Chile was governed by the same center–left coalition for twenty years, which has really been more center than anything. Although the right won the most recent election in early 2010, it is highly unlikely that the new Chilean government will represent a sharp break from recent political practice. The risk of political instability is quite low in Chile.

In Brazil there is more political uncertainty, but there is little danger of political meltdown as a consequence of global climate change. Brazilian politics is both robust and complex, with many cross-cutting currents. On the one hand, there is a great deal of electioneering. Politicians are quite aware that they are in a political arena that could prove extremely unforgiving if they did something wrong, so there is a good deal of healthy uncertainty, as is essential for any democracy. At the same time, there is a widespread sense that a political "oligarchy" runs things in Brazil, and there is some truth to this. Electoral success in Brazil has gradually moved from right to center to left, while the resulting governments have generally governed from the center. This is a familiar pattern in democratic states, and one that indicates that the Brazilian political system seems able to handle change fairly well, if only by refusing to overreact to demands for it. Whether these temperate habits will serve the country equally well in the face of long-term environmental challenges is difficult to say, but in the short run they bode well for continued political stability.

In Paraguay, there is more weakness and corruption in the state, and an appreciable degree of lawlessness in some parts of the country, especially in the triborder area with Brazil and Argentina.[20] Paraguay is accordingly the one country in the Southern Cone that could be viewed as vulnerable to state failure. Mitigating the risk of state failure is the interest of neighboring countries, which are likely to view state collapse in Paraguay as a significant threat to themselves and are thus likely to do something about it. There is some precedent for neighboring state intervention here; in 1996, when Paraguay experienced a military coup attempt, the other countries that belong to Mercosur (the "Southern Common Market," made up of Argentina, Brazil, Paraguay, and Uruguay) intervened to preserve the elected regime. Brazil played a particularly prominent role, as did the United States, in supporting the democratically elected regime.[21]

Brazilian Institutions and Climate Change

Returning again to the major player in South America, Brazil is less vulnerable than one might assume in the face of global climate change. It could well come to view itself as a winner in this arena. There is a great deal of environmental awareness in the country, and there is what one might describe as national pride when it comes to its own resources. There are many nongovernmental organizations in Brazil that organize around environmental issues, and a consciousness that Brazil can play a leading role when it comes to confronting potential environmental threats. There is pride, for example, in the large-scale production of sugarcane-based ethanol, and its widespread use in Brazilian automobiles. Brazilians view themselves as creative when it comes to confronting environmental issues, and they have a strong scientific infrastructure in the federal university system.

That said, there is also a popular fear in Brazil—stoked by the media, on occasion—that Brazil is vulnerable to an international "takeover" when it comes to environmental issues. The fear could be stated simply: that because of an impending environmental catastrophe, international actors (the United States, the UN, and others) will try to "internationalize" the Amazon River Basin and remove or limit Brazilian sovereignty over the region. Although it sounds rather absurd, and highly unlikely, it is a live fear in Brazil, to the point where cooperation with international organizations might be limited. Given the real global environmental threat, and the fact that environmental problems do not respect national boundaries, it is no less absurd to think that outside actors do *not* want to tell the Brazilians what they should be doing with their environment.

Brazilians will certainly resist this sort of outside control. And because of their desire to manage their relationship with the outside world when it comes to its environment, it is worth asking if Brazil is in fact capable, institutionally, of doing so. Brazilian politics is sometimes portrayed as chaotic, unstable, and difficult to manage. Although none of these adjectives should be discarded, this image is a legacy of the economic and political instability that occurred during and immediately after the country's transition to democracy more than two decades ago. Brazilian politics is now, if not orderly, at least more consistent and predictable than in the past.[22] Most observers of Brazilian politics in recent years would conclude that the country has become much more "normal" in political terms and that it is much more stable than many of its neighbors.

Another frequently cited reason for Brazil's supposed institutional incoherence is its federal system. The Constitution reserves substantial prerogatives for the states, and to the extent that central government policies can be undermined in the states, a national response to climate change or other environmental threats could be undermined. However, as José Cheibub and his colleagues have noted recently, the centrifugal tendencies in

Brazilian politics are more limited than commonly thought.[23] Presidents and the central government in Brazil retain a great deal of agenda-setting and law-making power, not least because of the constitutionally derived decree power that presidents preserve. Politics still focuses on the agenda of presidents.

Why does this matter in the context of global climate change? When considering a country's ability to respond effectively to the large-scale changes associated with climate change, institutional durability and stability will be important. In Brazil's case, its oft-maligned institutions might be viewed as inadequate and not up to the task of handling such threats. On the contrary, it is plausible to argue that Brazilian institutions are fairly robust, particularly when compared with those in many other middle- and low-income countries. Its institutions may be far from the most efficient, but its politics are reasonably functional, and many government agencies have a significant capacity, including some in the environmental area.[24] There is much to complain about with respect to the Brazilian state, to be sure, but there are also many effective agencies and a significant state capacity.

Finally, when considering the potential effects of climate change on Brazil, it is useful to think about the region's largest country in its own regional context. South America is one of the least likely regions on Earth to engage in war. There has been relative peace there for a long time, with no large-scale wars for decades. This is not to say that there has not been conflict between countries, including occasional outbreaks of violence; but these conflicts have almost always been resolved fairly quickly. There has been some recent conflict between Brazil and Bolivia over natural gas reserves (because of nationalizations carried out by the government of Evo Morales), for example, but it is hard to picture this kind of clash rising to the level of military violence, especially given the region's long tradition of the peaceful resolution of disputes.

Overall, then, Brazil is likely to set a reasonable pace in the region as the challenges of global climate change manifest themselves. There is no doubt that the country views itself as responsible when it comes to engaging the rest of the world, and this view would extend to issues related to global climate change. Brazil sees itself as a serious and important member of the international community, and given the long-standing role it has had as a significant player in world affairs, there is little danger of it going off the rails, as it were. In recent years, for example, Brazil has portrayed itself as a moderate counterweight to Venezuela in South America, and it has behaved that way, despite possessing a nominally "leftist" government, whose progressive credentials are mostly demonstrated in foreign policy and not domestically. Brazil's highly professional diplomatic corps—which is well trained and well educated—views foreign policy in a strategic way that seeks continued normal engagement with the rest of the world. The fact that its president, Luis Ignácio Lula da Silva (set to leave office at the beginning of 2011), managed to have a close relationship with U.S. president George W. Bush while also appealing to his more leftist base at home is a sign that whatever government takes control in Brazil, it will likely have a solid relationship with the United States. This does not imply agreement with the United States on all issues. Brazil can be expected to demonstrate its independence from the United States when it serves its interests; but it does imply a basic alignment within the West.

Conclusions

On the face of it, Brazil should be considered one of the world's most important countries when it comes to discussions of global climate change, and it is. Its role in global environmental politics is complex. It is an important actor on the international stage, and it has fully engaged environmental politics on a domestic level. At the same time its internal politics introduce a greater degree of uncertainty as to how it might come out on any particular environmental issue when acting as an agent.

Given the forecasts about the effects of global climate change, Brazil is somewhat less likely to face a catastrophic disruption of its ecosystems than many other countries and regions considered in this volume. For this, Brazilians should consider themselves lucky. Without a doubt, however, they also know that the policy choices they make will have an outsized effect on many other countries. To make the most effective choices in the environmental area, they will need strong and effective institutions. The good news is that Brazil does have a substantial state capacity in this regard. The bad news is that there is still much to be done. To get the most out of Brazil's still nascent institutional capacity, the international community, and wealthier countries in particular, will need to put significant financial and technological resources into the region.

Notes

1. A wide-ranging treatment of Brazil and its environmental politics is given by Kathryn Hochstetler and Margaret E. Keck, *Greening Brazil: Environmental Activism in State and Society* (Durham, NC: Duke University Press, 2007).

2. A good account of the initial "encounter" of Brazil, and the Portuguese reaction to it, is given by Jerry M. Williams, "Pero Vaz de Caminha: The Voice of the Luso-Brazilian Chronicle," *Luso-Brazilian Review* 28, no. 2 (1991): 59–72.

3. Eduardo Viola, "Brazil in the Politics of Global Governance and Climate Change, 1989–2003," University of Oxford Centre for Brazilian Studies Working Paper CBS-56-04, 2004, www.brazil.ox.ac.uk/__data/assets/pdf_file/0006/9366/Eduardo20Viola2056.pdf.

4. Gabriela Bielefeld Nardoto, Jean Pierre Henry Balbaud Ometto, James R. Ehleringer, Niro Higuchi, Mercedes Maria da Cunha Bustamante, and Luiz Antonio Martinelli, "Understanding the Influences of Spatial Patterns on N Availability within the Brazilian Amazon Forest," *Ecosystems* 11 (2008): 1235.

5. For a detailed account of Brazil's role and positions at the Stockholm Conference, see Roberto P. Guimarães, *The Economics of Development in the Third World: Politics and Environment in Brazil* (Boulder, CO: Lynne Rienner, 1991), chap. 6.

6. A prominent treatment of the exploitation of the Amazon as part of national economic policy is given by Stephen G. Bunker, *Underdeveloping the Amazon: Extraction, Unequal Exchange, and the Failure of the Modern State* (Urbana: University of Illinois Press, 1985).

7. See Thomas Skidmore, *The Politics of Military Rule in Brazil, 1964–85* (Oxford: Oxford University Press, 1988), 144–49.

8. For a discussion of the changing role of Itamaraty in Brazilian foreign policy, see Jeffrey Cason and Timothy J. Power, "Presidentialism, Pluralization, and the Rollback of Itamaraty: Explaining Change in Brazilian Foreign Policy Making in the Cardoso-Lula Era," *International Political Science Review* 30, no. 2 (2009): 117–40.

9. Hochstetler and Keck, *Greening Brazil*, 27–28.

10. Margaret E. Keck and Kathryn Sikkink, *Activists beyond Borders: Advocacy Networks in International Politics* (Ithaca, NY: Cornell University Press, 1998), 137–47.

11. A useful discussion of the continued relevance of inequality in the world system and how it affects relevant players in Brazil is given by Myanna Lahsen, "Transnational Locals: Brazilian Experiences of the Climate Regime," in *Earthly Politics: Local and Global in Environmental Governance*, edited by Sheila Jasanoff and Marybeth Long Martello (Cambridge, MA: MIT Press, 2004).

12. Fausto Brito discusses immigration trends in the latter half of the twentieth century, noting that the trend is now toward internal immigration from rural areas to medium sized cities. See "O deslocamento da população brasileira para as metrópoles," *Estudos Avançados* 20, no. 57 (2006): 221–36.

13. See Thomas Skidmore, "Brazil's Persistent Income Inequality: Lessons from History," *Latin American Politics and Society* 46, no. 2 (2004): 133–50.

14. See Rodolfo Hoffmann, "Mensuração da desigualdade e da pobreza no Brasil," in *Desigualdade e Pobreza no Brasil*, edited by Ricardo Henriques (Rio de Janeiro: IPEA, 2000), www.ipea.gov.br/sites/000/2/livros/desigualdadepobrezabrasil/capitulo03.pdf.

15. For an overview of the *bolsa família* program, see Mônica de Castro Maia Senna, Luciene Burlandy Campos de Alcântara, Giselle Lavinas Monnerat, Vanessa Schottz Rodrigues, and Rosana Magalhães, "Programa Bolsa Família: Nova institucionalidade no campo da política social brasileira?" *Revista Katálisys* 10, no. 1 (2007): 86–94.

16. These data are from *World Development Indicators*, www.worldbank.org/data/onlinedatabases/onlinedatabases.html.

17. See Arie Kacowicz, *Zones of Peace in the Third World: South America and West Africa in Comparative Perspective* (Albany: State University of New York Press, 1998).

18. On the MST and its interaction with the state, see Wendy Wolford, "Agrarian Moral Economies and Neoliberalism in Brazil: Competing Worldviews and the State in the Struggle for Land," *Environment and Planning* 37 (2005): 241–61.

19. For a recent discussion of Paraguay's political circumstances, see Paul C. Sondrol, "Paraguay: A Semi-Authoritarian Regime?" *Armed Forces & Society* 34, no. 1 (2007): 46–66.

20. For a discussion of the complications and potential for instability along the triborder region (with terrorism thrown into the mix), see Ana R. Sverdlick, "Terrorists and Organized Crime Entrepreneurs in the 'Triple Frontier' among Argentina, Brazil, and Paraguay," *Trends in Organized Crime* 9, no. 2 (2005): 84–93.

21. See Jeffrey Cason, "Democracy Looks South: Mercosul and the Politics of Brazilian Trade Strategy," in *Democratic Brazil: Actors, Institutions, and Processes*, edited by Peter R. Kingstone and Timothy J. Power (Pittsburgh: University of Pittsburgh Press, 2000), 204–16.

22. A good example of the literature focusing on the problems of Brazilian politics (and in many ways analyzing Brazil before greater stability had taken hold) is Scott Mainwaring, *Rethinking Party Systems in the Third Wave of Democratization* (Stanford, CA: Stanford University Press, 1999). For a more recent analysis of how political debate in Brazil has "centralized," see Timothy J. Power, "Centering Democracy? Ideological Cleavages and Convergence in the Brazilian Political Class," in *Democratic Brazil Revisited*, edited by Peter R. Kingstone and Timothy J. Power (Pittsburgh: University of Pittsburgh Press, 2008).

23. See José Antonio Cheibub, Argelina Figueiredo, and Fernando Limongi, "Political Parties and Governors as Determinants of Legislative Behavior in Brazil's Chamber of Deputies, 1988–2006," *Latin American Politics and Society* 51, no. 1 (2009): 1–30.

24. As Peter Evans notes in *Embedded Autonomy: States and Industrial Transformation* (Princeton, NJ: Princeton University Press, 1995), some parts of the Brazilian state work quite well, while others see rampant corruption. For more specific discussion of state actors in environmental policy, see Hochstetler and Keck, *Greening Brazil*.

21

Conclusion

The Politics of Uncertainty

Daniel Moran

As this book moved toward publication, it was pushed along in the final stage by the comments of three expert readers to whom the manuscript was sent by the publisher. All three offered incisive suggestions that have been incorporated throughout the text. All three also thought the book would benefit from a concluding chapter, in which the general implications of the project could be highlighted. It was a perfectly reasonable request; but as the moment to write these final observations approached, the task came to seem more difficult than it did at first.

As was discussed in the introduction, this volume was conceived in terms calculated to make theoretical generalizations about climate politics more difficult, by driving the discussion toward the concrete conditions of individual states and regions, and at least implicitly away from the kinds of broad conclusions that have so far dominated the literature on climate change. The result does not lack for common themes—questions of state capacity, social resilience, population movement, and the differential impact of climate change across the agricultural and industrial sectors of national economies weave their way through the volume like red threads. Yet their relative prominence from one chapter to the next varies a good deal, so that it is impossible to say, in general, how they should be weighed relative to each other. This, it must be insisted, is not a defect of the analysis but one of its merits. If there is a big picture here, it is by design a mosaic composed of lots of little pictures, each with its own story to tell.

Nevertheless, every mosaic needs its grout and glue. The most important source of cohesion among the contributors to this book is a shared sense that, whether or not the Earth's climate is palpably hotter in twenty years than it is now, the politics that surrounds climate almost certainly will be. This leads to a somewhat startling inference: Within the time frame that we are considering, it probably will not matter whether the predictions of contemporary earth science are correct. It is possible that the current scientific understanding of anthropogenic climate change is wrong, not just in detail (as is inevitable), but completely. Such things have happened before in the history of science, and all scientists know they will happen again. It is the essence of the scientific method to allow for such a possibility. At the same time, it is no less essential that a scientific consensus, once achieved, should not be capriciously overturned. It is thus most unlikely that any new insight will be achieved in the next twenty years that will falsify today's scientific consensus so decisively as to render the issue of climate change inconsequential to public life. It may be comforting to suppose that, some day, climate science will settle climate politics, one way or another. Ultimately, that day may come. But it is not going to come soon.[1]

It is also fair to say that all the studies presented here share a common reluctance to detach climate politics from politics in general. The kinds of actions that are most likely to have an impact on international security tend to be overdetermined in any case, and they tend to arise from the intersection of multiple, individually contingent factors that may well seem perfectly manageable in themselves. Anyone who has studied the origins of the world wars of the twentieth century will have seen this process unfold in its most disastrous form. Here again, climate politics resembles politics generally. Yet it also seems fair to say that climate change poses an especially insidious sort of challenge to policy, combining as it does the gradual accumulation of relatively subtle effects and an increasing tendency toward the kinds of dramatic events (flooding, droughts, storms, etc.) that are liable to galvanize public opinion at unexpected moments. Even the best governments have little relish for problems of this kind. They are likely to prove especially troublesome to authoritarian states whose institutions have been optimized (if that is the word) to deal with more traditional forms of opposition and social unrest.

It may, therefore, be easy to underestimate the threat that climate change poses to the stability of otherwise well-established regimes, whose true social base does not extend much beyond the armed forces and rent-seeking elites. Environmental challenges present themselves politically as problems demanding the evaluation of incomplete and inconsistent scientific information. They place a premium on a government's capacity to anticipate long-run second-order effects, and to educate and manage public opinion in ways that allow it to look beyond immediate experiences and short-term goals. These are not the core skills of the average state anywhere, and in the developing world especially. This was recently illustrated by Egypt's response to the spread of the H1N1A virus, known as "swine flu." The swine flu crisis was not, strictly speaking, an environmental issue—though changes in disease patterns are a widely anticipated effect of climate change—but it proved sufficiently similar in character to be instructive in our context.

The flu pandemic that began (or, more precisely, was first detected) in April 2009 was spread by an organism that combined genetic material from known strains of human, avian, and pig flu in a form that had not previously been encountered.[2] The disease was immediately dubbed "swine flu" by the American press, a phrase that obscured the fact that, while the genetic makeup of H1N1A owed something to viruses endemic to pigs, the disease itself was spread by direct human contact and respiratory droplets, not by contact with pigs or by eating pork.[3]

Most governments responded to the flu's appearance (initially in Mexico, though earlier cases were subsequently detected in Asia) with contained alarm, as anxiety about its unique genetic makeup was weighted against doubts about its true lethality (which in the end proved unexceptional). In late April, however, the World Health Organization declared that, under its guidelines, sufficient cases of the new flu had appeared in enough different countries to constitute an international public health emergency.[4] Egypt's leadership responded to this declaration by ordering the slaughter of several hundred thousand pigs.

The animals in question were the property of Coptic Christians, who make their living by collecting the garbage of Cairo, sorting though it in search of salvageable material, and then feeding the rest to pigs. The decision to kill the pigs was made before a single case of swine flu had occurred in Egypt, and with no particular plan for how to deal with the garbage of Cairo once the pigs were dead, nor to compensate their former owners for their loss, nor to provide them with an alternative future livelihood. The fact that Muslims do not eat pork caused some observers to suppose that the pig massacre was an act of religious persecution disguised as a public health measure; yet it seems to have been undertaken in good faith as a response to the flu, and from a desire not to be perceived as slow-footed in response to the World Health Organization's warning. Public opinion strongly favored slaughtering the pigs, and though the police professed surprise that their actions were met with rioting and rock throwing by the pig owners, they had no real difficulty executing the policy.[5]

It seems fair to wonder whether the Cairo pig massacre may not illustrate what the politics of climate change in the developing world will be like, at least at those moments when public anxiety reaches the point of demanding decisive action. Such episodes of misguided official caprice may occur in any policy context, of course, and the additional pressures arising from climate change may simply make them more frequent. Nevertheless, it is worth emphasizing that climate politics, whatever its eventual shape, will be complicated politics. It will challenge the decision-making capacities of even the most advanced societies and may prove especially puzzling for the well-ordered police states that still govern so much of humankind, even when they are motivated by a genuine desire to do the right thing.

Climate change will, without question, provide many opportunities for governments to embarrass themselves. Whether it will also create conditions that put the legitimacy of governments at risk is harder to say. None of the contributors to this volume is prepared to declare that the strains imposed by climate change are likely to lead to the outright failure of the states with which they are concerned, though a number regard such an outcome as possible. This collective reluctance is partly due to the fact that the current understanding of "state failure" has become a little too restrictive analytically and is now synonymous not merely with revolutionary political change but also with complete social collapse. It is the latter that the analyses assembled here regard as unlikely. One of the more prominent recurring themes of this book is the extent to which the most important sources of resilience to climate change reside within societies, rather than within governments. Conversely, it is easy to see why many governments may prove reluctant to take the kinds of measures required to tap into that resilience and mobilize social capacity, because doing so may loosen their grip on power.

The lion's share of space in nearly all the chapters above is devoted to analyzing the forms of social and political distress that disruptive environmental stress might cause—social violence, internal or external migration, military aggression, and so on. It is in this area where the intersection between climate change and local conditions becomes especially intricate, and where general statements about how things can go wrong at the macro level are least convincing to policymakers and scholars alike. It is of little use, for instance, to be told that "by 2020 between 75 million and 250 million [Africans] are projected to be exposed to increased water stress due to climate change," unless one also knows which Africans in particular are most at risk, and what kinds of governmental and other responses can be expected.[6] It is only when the analysis is driven down to the country level that such harbingers can yield insights that can support useful action.

Few contributors to this volume anticipate significant international violence as a consequence of climate change and associated resource scarcities, an outlook that is consistent with the majority of the recent literature on environmental security. Many, however, note the likelihood of climate-induced demographic shifts, both internally and across international frontiers, which may cause violent reactions depending on the social groups involved and the prevailing attitudes that are in play.[7] Climate change, as was noted in the introduction, is likely to increase social inequality within countries at almost every level of development, even if it does not produce many outright "winners" anywhere. It is also likely to heighten strains between urban and rural populations, a crucial fault line throughout the developing world, and one across which large-scale population movements are likely to be especially stressful. In this regard, one specific inference from the work presented here may be worth highlighting, which is that, in the period that concerns this study, the critical path connecting climate change to social and political failure lies less through rising temperatures or rising sea level than through the changing distribution of freshwater. If there is one piece of advice this study would appear to recommend to policymakers and strategic planners everywhere, it would surely be "follow the water."

With respect to the ultimate question of how climate change may affect relations between the United States and the larger world, the picture that emerges is ambiguous at best. This study includes all the current major producers of greenhouse gases apart from the United States itself, along with a number of countries for whom the imperatives of economic growth are likely to outweigh considerations of climate policy for some time to come—except, perhaps, when the latter are brought forcibly to the forefront of public attention by dramatic, climate-induced emergencies. In the developing world there is little doubt that, for the time being, adaptation to the negative consequences of carbon emissions will be more important as a policy priority than their mitigation, and also that, in the pursuit of effective adaptation, Western technological expertise and financial assistance will be essential. In this respect climate change may provide a framework of common interests that, if faced squarely and honestly, might bring North and South closer together.

It is also likely that, as public consciousness of climate change and its perils expands, so too will public awareness that the historical responsibility for these perils is not universally shared but lies at the feet of a handful of states. Some of these states are among the chief beneficiaries of the exploitative economic and colonial practices in the past. Others are among the main victims of those same practices. The largest producer of greenhouse gases today, as is well known, is China. It is less well known that Indonesia is now on the verge of outstripping the entirety of the European Union as a carbon emitter. India and Russia also rank high in the generation of greenhouse gases, and they appear destined to rise in the rankings. However, none of these nations is likely, in the period that concerns us, to surpass the United States on a per capita basis.

The United States' role as the proportionally dominant polluter may prove to be the most compelling metric of responsibility to the citizens of the developing world, whose economic aspirations have lately been subject to unwelcome scrutiny on account of environmental risks that can appear awfully remote when weighed against the needs of the moment and the inequities of the past. There is no question that such perceptions will play their part in determining whether climate change proves to be a source of strife or a means of reminding humankind that we really are all in this together. This volume is intended as a contribution to security studies, not to moral philosophy. Yet it is safe to say that, in managing the consequences of climate change, it will not be possible to leave questions of justice out of account.

Notes

1. As a policy issue, climate change stands out for the intractability (and circularity) of the controversy it inspires; on which see Mike Hulme, *Why We Disagree about Climate Change: Understanding Controversy, Inaction, and Opportunity* (Cambridge: Cambridge University Press, 2009).

2. Vladimir Trifonov, Hossein Khiabanian, and Raul Rabadan, "Geographic Dependence, Surveillance, and Origins of the 2009 Influenza A (H1N1) Virus," *New England Journal of Medicine* 61, no. 2 (July 9, 2009): 115–19.

3. The World Organization for Animal Health, an international association of veterinarians, attempted in vain to get the flu renamed "North American influenza," after its apparent point of origin. See Keith Bradsher, "The Naming of Swine Flu, a Curious Matter," *New York Times*, April 28, 2009, www.nytimes.com/2009/04/29/world/asia/29swine.html?_r=1.

4. The World Health Organization was later criticized for being excessively alarmist, and responded with an internal review to see whether its procedures were properly followed, or in need of revision. See Johnathan Lynn, "WHO to Review Its Handling of H1N1 Flu Pandemic," *Reuters*, 12 January 2010, www.reuters.com/article/idUSTRE5BL2ZT20100112.

5. "What a Waste," *Economist*, May 7, 2009, www.economist.com/world/mideast-africa/displaystory.cfm?story_id=13611723; Christopher Hitchens, "First They Came for the Pigs," *Slate*, September 28, 2009, www.slate.com/id/2229830/.

6. IPCC WGII, *Climate Change 2007*, 13.

7. Among the linkages connecting environmental stress to social and international conflict, migration has come in for especially careful study. This is largely because, in contrast to other hypothesized linkages between climate change and security, which may be rare or speculative in historical terms, enforced movements of population are a familiar phenomenon, on which a good deal of data exists. For a survey of the general issue, see Nils Petter Gleditsch, Ragnhild Nordås, and Idean Salehyan, *Climate Change and Conflict: The Migration Link*, Coping with Crisis Working Paper, International Peace Academy, May 2007, www.ipinst.org/media/pdf/publications/cwc_working_paper_climate_change.pdf.

Appendix A
Temperature Change and Freshwater Availability

Table A.1 summarizes the data and information provided to project participants with respect to anticipated changes in temperature, freshwater availability, and agricultural productivity in the countries included in this study. The columns headed "Temperature Vulnerability" and "Freshwater Availability" are based on material prepared for the National Intelligence Council by the Center for International Earth Science Information Network (CIESIN) at Columbia University.[1] The column headed "Agricultural Productivity Impact" is based on *Global Warming and Agriculture: Impact Estimates by Country*, by William Cline.[2] Additional details on the contents of individual columns follow.

Temperature Vulnerability

Three pieces of data are reported in the category "temperature vulnerability": an estimate of actual temperature change in degrees Celsius anticipated to occur by 2030 (the midpoint of a range that can be expected to vary by up to +/- 0.5 degree Celsius); an "aggregate vulnerability score," described below; and a verbal characterization of that score relative to global norms.

Temperature change is often used as a proxy for climate change generally. This is reasonable given the centrality of rising temperature to other environmental effects, provided the expression is recognized as a term of art and not as a claim about the direct impact of temperature change per se. "Temperature vulnerability" in turn is CIESIN's estimate of the degree to which a country can be expected to cope successfully with the social, political, and economic stresses that climate change may impose. Its work in this area is based in part upon, and summarizes, prior scholarship,[3] which has sought to take into account three traditional risk factors in estimating vulnerability: the frequency of wars and similar events in a given region, the frequency of internal political and social crises within a country,[4] and the historical capacity of the government to meet prior and current social and political challenges.[5]

It is obvious that this kind of estimate, which is reported here as a single number for each country ("aggregate vulnerability score"), presents a considerable risk of spurious precision and much difficulty in managing covariance. As a consequence, analyses arising from small estimated differences are certain to be fruitless. A verbal characterization of where

Table A.1 Anticipated Changes in Temperature, Freshwater, and Agriculture for the Countries in This Study

Region and Country		Temperature Vulnerability			Freshwater Availability			Agricultural Productivity Impact
		Aggregate Vulnerability Score	Relative Temperature Vulnerability	Temperature Change in Degrees Celsius	% of Population with <1,000 m³ per Year per Capita, 2000	% of Population with <1,000 m³ per Year per Capita, 2030	% Change	
Middle East and North Africa	Algeria	1.00	Average	0.96	86.2	94.4	8.2	Very Serious
	Egypt	0.90	Average	0.71	66.4	74.8	8.5	Positive
	Iran	0.96	Average	0.83	83.2	90.8	7.6	Serious
	Iraq	1.06	Average	0.74	31.5	50.1	18.6	Very Serious
	Morocco	1.22	Very Serious	0.91	98.8	99.5	0.7	Very Serious
	Saudi Arabia	0.78	Average	0.66	94.1	96.3	2.2	Moderate
	Tunisia	1.18	Very Serious	1.00	94.1	97.1	3.0	Serious
	Turkey	0.83	Average	0.78	34.7	54.5	19.8	Negligible
Russia and Central Asia	Russia	1.14	Serious	1.30	12.6	11.4	-1.2	Positive
	Kazakhstan	1.16	Serious	1.14	21.5	24.4	2.9	Positive
	Kyrgyzstan	0.87	Average	0.83	22.4	18.7	-3.7	Positive
	Tajikistan	1.02	Average	0.86	14.4	10.6	-3.8	Positive
	Turkmenistan	1.04	Average	0.91	61.6	63.5	1.9	Positive
	Uzbekistan	0.98	Average	0.91	33.2	45.4	12.2	Positive
South, Southeast, and East Asia	Bangladesh	1.18	Very Serious	0.77	61.8	64.4	2.6	Moderate
	China	1.03	Average	1.01	24.4	33.3	8.9	Positive
	India	0.86	Average	0.68	48.3	53.8	5.5	Very Serious
	Indonesia	0.55	Average	0.59	14.5	20.3	5.9	Negligible
	Pakistan	1.02	Average	0.79	67.2	72.4	5.2	Serious
	Philippines	0.59	Average	0.53	12.7	10.4	-2.3	Moderate
	Vietnam	0.52	Average	0.54	11.5	25.7	14.2	Negligible

Region	Country						
	Bolivia	1.06	Average	0.7	0.6	-0.1	Very Serious
	Brazil	0.73	Average	22.7	24.3	1.6	Negligible
South	Chile	0.98	Average	24.1	18.1	-5.9	Moderate
America	Colombia	0.49	Low	16.4	12.4	-4.0	Moderate
	Ecuador	0.55	Average	1.3	0.6	-0.7	Serious
	Paraguay	1.16	Serious	0.0	0.0	0.0	Very Serious
	Peru	0.74	Average	25.4	17.7	-7.7	Serious
	Côte d'Ivoire	1.12	Serious	0.0	22.7	22.7	Negligible
	Guinea	0.89	Average	0.0	0.0	0.0	Serious
	Guinea-Bissau	0.77	Average	0.0	0.0	0.0	Serious
	Lesotho	2.07	Very Serious	0.0	0.0	0.0	Very Serious
Sub-	Liberia	0.97	Average	0.0	0.0	0.0	Serious
Saharan	Nigeria	0.93	Average	26.1	45.3	19.2	Negligible
Africa	Senegal	0.83	Average	66.5	73.1	6.6	Very Serious
	Sierra Leone	0.90	Average	0.0	0.0	0.0	Serious
	South Africa	1.36	Very Serious	55.4	58.2	2.7	Serious
	Swaziland	1.33	Very Serious	0.0	0.0	0.0	Serious
	Zimbabwe	0.96	Average	44.9	56.3	11.3	Very Serious
Global average		0.80	Average	30.8	36.2	5.5	Negligible
United States (48 states)		0.60	Average	25.6	29.7	4.0	Positive

each country stands relative to world norms has been added by way of acknowledging and mitigating these difficulties. Five categories of "relative temperature vulnerability" were assigned: negligible, low, average, serious, and very serious. Of the 179 countries originally studied by CIESIN, 144 received temperature vulnerability scores that were within +/– 0.25 standard deviation (SD) of the global mean. All such countries are characterized on this chart as "average." Scores of "low" (< –0.25 and > –.50 SD) and "negligible" (< –.50 SD) reflect progressive deviation toward reduced vulnerability, while "serious" and "very serious" characterize states that are correspondingly more vulnerable.

Freshwater Availability

The data presented for the category "freshwater availability" estimate the percentage of each country's population that can be deemed to be short of water in 2000 and in 2030, along with a calculation of the percentage change. A population is deemed "short of water" for purposes of this study if its annual access to freshwater falls below 1,000 cubic meters per year, a widely used metric for estimating water stress.

Agricultural Productivity Impact

The information for the category "agricultural productivity impact" derives from William R. Cline's book *Global Warming and Agriculture: Impact Estimates by Country*. Cline's time frame is different from that employed in this study—his numerical estimates are for 2080 rather than 2030—which is why no actual data are provided. Instead, verbal characterizations have been employed to describe the relative trend of agricultural productivity that countries can expect to experience in the near term, based on Cline's estimates of conditions in 2080. Those trends will not be fully developed by 2030. Nevertheless, in cases where Cline's endpoint suggests very dramatic changes—some countries are estimated to have experienced productivity declines of more than 50 percent by 2080—it is reasonable to assume that the nature of the trend, if not its ultimate severity, will begin making itself felt in our period.

There are five possible entries in this column: "very serious," serious," "moderate," "negligible," and "positive." "Positive" entries mean that agricultural productivity in a given country may actually increase as a consequence of changes in temperature and precipitation patterns. For the most part, this hypothesized outcome is only possible if one accepts the reality of "carbon fertilization," a controversial hypothesis that seeks to take into account the fact that carbon dioxide, the principal human-made greenhouse gas, is an input to photosynthesis. Increasing concentrations of carbon dioxide in the atmosphere may benefit some crops under some conditions. Cline provides estimates for conditions in 2080 with and without carbon fertilization.[6] We have adopted the latter in keeping with our desire to avoid presenting what may appear to be worst-case scenarios. In this instance, however, it is arguable that we have presented something approaching a "best-case" scenario, which is in the nature of things equally unlikely. Almost without exception, if one were to eliminate the impact of carbon fertilization from the agricultural impact estimates presented above, each country would slip down at least one notch among our broad categories.

As in the calculation of relative temperature vulnerability, characterizations of agricultural productivity are made in relation to global norms. Countries are deemed to face a "negligible" agricultural impact in 2030 if their downward trend is no worse than 0.5 SD below the global average. This does not mean that the impact in 2080 will be negligible, merely that the impending decline may not be sufficiently apparent within our time frame to produce distinct social and political effects. "Moderate" impact is assigned to countries whose trend is between 0.5 and 1.0 SD worse than average. "Serious" impact applies to countries trending between 1.0 and 1.5 SDs below the global norm; and "very serious" conditions reflect anticipated decline at a rate more that 1.5 SDs faster than the world average.

Notes

1. The methodology by which the CIESIN data were prepared is described by Marc A. Levy, Bridget Anderson, Melanie Brickman, Chris Cromer, Brian Falk, Balazs Fekete, Pamela Green, et al., *Assessment of Climate Change Impacts on Select US Security Interests*, Center for International Earth Science Information Network Working Paper (New York: Columbia University, 2008), 30–48, www.ciesin.columbia.edu/documents/Climate_Security_CIESIN_July_2008_v1_0.ed070208_000.pdf.

2. William R. Cline, *Global Warming and Agriculture: Impact Estimates by Country* (Washington, DC: Peterson Institute for International Economics, 2007), 67–71.

3. CIESIN's work is based primarily upon that of G. Yohe, E. Malone, A. Brenkert, M. Schlesinger, H. Meij, X. Xing, and D. Lee, "Global Distributions of Vulnerability to Climate Change," *Integrated Assessment Journal* 6, no. 3 (2006): 35–44. Also see A. Brenkert E. and Malone, "Modeling Vulnerability and Resilience to Climate Change: A Case Study of India and Indian States," *Climatic Change* 72 (2005): 57–102; and Levy et al., *Assessment of Climate Change Impacts*, 30–33.

4. CIESIN estimates of regional instability and internal unrest are based on the public domain databases assembled by the Political Instability Task Force, http://globalpolicy.gmu.edu/pitf/pitfdata.htm.

5. CIESIN's concept of state capacity is based chiefly on the work of Daniel Kaufman, Aart Kraay, and their colleagues; see, inter alia, Daniel Kaufman, Aart Kraay, and Massimo Mastruzzi, *Governance Matters VI: Aggregate and Individual Governance Indicators: 1996–2006*, Policy Research Working Paper 428 (Washington, DC: World Bank, 2007). The World Bank maintains up-to-date indicators of governance effectiveness at http://info.worldbank.org/governance/wgi/index.asp. See also Robin Mearns and Andrew Norton, eds., *Social Dimensions of Climate Change: Equity and Vulnerability in a Warming World*, New Frontiers of Social Policy (Washington, DC: World Bank, 2010).

6. Cline, *Global Warming and Agriculture*, 67–71.

Appendix B
Sea-Level Rise

Table B.1 summarizes CIESIN estimates of population, territory, and productive infrastructure at risk from rising sea level in the countries discussed in this volume.[1] For nations without coastlines, the data cells are filled with "—." Data were not available for a few coastal nations. In those cases, the cells are marked "N.A."

Data are provided for low-elevation coastal zones of 1 and 3 meters. The effects of a 1-meter rise in sea level are difficult to estimate, because reliable measurements of what is often a narrow strip of coastal land are hard to obtain. Demographers and geographers tend to present a conflicting picture of what happens when the land meets the sea. If current demographic data are simply laid on top of current geographical knowledge, the result is that, for many countries, a fair number of people already appear to be living under water. The most conservative way to deconflict this discrepancy is to ignore people who are already living, statistically speaking, in the ocean. That is what has been done here. The population figures do not include people whose current nominal location already places them in the sea. These data therefore represent a conservative estimate of likely populations affected by a 1-meter rise in sea level.

A 3-meter rise in sea level will inundate a much larger area than a 1-meter rise. In scientific terms this means that more accurate measurements are possible, because the margin of potential error, relative to the size of what is being measured, is smaller. For the 3-meter low-elevation coastal zone, absolute and percentage figures for the population, area, and GDP that would be put at risk are included. Note that population and GDP estimates are based on figures from 2004 to 2007, depending on the country in question, and do not take account of future economic or population growth. Given that coastal regions are often among the most economically productive parts of coastal states, these figure are accordingly at the conservative end of realistic future estimates of affected populations.

Note

1. See Marc A. Levy, Bridget Anderson, Melanie Brickman, Chris Cromer, Brian Falk, Balazs Fekete, Pamela Green, et al., *Assessment of Climate Change Impacts on Select US Security Interests*, Center for International Earth Science Information Network Working Paper (New York: Columbia University, 2008), 2–29, www.ciesin.columbia.edu/documents/Climate_Security_CIESIN_July_2008_v1_0.ed070208_000.pdf.

Table B.1 Estimates of Population, Territory, and Productive Infrastructure at Risk from Rising Sea Level for the Countries in This Study

| | | | | | Sea-Level Rise | | | | | | |
| | | | | 1-Meter Low-Elevation Coastal Zone | | 3-Meter Low-Elevation Coastal Zone | | | | | |
Region and Country	Total Population	Total Area (square kilometers)	Total GDP	Population	%	Population	%	Area	%	GDP	%
Middle East and North Africa											
Algeria	30,309,115	2,307,956	185,239,136,256	48,255	0.2	341,943	1.1	920	0.0	1,959,715,202	1.1
Egypt	67,280,560	970,455	212,201,467,904	6,061,429	9.0	10,246,438	15.2	10,839	1.1	4,592,876,337	2.2
Iran	70,016,722	1,590,342	363,445,796,864	177,946	0.3	597,918	0.9	7,091	0.4	264,334,481	0.1
Iraq	22,946,214	430,575	57,745,177,088	146,041	0.6	745,896	3.3	7,469	1.7	900,446,573	1.6
Morocco	29,511,366	670,467	98,004,045,824	67,623	0.2	701,221	2.4	1,779	0.3	778,329,902	0.8
Saudi Arabia	20,346,063	1,940,804	278,807,078,400	N.A.	N.A.	N.A.	N.A.	N.A.	N.A.	N.A.	N.A.
Tunisia	9,478,949	148,780	57,871,109,120	290,128	3.1	710,711	7.5	2,555	1.7	2,667,636,028	4.6
Turkey	66,667,710	768,751	403,124,543,488	180,357	0.3	1,033,691	1.6	3,625	0.5	777,493,705	0.2
Russia and Central Asia											
Russia	145,377,748	16,564,764	966,550,421,504	1,121,908	0.8	1,816,859	1.2	91,598	0.6	887,512,646	0.1
Kazakhstan	16,170,689	2,648,888	86,736,797,696	—	—	—	—	—	—	—	—
Kyrgyzstan	4,920,256	185,144	13,653,171,200	—	—	—	—	—	—	—	—
Tajikistan	6,086,855	130,295	7,067,394,304	—	—	—	—	—	—	—	—
Turkmenistan	4,737,194	460,407	19,580,956,672	—	—	—	—	—	—	—	—
Uzbekistan	24,799,171	421,509	58,326,272,000	—	—	—	—	—	—	—	—
South, Southeast, and East Asia											
Bangladesh	137,232,248	136,305	201,968,775,168	1,370,973	1.0	5,115,983	3.7	5,806	4.3	7,521,027,674	3.7
China	1,262,333,728	9,198,069	4,697,121,488,896	16,951,343	1.3	40,078,333	3.2	58,045	0.6	119,358,133,247	2.5
India	1,007,874,208	3,213,908	2,728,495,841,280	25,991,741	2.6	66,546,379	6.6	25,294	0.8	31,657,719,729	1.2
Indonesia	212,067,840	1,898,677	598,999,433,216	2,260,524	1.1	13,482,941	6.4	37,913	2.0	18,588,170,182	3.1
Pakistan	141,255,616	785,319	258,012,872,704	232,220	0.2	699,444	0.5	6,765	0.9	2,348,410,915	0.9
Philippines	75,289,646	295,408	278,108,258,304	2,997,929	4.0	16,111,268	21.4	6,173	2.1	6,021,920,889	2.2
Vietnam	78,136,408	328,535	154,137,812,992	7,024,774	9.0	23,039,073	29.5	35,764	10.9	24,232,196,735	15.7

Region	Country											
	Bolivia	8,328,107	1,087,730	19,369,255,424	—	—	—	—	—	—	—	—
	Brazil	170,313,088	8,480,403	1,215,743,524,864	565,076	0.3	4,145,528	2.4	52,844	0.6	80,967,690,385	6.7
South	Chile	15,210,178	721,229	134,686,474,240	12,920	0.1	177,538	1.2	17,856	2.5	1,904,228,314	1.4
America	Colombia	42,103,731	1,141,569	295,367,606,272	67,726	0.2	289,198	0.7	3,765	0.3	2,285,355,760	0.8
	Ecuador	12,645,020	247,404	39,432,658,432	127,440	1.0	348,618	2.8	2,128	0.9	637,401,567	1.6
	Paraguay	5,496,268	395,886	27,771,606,016	—	—	—	—	—	—	—	—
	Peru	25,596,223	1,289,526	112,633,380,864	31,017	0.1	384,214	1.5	3,807	0.3	2,250,879,089	2.0
	Côte d'Ivoire	16,734,756	318,979	24,557,228,032	273,878	1.6	549,297	3.3	589	0.2	116,811,516	0.5
	Guinea	8,433,902	245,401	14,073,371,456	39,088	0.5	350,817	4.2	1,415	0.6	443,022,090	3.1
	Guinea-Bissau	1,365,579	33,528	1,159,567,136	9,484	0.7	42,500	3.1	1,085	3.2	51,927,932	4.5
	Lesotho	1,779,168	30,431	5,024,486,840	—	—	—	—	—	—	—	—
Sub-	Liberia	3,065,385	96,056	3,325,665,488	17,368	0.6	92,516	3.0	132	0.1	7,979,598	0.2
Saharan	Nigeria	117,581,548	898,795	108,906,312,704	174,872	0.1	1,628,142	1.4	1,996	0.2	811,703,603	0.7
Africa	Senegal	10,288,773	195,472	13,178,484,992	115,785	1.1	834,477	8.1	3,839	2.0	1,104,890,143	8.4
	Sierra Leone	4,508,800	72,086	2,250,160,480	14,124	0.3	88,296	2.0	713	1.0	25,141,277	1.1
	South Africa	45,569,236	1,216,706	467,851,427,840	14,636	0.0	36,428	0.1	381	0.0	369,736,508	0.1
	Swaziland	1,021,190	17,350	5,003,625,792	—	—	—	—	—	—	—	—
	Zimbabwe	12,595,012	388,205	32,017,589,248	—	—	—	—	—	—	—	—

Note: N.A. = not available.
Source: CIESIN data.

Contributors

Ibrahim Al-Marashi is associate dean of international relations at IE University, Segovia.

Linda J. Beck is associate professor of political science at the University of Maine at Farmington.

Chad M. Briggs is senior fellow at the Institute for Environmental Security, and director of GlobalINT LLC.

Jeffrey Cason is professor of political science and dean of international programs at Middlebury College.

Kent Eaton is professor of politics at the University of California, Santa Cruz.

Dennis Galvan is associate professor of political science and international studies and co-director of the Global Oregon Initiative at the University of Oregon.

Brian Guy is a doctoral candidate in political science at the University of Oregon.

Paul D. Hutchcroft is a professor in the Department of Political and Social Change and director of the School of International, Political, and Strategic Studies at the Australian National University's College of Asia and the Pacific.

Joanna I. Lewis is assistant professor of science, technology, and international affairs in the Edmund A. Walsh School of Foreign Service at Georgetown University.

Michael S. Malley is assistant professor of national security affairs at the Naval Postgraduate School.

Daniel Markey is senior fellow for India, Pakistan, and South Asia at the Council on Foreign Relations.

Daniel Moran is professor of national security affairs at the Naval Postgraduate School.

Ngonidzashe Munemo is assistant professor of political science at Williams College.

T. V. Paul is James McGill Professor of International Relations in the Department of Political Science and director of the Center for International Peace and Security Studies at McGill University.

E. Mark Pires is associate professor of geography in the Earth and Environmental Science Department at Long Island University in Brookville, New York.

Ali Riaz is professor and chair of the Department of Politics and Government at Illinois State University.

James A. Russell is associate professor of national security affairs at the Naval Postgraduate School.

Edward Schatz is associate professor of political science at the University of Toronto.

Carlyle A. Thayer is professor emeritus in the School of Humanities and Social Sciences, University College, University of New South Wales, at the Australian Defence Force Academy.

Stacy D. VanDeveer is associate professor of political science at the University of New Hampshire.

Celeste A. Wallander is associate professor in the School of International Service at American University, on leave since May 2009 as deputy assistant secretary of defense for Russia, Ukraine, and Eurasia Policy.

Gregory W. White is professor of government at Smith College.

Index